T0235150

Foundations of Quantitative Finance

Chapman & Hall/CRC Financial Mathematics Series

Series Editors

M.A.H. Dempster
Centre for Financial Research
Department of Pure Mathematics and Statistics
University of Cambridge, UK

Dilip B. Madan
Robert H. Smith School of Business
University of Maryland, USA

Rama Cont
Department of Mathematics
Imperial College, UK

Robert A. Jarrow
Ronald P. & Susan E. Lynch Professor of Investment
ManagementSamuel Curtis Johnson Graduate School
of Management Cornell University

Recently Published Titles

For more information about this series please visit: https://www.crcpress.com/Chapman-and-HallCRC-Financial-Mathematics-Series/book series/CHFINANCMTH

Foundations of Quantitative Finance

Book III: The Integrals of Riemann, Lebesgue and (Riemann-)Stieltjes

Robert R. Reitano

Brandeis International Business School
Waltham, MA

CRC Press
Taylor & Francis Group
Boca Raton London New York

CRC Press is an imprint of the
Taylor & Francis Group, an **informa** business

A CHAPMAN & HALL BOOK

First edition published 2023
by CRC Press
6000 Broken Sound Parkway NW, Suite 300, Boca Raton, FL 33487-2742

and by CRC Press
4 Park Square, Milton Park, Abingdon, Oxon, OX14 4RN

CRC Press is an imprint of Taylor & Francis Group, LLC

© 2023 Robert R. Reitano

Reasonable efforts have been made to publish reliable data and information, but the author and publisher cannot assume responsibility for the validity of all materials or the consequences of their use. The authors and publishers have attempted to trace the copyright holders of all material reproduced in this publication and apologize to copyright holders if permission to publish in this form has not been obtained. If any copyright material has not been acknowledged please write and let us know so we may rectify in any future reprint.

Except as permitted under U.S. Copyright Law, no part of this book may be reprinted, reproduced, transmitted, or utilized in any form by any electronic, mechanical, or other means, now known or hereafter invented, including photocopying, microfilming, and recording, or in any information storage or retrieval system, without written permission from the publishers.

For permission to photocopy or use material electronically from this work, access www.copyright.com or contact the Copyright Clearance Center, Inc. (CCC), 222 Rosewood Drive, Danvers, MA 01923, 978-750-8400. For works that are not available on CCC please contact mpkbookspermissions@tandf.co.uk

Trademark notice: Product or corporate names may be trademarks or registered trademarks and are used only for identification and explanation without intent to infringe.

Library of Congress Cataloging-in-Publication Data
Names: Reitano, Robert R., 1950- author.
Title: Foundations of quantitative finance. III, The integrals of Riemann, Lebesgue, and (Riemann-)Stieltjes / Robert R. Reitano.
Other titles: Integrals of Riemann, Lebesgue, and (Riemann-)Stieltjes
Description: Boca Raton : CRC Press, [2023] | Includes bibliographical references.
Identifiers: LCCN 2022054958 (print) | LCCN 2022054959 (ebook) | ISBN 9781032206547 (paperback) | ISBN 9781032206561 (hardback) | ISBN 9781003264590 (ebook)
Subjects: LCSH: Finance--Mathematical models. | Integrals.
Classification: LCC HG106 .R4483 2023 (print) | LCC HG106 (ebook) | DDC 332.01/5195--dc23/eng/20230113
LC record available at https://lccn.loc.gov/2022054958
LC ebook record available at https://lccn.loc.gov/2022054959

ISBN: 978-1-032-20656-1 (hbk)
ISBN: 978-1-032-20654-7 (pbk)
ISBN: 978-1-003-26459-0 (ebk)

DOI: 10.1201/9781003264590

Publisher's note: This book has been prepared from camera-ready copy provided by the authors.

To Michael, David, and Jeffrey

Contents

Preface

The idea for a reference book on the mathematical foundations of quantitative finance has been with me throughout my professional and academic careers in this field, but the commitment to finally write it didn't materialize until completing my first "introductory" book in 2010.

My original academic studies were in "pure" mathematics in a field of mathematical analysis, and neither applications generally nor finance in particular were then even on my mind. But on completion of my degree, I decided to temporarily investigate a career in applied math, becoming an actuary, and in short order became enamored with mathematical applications in finance.

One of my first inquiries was into better understanding yield curve risk management, ultimately introducing the notion of partial durations and related immunization strategies. This experience led me to recognize the power of greater precision in the mathematical specification and solution of even an age-old problem. From there my commitment to mathematical finance was complete, and my temporary investigation into this field became permanent.

In my personal studies, I found that there were a great many books in finance that focused on markets, instruments, models and strategies, and which typically provided an informal acknowledgment of the background mathematics. There were also many books in mathematical finance focusing on more advanced mathematical models and methods, and typically written at a level of mathematical sophistication requiring a reader to have significant formal training and the time and motivation to derive omitted details.

The challenge of acquiring expertise is compounded by the fact that the field of quantitative finance utilizes advanced mathematical theories and models from a number of fields. While there are many good references on any of these topics, most are again written at a level beyond many students, practitioners, and even researchers of quantitative finance. Such books develop materials with an eye to comprehensiveness in the given subject matter, rather than with an eye toward efficiently curating and developing the theories needed for applications in quantitative finance.

Thus the overriding goal I have for this collection of books is to provide a complete and detailed development of the many foundational mathematical theories and results one finds referenced in popular resources in finance and quantitative finance. The included topics have been curated from a vast mathematics and finance literature for the express purpose of supporting applications in quantitative finance.

I originally budgeted 700 pages per book, in two volumes. It soon became obvious this was too limiting, and two volumes ultimately turned into ten. In the end, each book was dedicated to a specific area of mathematics or probability theory, with a variety of applications to finance that are relevant to the needs of financial mathematicians.

My target readers are students, practitioners, and researchers in finance who are quantitatively literate, and recognize the need for the materials and formal developments presented. My hope is that the approach taken in these books will motivate readers to navigate these details and master these materials.

Most importantly for a reference work, all ten volumes are extensively self-referenced. The reader can enter the collection at any point of interest, and then using the references cited, work backward to prior books to fill in needed details. This approach also works for a course on a given volume's subject matter, with earlier books used for reference, and for both course-based and self-study approaches to sequential studies.

The reader will find that the developments herein are at a much greater level of detail than most advanced quantitative finance books. Such developments are of necessity typically longer, more meticulously reasoned, and therefore can be more demanding on the reader. Thus before committing to a detailed line-by-line study of a given result, it is always more efficient to first scan the derivation once or twice to better understand the overall logic flow.

I hope the additional details presented will support your journey to better understanding.

I am grateful for the support of my family: Lisa, Michael, David, and Jeffrey, as well as the support of friends and colleagues at Brandeis International Business School.

Robert R. Reitano
Brandeis International Business School

Author

Robert R. Reitano is Professor of the Practice of Finance at the Brandeis International Business School where he specializes in risk management and quantitative finance. He previously served as MSF Program Director and Senior Academic Director. He has a PhD in mathematics from MIT, is a fellow of the Society of Actuaries, and is a Chartered Enterprise Risk Analyst. Dr. Reitano consults in investment strategy and asset/liability risk management, and previously had a 29-year career at John Hancock/Manulife in investment strategy and asset/liability management, advancing to Executive Vice President and Chief Investment Strategist. His research papers have appeared in a number of journals and have won both the Annual Prize of the Society of Actuaries as well as two F.M. Redington Prizes of the Investment Section of the Society of the Actuaries. Dr. Reitano serves on various not-for-profit boards and investment committees.

Introduction

Foundations of Quantitative Finance is structured as follows:

    Book I: *Measure Spaces and Measurable Functions*

    Book II: *Probability Spaces and Random Variables*

    Book III: *The Integrals of Riemann, Lebesgue, and (Riemann-)Stieltjes*

    Book IV: *Distribution Functions and Expectations*

    Book V: *General Measure and Integration Theory*

    Book VI: *Densities, Transformed Distributions, and Limit Theorems*

    Book VII: *Brownian Motion and Other Stochastic Processes*

    Book VIII: *Itô Integration and Stochastic Calculus 1*

    Book IX: *Stochastic Calculus 2 and Stochastic Differential Equations*

    Book X: *Classical Models and Applications in Finance*

The series is logically sequential. Books I, III, and V develop foundational mathematical results needed for the probability theory and finance applications of Books II, IV, and VI, respectively. Then Books VII, VIII, and IX develop results in the theory of stochastic processes. While these latter three books introduce ideas from finance as appropriate, the final realization of the applications of these stochastic models to finance is deferred to Book X.

This Book III, *The Integrals of Riemann, Lebesgue, and (Riemann-)Stieltjes,* reflects my objective of developing the Riemann constructions, and then generalizing to two models for integration that in different ways expand the ideas of the Riemann approach.

The Lebesgue integration (and differentiation) theory will be seen to generalize familiar Riemann results using the measure theoretic tools of Book I. While the Riemann theory requires the integrand $f(x)$ to be more or less definable pointwise and smoothly on a given interval, the Lebesgue theory requires only that intervals in the range of $f(x)$ be obtainable by measurable sets in the domain. This perspective will also provide an introduction to the measure theoretical approach to integration that will see its fullest realization in Book V, and then generalized in yet another direction in Book VIII.

The Stieltjes modification of the Riemann integral, producing what is now known as the Riemann-Stieltjes integral, again requires a certain regularity in the pointwise definition of the integrand. But the innovation here is that it generalizes the Riemann integral by changing the "length" of a given interval $[a, b]$ from $b - a$ to $F(b) - F(a)$, for a properly specified integrator function $F(x)$. This theory will be useful in Books IV and VI when we return to probability theory and expectations of random variables, with $F(x)$ there as a distribution function of a random variable or vector. Such integrals will also reappear within the integrals of stochastic processes studied in Book VIII.

Chapter 1 begins with the classical definition of the Riemann integral of $f(x)$ over a bounded interval $[a, b] \subset \mathbb{R}$ in terms of Riemann sums, and then the Darboux sums framework for this integral is introduced. The first questions investigated are the equivalence of these definitions, and the existence of this integral. Continuity of $f(x)$ is seen to be more than necessary, and the classical Lebesgue existence theorem, which characterizes the result for bounded functions, is derived.

Properties of the Riemann integral are then developed, including approaches to its evaluation using the two classical versions of the fundamental theorem of calculus. The Leibniz integral rule for differentiating a function defined parametrically by an integral is also derived. The chapter then turns to integrals of convergent function sequences, illustrating both positive results and the difficulties that exist in generalizing such results within the Riemann framework.

The Riemann integral over rectangles in $\mathbb{R}^n$ is then defined and questions of existence addressed, including the generalized Lebesgue existence theorem. After properties of such integrals are developed, the chapter ends with results on evaluations of integrals using so-called iterated integrals.

The definition of the Lebesgue integral of a simple function on $\mathbb{R}^n$ is the starting point for Chapter 2, and due to its generality, the question of well-definedness is addressed, as are basic properties. The process of generalization to other functions then proceeds in steps to bounded, then nonnegative, and finally general measurable functions. Each step uses the prior results to both define the integral and derive the associated properties.

For bounded functions on bounded rectangles, existence is characterized by Lebesgue measurability of the integrand, and it is then proved that Riemann integrability implies Lebesgue integrability for such functions and rectangles. Properties of this integral, and a first result on convergent function sequences, are derived, one of several so-called "integration to the limit" results of the chapter.

This pattern is repeated in the other steps, addressing definition, existence, properties, and several all-important classical results on the integrals of a convergent function sequence. The link between Riemann and Lebesgue integrals is then generalized to integrals over $\mathbb{R}^n$. The chapter ends with a summary of the integration to the limit results of Lebesgue theory, and a discussion of the Chapter 1 positive and negative results on Riemann integrals.

Fundamental to most valuation approaches to the Riemann integral are the aptly named fundamental theorems of calculus. One theorem addresses the integral of a derivative, the other the derivative of an integral with a variable limit of integration. Since the existence and value of Lebesgue integrals are independent of many pointwise redefinitions of the integrand, the Riemann results cannot generalize without further restrictions. Chapter 3 investigates these generalizations.

For a result on the differentiability of a Lebesgue integral, a very general and surprising result is first derived with the aid of the Vitali covering lemma. The result is: Every monotonic function is differentiable almost everywhere. A study of functions of bounded variation is then undertaken, and this provides the necessary tools to prove this version of the fundamental theorem within the Lebesgue framework.

The complementary result on the Lebesgue integral of a derivative is addressed next, starting with an investigation into what it means for a function to be expressible as the integral of its derivative. Using a variety of examples, this question is seen to be deeper than it first appears. Indeed, continuous, increasing functions exist with derivative equal to zero almost everywhere! Certainly, integrating such derivatives cannot restore the original functions.

The final resolution requires the notion of absolute continuity, a property possessed by any Lebesgue integral with variable upper limit of integration. It is then seen that any absolutely continuous function is differentiable almost everywhere, and that this derivative then integrates back to the function everywhere. Lebesgue integration by parts is then addressed.

The final Chapter 4 addresses the Stieltjes innovation to integration, beginning with an introduction that applies Stieltjes' approach within a Riemann context, and also the generalization of this idea to Lebesgue integration. The chapter then focuses on the former integral, known as the Riemann-Stieltjes integral, with the study of the Lebesgue-Stieltjes integral deferred to Book V.

As in Chapter 1, we begin with a development on $\mathbb{R}$, using both the Riemann-Stieltjes sums and the Darboux sums approaches for its definition, and addressing both increasing and bounded variation integrators. Equivalence of approaches and existence results for increasing integrators are derived, as is a Riemann-Stieltjes integration by parts formula. Existence results for bounded variation integrators are then derived, and these integrals are seen to be decomposable into a difference of integrals with increasing integrators. Properties of this integral, an integration to the limit result, and evaluation approaches are then developed.

The next investigation addresses a very general existence result for Riemann-Stieltjes integration on $\mathbb{R}$. It requires the development of several results on Banach spaces and bounded linear operators, results that will reappear and be expanded upon in later books. The key result is summarized to say that given a bounded integrator function $F(x)$, if the Riemann-Stieltjes integral exists for all continuous integrands, then $F(x)$ must be of bounded variation. This insight will be fundamental in Book VIII, and will motivate the need for a new notion of integration developed there.

The chapter then turns to Riemann-Stieltjes integrals on $\mathbb{R}^n$, with the notions of "increasing" and "of bounded variation" appropriately generalized. As in the 1-dimensional case, the development addresses definition, existence, properties, and finally evaluation.

I hope this book and the other books in the collection serve you well.

Notation 0.1 (Referencing within FQF Series) *To simplify the referencing of results from other books in this series, we will use the following convention.*

A reference to "Proposition I.3.33" is a reference to Proposition 3.33 of Book I, while "Chapter III.4" is a reference to Chapter 4 of Book III, and "II.(8.5)" is a reference to formula (8.5) of Book II, and so forth.

1

The Riemann Integral

In this chapter, we introduce constructions of the Riemann integral and review many of this integral's properties. Beginning in 1-dimension, both the classical Riemann sums approach and the Darboux sums approach are introduced and shown to be equivalent. While the former will be more familiar to the reader, variations in the latter approach will be seen in later chapters and in Books V and VIII. The question of existence of this integral is then investigated, as are various properties and evaluation approaches.

The Riemann approach provides a beautiful theory of integration, along with its intimate connections with differentiation, for appropriately specified functions. It also enjoys a wide variety of applications in finance, probability theory, and the various sciences, where functions of interest often satisfy the necessary requirements.

There are also limitations to the Riemann approach to the integral that motivate the need for the Lebesgue development of the next two chapters. These limitations include the necessary restrictions on the integrated function and the domains over which such functions can be integrated. In addition, this integral does not generally work well with convergent sequences of functions. It will be seen by example that neither the integrability of the limit function nor the value of this integral when so integrable, are generally predictable.

The final section develops the Riemann integral in n-dimensions and its evaluation by iterated integrals.

1.1 The Integral in $\mathbb{R}$

In this section, we develop the theory of the Riemann integral on $\mathbb{R}$, beginning with the classical Riemann approach. The Darboux criterion for existence of an integral is then presented; and these criteria are proved to be equivalent and indeed equivalent to a third Cauchy criterion.

We then turn to existence of Riemann integrals for continuous functions and derive the Lebesgue existence theorem, which characterizes bounded functions that are so integrable. Properties of Riemann integrals are then summarized, and then evaluation of integrals is addressed, investigating the fundamental theorems of calculus and related integration results.

1.1.1 The Riemann Approach

The theory of Riemann integration was introduced in 1854 by **Bernhard Riemann** (1826–1866), though it was not published until 1868. It begins with an arbitrary **partition** $\{x_i\}_{i=0}^n$ of the interval $[a, b]$ into subintervals $\{[x_{i-1}, x_i]\}_{i=1}^n$:

$$a = x_0 < x_1 < \cdots < x_{n-1} < x_n = b, \tag{1.1}$$

DOI: 10.1201/9781003264590-1

with **mesh size** μ defined by:

$$\mu \equiv \max_{1 \le i \le n} \{x_i - x_{i-1}\}. \tag{1.2}$$

The Riemann integral of a function $f(x)$ is first estimated with a **Riemann sum**:

$$\sum_{i=1}^{n} f(\widetilde{x}_i)\Delta x_i, \tag{1.3}$$

where $\Delta x_i = x_i - x_{i-1}$ and $\widetilde{x}_i \in [x_{i-1}, x_i]$. The point $\widetilde{x}_i$ is often called an **interval tag**, and the collection $\{\widetilde{x}_i, [x_{i-1}, x_i]\}_{i=1}^{n}$ is referred to as a **tagged partition of** $[a, b]$.

To understand to what extent this summation is an "estimate" of something of interest and to identify circumstances under which this estimate converges as $\mu \to 0$, we can sometimes identify useful bounds:

$$m(b-a) \le \sum_{i=1}^{n} m_i \Delta x_i \le \sum_{i=1}^{n} f(\widetilde{x}_i)\Delta x_i \le \sum_{i=1}^{n} M_i \Delta x_i \le M(b-a). \tag{1.4}$$

In this expression, m_i denotes the **greatest lower bound** or **infimum** of $f(x)$, and M_i the **least upper bound** or **supremum** of $f(x)$, both defined on the subinterval $[x_{i-1}, x_i]$:

$$m_i = \inf_{[x_{i-1}, x_i]} f(x); \quad M_i = \sup_{[x_{i-1}, x_i]} f(x). \tag{1.5}$$

Similarly, m and M are defined with respect to $[a, b]$.

Of course, these bounds are only meaningful if the function f is bounded on $[a, b]$, meaning $|f(x)| \le c < \infty$.

Exercise 1.1 (On (1.4)) *Given a partition $\{x_i\}_{i=0}^{n}$ of $[a, b]$, let $\{x_i'\}_{i=0}^{m}$ be any* **refinement,** *meaning a partition with $\{x_i\}_{i=0}^{n} \subset \{x_i'\}_{i=0}^{m}$. Prove that for a bounded function:*

$$\sum_{i=1}^{n} m_i \Delta x_i \le \sum_{i=1}^{m} m_i' \Delta x_i' \le \sum_{i=1}^{n} M_i' \Delta x_i' \le \sum_{i=1}^{n} M_i \Delta x_i. \tag{1.6}$$

In other words, refinements always improve the bounds in (1.4).

The goal of this construction is to have these Riemann sums converge as we refine these partitions, by which we mean, converge as the mesh size $\mu \to 0$. It will be seen below that boundedness will not be enough to make this construction achieve its ultimate goal.

Definition 1.2 (Riemann integral of f over $[a, b]$) *The* **Riemann integral** *of $f(x)$* **over bounded** *$[a, b] \subset \mathbb{R}$ is defined by:*

$$(\mathcal{R}) \int_a^b f(x)dx = \lim_{\mu \to 0} \sum_{i=1}^{n} f(\widetilde{x}_i)\Delta x_i, \tag{1.7}$$

when this limit exists, is finite, and has value that is independent of $\widetilde{x}_i \in [x_{i-1}, x_i]$.

That is, for any $\epsilon > 0$ there is a δ, so that:

$$\left| \sum_{i=1}^{n} f(\widetilde{x}_i)\Delta x_i - (\mathcal{R}) \int_a^b f(x)dx \right| < \epsilon,$$

for all **tagged partitions** *$\{\widetilde{x}_i, [x_{i-1}, x_i]\}_{i=1}^{n}$ of mesh size $\mu \le \delta$.*

The function $f(x)$ is called the **integrand;** *and the constants a and b are the* **limits of integration** *of the Riemann integral.*

This definition involves a complicated notion within the limit denoted $\mu \to 0$. That is, for all μ we need to contemplate all possible partitions with this mesh size, as well as all possible interval tags for those partitions. In both cases, we have uncountably many choices.

To appreciate what can go wrong within the definition of the Riemann integral, consider the following.

Exercise 1.3 *Evaluate Riemann integrability of the following.*

1. *A classical example is the **Dirichlet function** $d(x)$, named after its discoverer, **J. P. G. Lejeune Dirichlet** (1805–1859):*

$$d(x) = \left\{ \begin{array}{ll} 0, & x \in [0,1] \ irrational, \\ 1, & x \in [0,1] \ rational. \end{array} \right.$$

This is a bounded function that obtains all $m_i = 0$ and $M_i = 1$ in (1.4), but $(\mathcal{R}) \int_0^1 d(x)dx$ does not exist.

2. *Let $f(x) = x^2$ on $[0,1]$. This function is bounded and is Riemann integrable.*

3. *Let $f(x) = 1/x$ on $(0,1]$, and $f(0) = 0$. This function is not bounded and is not Riemann integrable.*

4. *Let $f(x) = 1/x^2$ on $(0,1]$, and $f(0) = 0$. This function is not bounded but is Riemann integrable.*

1.1.2 The Darboux Approach

The bounding summations in (1.4) are in fact **upper and lower Darboux sums**, reflecting an approach to the integral introduced by **Jean-Gaston Darboux** (1842–1917). The resulting **Darboux integral** will be seen to exist if and only if the Riemann integral exists (Proposition 1.11). But the framework that Darboux introduced provided an important conceptual bridge to the frameworks used in more general approaches to integration, starting with Lebesgue's approach in the next chapter.

The conceptual leap is that these Darboux sums can be interpreted as the Riemann integrals of upper and lower bounding step functions.

Definition 1.4 (Step function) *Given the partition $\{x_i\}_{i=0}^n$ of $[a,b]$ in (1.1), and $A_i \equiv [x_{i-1}, x_i]$ for $1 \leq i \leq n$, let the **characteristic function** $\chi_{A_i}(x)$ be defined:*

$$\chi_{A_i}(x) \equiv \left\{ \begin{array}{ll} 1, & x \in A_i, \\ 0, & x \notin A_i. \end{array} \right. \tag{1.8}$$

*A **step function** on $[a,b]$ is defined as:*

$$\varphi(x) = \sum_{i=1}^n a_i \chi_{A_i}(x), \tag{1.9}$$

where $\{a_i\}_{i=1}^n \subset \mathbb{R}$.

Exercise 1.5 (Riemann integral of a step function) *Show that the Riemann integral of a step function $\varphi(x)$ exists and is given by:*

$$(\mathcal{R}) \int_a^b \varphi(x)dx = \sum_{i=1}^n a_i \Delta x_i. \tag{1.10}$$

This can also be expressed in terms of m, the Lebesgue measure of A_i, since $m(A_i) \equiv \Delta x_i$ (Proposition I.2.28 and Definition I.2.40).

Note that this result is not affected by the fact that $\varphi(x_i) = a_i + a_{i+1}$ at the interior partition points $\{x_i\}_{i=1}^{n-1}$.

Remark 1.6 (On partition intersections) *There is a little informality in the above definition that is allowed due to the robustness of the Riemann integral relative to isolated points. In general, one wants step functions, and more generally **simple functions** of Definition 2.2, to be defined over **disjoint partitions** of the given interval or set. The above collection is not quite disjoint, but had $\varphi(x)$ been defined relative to $A_i \equiv (x_{i-1}, x_i]$, then this would have been such a disjoint decomposition of $(a, b]$.*

For the Riemann integral, nondisjointness of A_i at the interior partition points $\{x_i\}_{i=1}^{n-1}$ does not affect the value of the integral as seen in Exercise 1.5. This would also be true using any other measure with $\mu[\{x\}] = 0$ for all x. However, we will need to be more careful for more general integration theories, because such intersections can change the value of the integral materially.

The following framework for the definition of the Darboux integral will be seen again and again in the context of other integration theories in these books and elsewhere. We will see in Proposition 1.11 that a bounded function is Darboux integrable if and only if it is Riemann integrable, and that when integrable, the integrals agree. Hence, the following definition is sometimes used as the definition of a Riemann integral.

Definition 1.7 (Darboux integral of f over $[a, b]$) *Let f be a bounded function defined on $[a, b]$ with:*

$$\sup_{\varphi_\mu \leq f} \int_a^b \varphi_\mu(x)dx = I = \inf_{f \leq \psi_\mu} \int_a^b \psi_\mu(x)dx. \tag{1.11}$$

Here $\varphi_\mu(x)$, $\psi_\mu(x)$ denote step functions with $\varphi_\mu(x) \leq f(x) \leq \psi_\mu(x)$, defined relative to arbitrary partitions of $[a, b]$ of mesh size μ, with integrals as given by (1.10).

*Then define the **Darboux integral of $f(x)$ over $[a, b]$** by:*

$$(\mathcal{D}) \int_a^b f(x)dx = I. \tag{1.12}$$

If the extrema in (1.11) are unequal, the Darboux integral of $f(x)$ over $[a, b]$ does not exist.

Remark 1.8 (On the Darboux integral) *Two comments on the above definition:*

1. ***Best step functions:*** *Given any partition of mesh size μ, one can without loss of generality in (1.11) consider only φ_μ and ψ_μ defined by:*

$$\varphi_\mu(x) \equiv \sum_{i=1}^n m_i \chi_{A_i}(x), \quad \psi_\mu(x) \equiv \sum_{j=1}^m M_j \chi_{A_j'}(x), \tag{1.13}$$

whereas in (1.5), $m_i \equiv \inf\{f(x)|x \in A_i\}$ and $M_j \equiv \sup\{f(x)|x \in A_j'\}$. This is the usual approach taken for the Darboux integral, and then (1.11) is stated:

$$\sup_{\varphi_\mu \leq f} \sum_{i=1}^n m_i \Delta x_i = I = \inf_{f \leq \psi_\mu} \sum_{i=1}^n M_i \Delta x_i. \tag{1.14}$$

*Given any partition, the summations in (1.14) are called the **lower and upper Darboux sums** of $f(x)$, respectively.*

2. On $\mu \to 0$: *It follows from Exercise 1.1 that:*

$$\sup_{\varphi_\mu \leq f} \int_a^b \varphi_\mu(x)dx = \lim_{\mu \to 0} \sup_{\varphi_\mu \leq f} \int_a^b \varphi_\mu(x)dx, \tag{1.15}$$

where the supremum on the right is over all φ_μ with given μ. The same is true with infima of $\psi_\mu(x)$-integrals.

In other words, these extrema are necessarily achieved as the mesh size decreases to zero, since refining any partition improves the bounds.

Remark 1.9 (On the Darboux framework) *As presented above, the Darboux approach appears to provide only a slight modification of the original idea with Riemann sums. This modification is to focus on the integrals of bounding step functions rather than on Riemann sum approximations using tags and function values. The representation in (1.14) makes this connection even closer.*

But this perspective opens a few significant doors to future generalizations:

1. *The approach in (1.11) suppresses the actual pointwise values of the function $f(x)$ and replaces these with bounding step functions. This opens the door to definitions of the integral of a function that has far less regularity than continuity, and for which pointwise values are less informative.*

Also, while we derived the value of the integral of a step function in (1.10) using the Riemann framework, one can imagine that this could be generalized using measures other than the Lebesgue measure to evaluate the measure of the subintervals in the partition.

2. *The above development used subinterval partitions of an interval $[a, b]$ to define the integral over this interval. Again, one could contemplate partitioning this interval into more general sets as long as these were measurable, disjoint, and unioned to $[a, b]$. Then the step function would be called a **simple function** in the terminology of Definition 2.2, notationally appearing as in (1.9) but with non-interval A_i-sets. Indeed, the same approach could be used to define the integral over any measurable set A, by decomposing this set into disjoint measurable subsets.*

3. *Finally, the real leap of imagination comes from looking again at (1.10). In the above development, we **derived** the value of this step function integral from the Riemann framework that was already in place. More generally, could we **define** the integral of a step function to be the result in (1.10)? Or in the most general case with a measure μ, could we **define** the integral of the function in (1.9) by:*

$$\int_A \varphi(x)d\mu \equiv \sum_{i=1}^n a_i \mu(A_i). \tag{1}$$

This definition would seem to make sense if $A = \bigcup_{j=1}^m A_i$ as a disjoint union. We could then proceed to define integrals of other functions as in (1.11).

*One detail will need to be addressed if (1) is to be considered a definition. Is this result **well-defined** given that step/simple functions can be expressed in multiple ways?*

Example 1.10 (Well-definedness of (1)) *For simplicity, we investigate well-definedness of (1) of Remark 1.9 in the context of step functions for which partitions are disjoint. See Proposition 2.7 for a more general derivation.*

Assume that $\varphi(x)$ can also be expressed by:

$$\varphi(x) = \sum\nolimits_{j=1}^{m} a'_j \chi_{A'_j}(x),$$

where $A'_j = (x'_{j-1}, x'_j]$ using the partition:

$$a = x'_0 < x'_1 < \cdots < x'_{m-1} < x'_m = b.$$

It must be verified that the integrals agree as defined in (1):

$$\sum\nolimits_{i=1}^{n} a_i \mu(A_i) = \sum\nolimits_{j=1}^{m} a'_j \mu(A'_j).$$

To prove this, consider the combined partition including both $\{x_i\}_{i=1}^{n}$ and $\{x'_j\}_{j=1}^{m}$, and associated intervals $B_{ij} \equiv A_i \cap A'_j$, noting that many such intervals will be empty. Define:

$$\bar{\varphi}(x) = \sum\nolimits_{j=1}^{m} \sum\nolimits_{i=1}^{n} b_{ij} \chi_{B_{ij}}(x),$$

where on any non-empty B_{ij} define $b_{ij} = a_i = a'_j$. Thus by (1):

$$(\mathcal{R}) \int_a^b \bar{\varphi}(x) dx = \sum\nolimits_{j=1}^{m} \sum\nolimits_{i=1}^{n} b_{ij} \mu(B_{ij}).$$

But $A_i = \bigcup_{j=1}^{m} B_{ij}$ and $A'_j = \bigcup_{i=1}^{n} B_{ij}$ as disjoint unions, and this obtains by finite additivity of measure μ that $\mu(A_i) = \sum_{j=1}^{m} \mu(B_{ij})$ and $\mu(A'_j) = \sum_{i=1}^{n} \mu(B_{ij})$.

This double summation can then be reordered to conclude that:

$$\sum\nolimits_{j=1}^{m} a'_j \mu(A'_i) = \sum\nolimits_{j=1}^{m} \sum\nolimits_{i=1}^{n} b_{ij} \mu(B_{ij}) = \sum\nolimits_{i=1}^{n} a_i \mu(A_i).$$

Consequently, the integral of a step function would be well-defined by (1).

While many generalizations will be seen to flow from the Darboux framework, perhaps the most important is the possibility of taking an axiomatic approach to integration theory as implied by item 3 of Remark 1.9. In this approach, we define the integral of step or simple functions as above, and then investigate how this definition extends to a broader class of functions. Specifically, we can investigate:

When does convergence of step/simple functions $\varphi_n(x) \to f(x)$ imply convergence of integrals $\int_A \varphi_n(x) d\mu \to \int_A \varphi(x) d\mu$?

And in this extension, we may even want to consider various definitions of convergence.

The Lebesgue and Riemann-Stieltjes integrals of this book, and the Lebesgue-Stieltjes integrals of Book V, will use this framework and the ordinary notion of convergence. Then in Book VIII, we will use this framework and also change the notion of convergence for Itô and other stochastic integrals.

1.1.3   Riemann and Darboux Equivalence

In this section, we prove that the Riemann sums Definition 1.2 and Darboux sums Definition 1.7 provide equivalent criteria for the existence of the integral of a bounded function. In other words, we prove that a bounded function is Riemann integrable if and only if it is Darboux integrable, and then the integrals agree.

We do this by showing that each is equivalent to (1.16), which is known as the **Cauchy criterion for the existence of the Riemann integral** and named after **Augustin-Louis Cauchy** (1789–1857). As might be appreciated given the simplicity of its statement, the Cauchy criterion is often the easiest to use when proving integrability. See Proposition 1.15 as an example.

Proposition 1.11 (Cauchy criterion for the Riemann integral) *A bounded function $f(x)$ is Riemann integrable on $[a, b]$ by Definition 1.2 if and only if it is Darboux integrable on this interval by Definition 1.7, and then the integrals agree.*

In either case, $f(x)$ is so integrable if and only for any $\epsilon > 0$ there is a partition $\{x_i\}_{i=0}^n$ of $[a, b]$, so that:

$$0 \leq \sum_{i=1}^n M_i \Delta x_i - \sum_{i=1}^n m_i \Delta x_i < \epsilon, \tag{1.16}$$

where $m_i \equiv \inf\{f(x) | x \in [x_{i-1}, x_i]\}$, $M_i \equiv \sup\{f(x) | x \in [x_{i-1}, x_i]\}$.

Further, (1.16) is then also satisfied for every refinement of $\{x_i\}_{i=0}^n$.

Proof. *As noted above, we show that the requirement of either definition of integrability is equivalent to the above Cauchy criterion, and thus each is equivalent to the other.*

Given $\epsilon > 0$, assume that (1.16) is satisfied. It follows by the definition of infimum and supremum that:

$$\sum_{i=1}^n m_i \Delta x_i \leq \sup_{\varphi_\mu \leq f} \int_a^b \varphi_\mu(x) dx \leq \inf_{f \leq \psi_\mu} \int_a^b \psi_\mu(x) dx \leq \sum_{i=1}^n M_i \Delta x_i.$$

Thus:

$$0 \leq \inf_{f \leq \psi_\mu} \int_a^b \psi_\mu(x) dx - \sup_{\varphi_\mu \leq f} \int_a^b \varphi_\mu(x) dx < \epsilon, \tag{1}$$

and since this is true for all ϵ, f is Darboux integrable on $[a, b]$ by Definition 1.7.

Now (1.15) implies that there exists μ' so that for every step function $\varphi'_{\mu_1}(x) = \sum_{i=1}^n m_i \Delta x_i$ with partition mesh size $\mu_1 \leq \mu'$:

$$\sup_{\varphi_\mu \leq f} \int_a^b \varphi_\mu(x) dx \leq \int_a^b \varphi'_{\mu_1}(x) dx + \epsilon/2.$$

Similarly, there exists μ'' so that for every step function $\psi''_{\mu_2}(x) = \sum_{i=1}^n M_i \Delta x_i$ with partition mesh size $\mu_2 \leq \mu''$:

$$\inf_{f \leq \psi_\mu} \int_a^b \psi_\mu(x) dx \geq \int_a^b \psi''_{\mu_2}(x) dx - \epsilon/2.$$

So by (1.16) and (1), if $\mu_3 \leq \min\{\mu', \mu''\}$:

$$0 \leq \int_a^b \psi''_{\mu_3}(x) dx - \int_a^b \varphi'_{\mu_3}(x) dx < 2\epsilon.$$

Thus given any such partition and step function, then for any interval tags $\{\tilde{x}_i\}_{i=1}^n$:

$$\int_a^b \varphi''_{\mu_3}(x) dx \leq \sum_{i=1}^n f(\tilde{x}_i) \Delta x_i \leq \int_a^b \varphi''_{\mu_3}(x) dx + 2\epsilon. \tag{2}$$

As (2) is satisfied for every partition with mesh size $\mu_3 \leq \min\{\mu', \mu''\}$, it is thus true taking a limit as $\mu_3 \to 0$. Since ϵ is arbitrary, f is Riemann integrable on $[a, b]$ by Definition 1.2.

Further, by (2) and (1), the Darboux and Riemann integrals agree.

Conversely, if f is Darboux integrable by Definition 1.7 and $\epsilon > 0$ is given, then again *by definition of extrema there exists partitions of $[a, b]$ of mesh sizes μ_1 and μ_2, and step functions $\varphi_{\mu_2} \leq f \leq \psi_{\mu_1}$, so that:*

$$0 \leq \int_a^b \psi_{\mu_1}(x)dx - (\mathcal{D})\int_a^b f(x)dx < \epsilon/2,$$

$$0 \leq (\mathcal{D})\int_a^b f(x)dx - \int_a^b \varphi_{\mu_2}(x)dx < \epsilon/2.$$

Thus:

$$0 \leq \int_a^b \psi_{\mu_1}(x)dx - \int_a^b \varphi_{\mu_2}(x)dx < \epsilon. \qquad (3)$$

It then follows from (1.6) that (3) is satisfied with both $\psi(x)$ and $\varphi(x)$ defined relative to the common refinement of these partitions, and this is (1.16).

Similarly, if f is Riemann integrable by Definition 1.2 and $\epsilon > 0$ is given, then there is a δ so that for all tagged partitions $\{\widetilde{x}_i, [x_{i-1}, x_i]\}_{i=1}^n$ of mesh size $\mu \leq \delta$:

$$\left| \sum_{i=1}^n f(\widetilde{x}_i)\Delta x_i - (\mathcal{R})\int_a^b f(x)dx \right| < \epsilon/3.$$

Thus for any such partition and any two sets of tags $\{\widetilde{x}_i\}_{i=1}^n$ and $\{\widetilde{x}_i'\}_{i=1}^n$:

$$\left| \sum_{i=1}^n f(\widetilde{x}_i)\Delta x_i - \sum_{i=1}^n f(\widetilde{x}_i')\Delta x_i \right| < 2\epsilon/3.$$

By definition of M_i and m_i, we can choose such tags so that these Riemann sums are within $\epsilon/6$ of $\sum_{i=1}^n M_i\Delta x_i$ and $\sum_{i=1}^n m_i\Delta x_i$, and this obtains (1.16).

Finally, if (1.16) is satisfied for $\{x_i\}_{i=0}^n$, it is also satisfied for every refinement by (1.6). ∎

1.1.4   On Existence of the Integral

The next question is one of existence. What types of functions are Riemann integrable? Intuitively, continuity of $f(x)$ on $[a, b]$ would seem sufficient to assure existence, and we will prove this. Again intuitively, one expects that continuity is not necessary. As an example, a single or even finite number if isolated discontinuities would likely not cause any trouble, and this result is assigned as an exercise. But then, how much discontinuity can be tolerated? We will see with Lebesgue's existence theorem.

The intuition for continuity is often supported by familiar graphical representations. If $f(x)$ is continuous and nonnegative, the Riemann sum is seen to approximate the area between the graph of $f(x)$ and the x-axis over the interval $[a, b]$, using a collection of rectangles. The upper and lower Darboux sums provide upper and lower estimates to this area.

For general continuous $f(x)$, these sums approximate a "signed" area between the graph of $f(x)$ and the x-axis over the interval $[a, b]$. Here, areas above the x-axis are positive, areas below are negative, and the sums approximate the net signed area.

Continuity of $f(x)$ would seem to assure that these areas are well defined, and that approximations and bounds will improve as $\mu \to 0$.

To proceed toward proofs, we first recall some definitions.

Definition 1.12 (Continuous function) *The function $f : \mathbb{R} \to \mathbb{R}$ is **continuous** at x_0 if:*

$$\lim_{x \to x_0} f(x) = f(x_0).$$

That is, given $\epsilon > 0$ there is a $\delta \equiv \delta(x_0, \epsilon) > 0$, so that:

$$|f(x) - f(x_0)| < \epsilon \ \text{whenever} \ |x - x_0| < \delta. \tag{1.17}$$

*A function is said to be **continuous on an interval** $[a, b]$ if it is continuous at each $x_0 \in (a, b)$, and also continuous at $x_0 = a$ and $x_0 = b$, where the bounds in (1.17) are understood as one-sided, meaning for $x < b$ or $x > a$. A function is said to be **continuous** if it is continuous everywhere on its domain.*

*The function $f : \mathbb{R} \to \mathbb{R}$ is **uniformly continuous on** $[a, b]$ if given $\epsilon > 0$ there is a $\delta \equiv \delta(\epsilon) > 0$ so that (1.17) is true for all $x, x_0 \in [a, b]$. A function is said to be **uniformly continuous** if this definition applies for all x, x_0 in its domain.*

The same definitions apply to a function $f : \mathbb{R}^n \to \mathbb{R}$, where $|x - x_0|$ is interpreted in terms of the standard distance function on $\mathbb{R}^n$:

$$|x - y| \equiv \left[\sum_{i=1}^{n} (x_i - y_i)^2 \right]^{1/2}. \tag{1.18}$$

Exercise 1.13 (Intermediate value theorem) *Assume that $f(x)$ is continuous on $[a, b]$, and let m, M be defined as in (1.5) relative to this interval. Prove that there exists $x_m, x_M \in [a, b]$ so that $f(x_m) = m$ and $f(x_M) = M$. Further, show that if $x_m < y < x_M$, then there exists $x_y \in [a, b]$ so that $f(x_y) = y$. This last result is known as the **intermediate value theorem**. Hint: By definition you can get arbitrarily close to m and M with points in $[a, b]$. Prove that these points must converge to a point in this closed interval, and now use continuity. For the other statement, consider the collection $\{x | f(x) \le y\}$, and let x_y equal the least upper bound. Note that $x_y \in [a, b]$ and thus f is continuous at this point.*

Generalize this result to $f(x)$ continuous on $\prod_{i=1}^{n} [a_i, b_i] \subset \mathbb{R}^n$.

Exercise 1.14 (When continuity $\Rightarrow$ uniform continuity) *Prove that if $f(x)$ is continuous on a closed and bounded interval $[a, b]$, then it is uniformly continuous and bounded on this interval. Hint: This is actually a deep result that requires the Heine-Borel theorem of Proposition I.2.27. Given $\epsilon > 0$, for each $x_0 \in [a, b]$ determine δ_{x_0} and the interval $(x_0 - \delta_{x_0}, x_0 + \delta_{x_0})$ on which $|f(x) - f(x_0)| < \epsilon$. These intervals are an open cover of $[a, b]$.*

Then show by example that uniform continuity does not follow if $f(x)$ is continuous on bounded (a, b), or continuous on an unbounded interval, say (a, ∞) or $[a, \infty)$.

Generalize: If $f(x)$ is continuous on a closed and bounded rectangle $R = \prod_{j=1}^{n} [a_j, b_j]$, then it is uniformly continuous and bounded on this rectangle. By example show that this conclusion does not follow on $R' \equiv \prod_{j=1}^{n} (a_j, b_j)$ or an unbounded rectangle such as $R' \equiv \prod_{j=1}^{n} (a_j, \infty)$. Hint: Let $f(x) = \prod_{j=1}^{n} f_j(x_j)$.

We are now ready for a fundamental existence result on Riemann integrability. Using the Cauchy criterion in (1.16), the proof is remarkably simple. The reader may be interested to verify that this proof is much less simple using the Riemann or Darboux criteria of Definitions 1.2 and 1.7.

Proposition 1.15 (Continuous $\Rightarrow$ Riemann integrable) *If $f(x)$ is continuous on $[a, b]$, then $(\mathcal{R}) \int_a^b f(x) dx$ exists.*

Proof. Let $\epsilon > 0$ be given. As $f(x)$ is uniformly continuous by Exercise 1.14, there exists $\delta > 0$ so that (1.17) is satisfied for all $x, x_0 \in [a, b]$. Thus given a partition $\{x_i\}_{i=0}^{n}$ with *mesh size $\mu < \delta(\epsilon)$, it follows that $|f(x_i'') - f(x_i')| < \epsilon$ for any $x_i', x_i'' \in [x_{i-1}, x_i]$. By Exercise 1.13, $M_i - m_i \leq \epsilon$ for all i, and so:*

$$0 \leq \sum_{i=1}^{n} M_i \Delta x_i - \sum_{i=1}^{n} m_i \Delta x_i \leq \epsilon(b - a),$$

which is (1.16). ∎

The continuity assumption in Proposition 1.15 is quite strong. For the next exercise, note the importance of assuming boundedness once we stray from continuity.

Exercise 1.16 (Bounded and finitely many discontinuities) *Prove that if $f(x)$ is bounded on $[a, b]$ and continuous except at $\{c_j\}_{j=1}^{m} \subset [a, b]$, then $(\mathcal{R}) \int_a^b f(x)dx$ exists. Hint: It is enough to prove the case $m = 1$, and it will be seen that nothing changes except notational complexity with finitely many discontinuities. If $c_1 \in (a, b)$, verify (1.16) by choosing partitions $\{x_i\}_{i=0}^{n}$ with $c_1 \in (x_{i-1}, x_i)$ for some i. This function is now uniformly continuous on $[a, x_{i-1}]$ and $[x_i, b]$, is bounded, and we can choose $x_i - x_{i-1}$ as small as we want. Now consider the case where c_1 is an endpoint.*

The proof in Exercise 1.16 falls apart when there are infinitely many discontinuities. For one thing, these discontinuities will of necessity have at least one accumulation point in this compact interval, and thus isolating them as in the Hint will be more challenging.

Now recall that there are different orders of infinity. The Dirichlet function, which is nowhere continuous, is not Riemann integrable as seen in Exercise 1.3. But there is a deceptively small modification of this function that makes the discontinuities infinite but countable, and also makes this function Riemann integrable as a result of Lebesgue's existence theorem below.

Example 1.17 (Thomae's function) *A function defined on $[0, 1]$ with a countable number of discontinuities is **Thomae's function**, named after **Carl Johannes Thomae** (1840–1921):*

$$f(x) = \begin{cases} 1, & x = 0, \\ 1/n, & x = m/n \text{ in lowest terms,} \\ 0, & x \text{ irrational.} \end{cases}$$

*This function modifies the Dirichlet function only on the rationals, but is now continuous except on the rationals. Recalling Definition I.3.17, Thomae's function is continuous m-a.e., or **almost everywhere relative to Lebesgue measure**.*

To prove this, we show that $\lim_{x \to x_0} f(x) = f(x_0) = 0$ for all irrational x_0. For any integer N, choose δ_N, so that:

$$\delta_N < \min\{|x_0 - m/n| \, | n \leq N\}.$$

This minimum exists because only finitely many rationals need be considered. Further, this minimum is greater than 0 because x_0 is irrational. Then since $f(x) = 0$ for irrational x and $f(x) < 1/N$ for rationals in this interval by construction, it follows that $|f(x) - f(x_0)| = f(x) < 1/N$ for all x with $|x - x_0| < \delta$.

That $f(x)$ is discontinuous on the rationals follows from the observation that given m/n and any δ, there are infinitely many irrationals x with $|m/n - x| < \delta$, and thus $|f(m/n) - f(x)| = 1/n$. Choosing $\epsilon < 1/n$ obtains the conclusion.

It turns out that $(\mathcal{R}) \int_a^b f(x)dx$ exists and equals 0, but to prove this is to implement the general construction for Proposition 1.22 below.

The **Lebesgue existence theorem for the Riemann integral,** named after **Henri Lebesgue** (1875–1941) who first proved it, states that a bounded function $f(x)$ is Riemann integrable over a compact interval $[a, b]$ if and only if it is continuous except on a set of Lebesgue measure 0.

For example, this result applies to any bounded function with at most countably many discontinuities, since every countable set has Lebesgue measure 0. But it should be noted that this result does not require that the set of discontinuities be countable, only that this set has measure 0.

While Lebesgue measure is developed in Chapter I.2, sets of Lebesgue measure 0 can be characterized simply as in the definition below. Comparing with Definition I.2.25, the set E below is said to have Lebesgue measure 0 if $m^*(E) = 0$, with m^* denoting Lebesgue outer measure of this definition. However, $m \equiv m^*$ on the complete sigma algebra $\mathcal{M}_L(\mathbb{R})$ by Definition I.2.40, and $\mathcal{M}_L(\mathbb{R})$ contains every set with $m^*(E) = 0$ by Proposition I.2.35. Thus $m(E) = 0$ if and only if $m^*(E) = 0$.

Definition 1.18 (Set of measure 0 **in** $\mathbb{R}$**)** *A set $E \subset \mathbb{R}$ has **measure 0,** or more formally **Lebesgue measure 0,** if for any $\epsilon > 0$ there is a finite or countable collection of open intervals $\{G_i\}$, with $G_i = (a_i, b_i)$, so that $E \subset \bigcup_i G_i$ and $\sum_i m(G_i) < \epsilon$, where $m(G_i) \equiv b_i - a_i$ is the Lebesgue measure of (a_i, b_i).*

To obtain a characterization of continuity that is more usable for Lebesgue's existence result, we require the notion of the **oscillation of a function** at a point x.

Definition 1.19 (Oscillation of $f(x)$ **at** x**)** *Given an open interval $I = (x_{i-1}, x_i)$, the **oscillation of** $f(x)$ **on** I, denoted $\omega(I)$, is defined:*

$$\omega(I) = M_i - m_i,$$

where M_i and m_i are respectively defined as in (1.5), as the least upper bound (l.u.b.) and greatest lower bound (g.l.b.) of $f(x)$ on the interval I.

*The **oscillation of** $f(x)$ **at** x, denoted $\omega(x)$, is defined:*

$$\omega(x) = \inf\{\omega(I) | x \in I\}.$$

Exercise 1.20 (Characterizing the discontinuity set) *Define:*

$$E_N = \{x | \omega(x) \geq 1/N\}, \qquad E \equiv \bigcup_{N \geq 1} E_N = \{x | \omega(x) > 0\}.$$

Using Definition 1.19, show that x is a discontinuity point of $f(x)$ if and only if $\omega(x) > 0$, and thus E is the collection of discontinuities of $f(x)$.

Prove that E_N is closed for all N, and thus E is a Borel set, $E \in \mathcal{B}(\mathbb{R})$, and specifically an $\mathcal{F}_\sigma$-set (Notation I.2.16). Hint: Verify that a set F is closed (Definition I.2.10) if and only if every convergent sequence in F converges to a point in F. Now show that if $\omega(x_n) \geq 1/N$ for all n and $x_n \to x$, then $\omega(x) \geq 1/N$.

The final step is to connect the oscillation of $f(x)$ with the difference between the bounding summations in (1.16). This result provides another proof of Proposition 1.15, that every continuous function satisfies Cauchy's criterion on compact intervals, and is thus Riemann integrable by Proposition 1.11.

Proposition 1.21 (Oscillation and the Cauchy criterion) *If $\omega(x) < k$ for all $x \in [c, d]$, then for some partition:*

$$\sum_{i=1}^{n} (M_i - m_i) \Delta y_i < k(d - c). \tag{1.19}$$

Proof. *Given $x \in [c, d]$, choose an open interval I_x so that $\omega(I_x) < k$, which is possible by definition of $\omega(x)$. As $\{I_x\}_{x \in [c,d]}$ is an open cover of compact $[c, d]$, the Heine-Borel theorem of Proposition I.2.27 obtains a finite subcover $\{I_{x_j}\}_{j=1}^{n}$, where the collection $\{x_j\}_{j=1}^{n}$ can be assumed to be in increasing order, so $y_0 \equiv c \in I_{x_1}$ and $y_n \equiv d \in I_{x_n}$.*

Now $I_{x_j} \bigcap I_{x_{j+1}}$ is an open interval for each $j = 1, ..., n-1$, so choose $y_j \in I_{x_j} \bigcap I_{x_{j+1}}$, for example choose the midpoints of such intervals. Then $\{[y_{j-1}, y_j]\}_{j=1}^{n}$ is a partition of $[c, d]$, and since $[y_{j-1}, y_j] \subset I_{x_j}$ for all j, it follows that $M_j - m_j \leq \omega(I_{x_j}) < k$. Then (1.19) follows from $\sum_{j=1}^{n} \Delta y_j = d - c$. ∎

We now have the tools to prove Lebesgue's existence result, which incidentally proves the Riemann integrability of Thomae's function of Example 1.17.

Proposition 1.22 (Lebesgue's existence theorem) *If $f(x)$ is a bounded function on the finite interval $[a, b]$, then:*

$$(\mathcal{R}) \int_a^b f(x) dx$$

exists if and only if $f(x)$ is continuous almost everywhere. That is, if $f(x)$ is continuous except on a set of points $E \equiv \{x_\alpha\}$ of Lebesgue measure 0.

Proof. *If this integral exists, we show that E_N has measure 0 for each N, and thus as a countable union, E has measure 0 by countable subadditivity of Lebesgue measure.*

By boundedness of $f(x)$ and the Cauchy criterion of Proposition 1.11, for any $\epsilon > 0$ and integer N there exists a partition of $[a, b]$, so that:

$$\sum_{i=1}^{n} (M_i - m_i) \Delta x_i < \frac{\epsilon}{N}.$$

If any of the discontinuity points $x \in E_N$ of $f(x)$ are among the partition endpoints, these points have measure 0 and can be ignored.

Focusing on the open intervals defined as the interiors of this partition's intervals, we denote by $\{I_j\}_{j=1}^{m}$, the subset of open intervals that have at least one discontinuity point. On any such interval, $M_j - m_j \geq 1/N$ by definition of E_N, since $1/N$ is the greatest lower bound over all open intervals that contain these discontinuities. Thus as a subset of the partition, recalling that $m(I_j) = \Delta x_j$:

$$\frac{1}{N} \sum_{j=1}^{m} m(I_j) \leq \sum_{i=1}^{n} (M_i - m_i) \Delta x_i < \frac{\epsilon}{N}. \tag{1}$$

By (1), the discontinuities in E_N can be covered by intervals of arbitrarily small total interval length, and so $m(E_N) = 0$ by Definition 1.18.

Needed below, the inequality in (1) is also satisfied using the closed intervals $\{\bar{I}_j\}_{j=1}^{m}$ since $m(I_j) = m(\bar{I}_j) = \Delta x_j$.

Now assume that $f(x)$ is bounded on $[a, b]$ and that $m(E) = 0$. Then $m(E_N) = 0$ for any N, and so for any $\epsilon > 0$ there exists an at most countable collection of open intervals $\{I_j\}$ with $E_N \subset \bigcup I_j$ and $\sum m(I_j) < \epsilon$. As E_N is closed by Exercise 1.20 and bounded, it is compact.

By the Heine-Borel theorem of Proposition I.2.27, there exists a finite subcollection $\{I_j\}_{j=1}^{m}$ with $E_N \subset \bigcup_{j=1}^{m} I_j$, and:

$$\sum_{j=1}^{m} m(\bar{I}_j) < \epsilon, \tag{2}$$

using the closed intervals $\{\bar{I}_j\}_{j=1}^{m}$ since $m(I_j) = m(\bar{I}_j) = \Delta x_j$.

Now $[a, b] - \bigcup_{j=1}^{m} I_j$ is a union of at most $m + 1$ closed disjoint intervals, say $\{K_i\}_{i=1}^{m+1}$, *and $\omega(x) < 1/N$ on these intervals since these are in the complement of E_N. By Proposition 1.21, there is a partition of each such K_i for which (1.19) is satisfied with $k = 1/N$, and thus with m' the total number in all partitions of $\{K_i\}_{i=1}^{m+1}$:*

$$\sum_{i=1}^{m'} (M_i - m_i)\, \Delta x_i < \frac{1}{N} \sum_{i=1}^{m'} m(K_i). \tag{3}$$

Partition $[a, b]$ with $\{\bar{I}_j\}_{j=1}^m$ and the above partitions of $\{K_i\}_{i=1}^{m+1}$. Noting that by boundedness of $f(x)$ that $M_i - m_i \leq M - m$ for all i with m, M defined on $[a, b]$, it follows from (2) and (3):

$$\sum (M_i - m_i)\, \Delta x_i < \frac{1}{N} \sum_{i=1}^{m'} m(K_i) + (M - m)\, \epsilon.$$

As $\epsilon > 0$ and N are arbitrary, this proves the existence of $(\mathcal{R}) \int_a^b f(x)dx$ by the Cauchy criterion of Proposition 1.11. ∎

1.1.5   Properties of the Integral

The Lebesgue existence theorem for the Riemann integral of Proposition 1.22 states that a bounded function is Riemann integrable over a compact interval $[a, b]$ if and only if it is continuous almost everywhere. Knowing that this integral exists then implies by Proposition 1.11 that it exists by any and all criteria as specified in that result. For the proofs below, the Riemann criterion of Definition 1.2 often provides the simplest approach. We sketch the proof with details left as an exercise.

Note that item 5 of this result cannot, in general, be extended to infinite disjoint unions of intervals. See the discussion following.

Proposition 1.23 (Properties of the Riemann integral) *If $f(x)$ and $g(x)$ are bounded and Riemann integrable on $[a, b]$, then suppressing the $(\mathcal{R})$-notation:*

1. **Triangle inequality:** *The integral $\int_a^b |f(x)|\, dx$ exists, and if $|f(x)| \leq c$ on $[a, b]$:*

$$\left| \int_a^b f(x)dx \right| \leq \int_a^b |f(x)|\, dx \leq c(b - a).$$

   Further, if $c_1 \leq f(x) \leq c_2$:

$$c_1(b - a) \leq \int_a^b f(x)dx \leq c_2(b - a).$$

2. *For any constants k_1 and k_2, $k_1 f(x) + k_2 g(x)$ is Riemann integrable and:*

$$\int_a^b (k_1 f(x) + k_2 g(x))\, dx = k_1 \int_a^b f(x)dx + k_2 \int_a^b g(x)dx.$$

3. *If $[c, d] \subset [a, b]$, then $\int_c^d f(x)dx$ exists.*

4. *If $a < c < b$, then $\int_a^c f(x)dx$ and $\int_c^b f(x)dx$ exist, and:*

$$\int_a^b f(x)dx = \int_a^c f(x)dx + \int_c^b f(x)dx.$$

5. *For a finite disjoint union $E \equiv \bigcup_{i=1}^{n} [c_j, d_j] \subset [a, b]$, the integral $\int_E f(x) dx$ is well defined by:*

$$\int_E f(x) dx \equiv \int_a^b \chi_E(x) f(x) dx,$$

*with the **characteristic function** of E:*

$$\chi_E(x) \equiv \begin{cases} 1, & x \in E, \\ 0, & x \notin E. \end{cases} \tag{1.20}$$

Further:

$$\int_E f(x) dx = \sum_{i=1}^{n} \int_{c_i}^{d_i} f(x) dx.$$

Proof. *By assumption and the Lebesgue existence theorem, $f(x)$ and $g(x)$ are bounded and continuous almost everywhere. The proof of integrability of $|f(x)|$ in item 1 follows from the observation that a continuity point of f must be a continuity point of $|f|$, so $|f|$ is also bounded and continuous almost everywhere. The bounds for this integral then follow from a consideration of Riemann sums from Definition 1.2, and the triangle inequality.*

For item 2, $k_1 f(x) + k_2 g(x)$ is Riemann integrable since the integral identity stated is true for every Riemann sum, and thus this identity is preserved in the limit. A detail to address is that the Riemann sums for the integrals on the right need not have the same partitions. So given two such partitions, derive the combined partition, and note that each of these Riemann sums converges to the same integrals starting with the original or combined partitions.

For item 3, with $\chi_{[c,d]}(x)$ defined as in (1.20), the function $\chi_{[c,d]}(x) f(x)$ is bounded and has at most two more discontinuities than does $f(x)$ on $[a, b]$. Thus this function is Riemann integrable, and by Riemann sums:

$$\int_c^d f(x) dx = \int_a^b \chi_{[c,d]}(x) f(x) dx.$$

Item 4 is derived similarly, using $\chi_{[a,c]}(x)$ and $\chi_{[c,b]}(x)$, while item 5 uses $\chi_E(x) = \sum_{i=1}^{n} \chi_{[c_j, d_j]}(x)$ and item 2. ∎

Item 1 above has a simple yet useful corollary. Continuity is essential for this result, and the reader is invited to provide examples of Riemann integrable functions for which this result fails.

Corollary 1.24 (Mean value theorem for integrals) *If $f(x)$ is continuous on $[a, b]$, then there exists $c \in [a, b]$ so that:*

$$\int_a^b f(x) dx = f(c)(b - a). \tag{1.21}$$

Proof. *Continuity of $f(x)$ obtains by Exercise 1.13 that $m \leq f(x) \leq M$ for all $x \in [a, b]$, and $f(x)$ achieves these extremes. Then by item 1 of Proposition 1.23:*

$$m \leq \frac{1}{b - a} \int_a^b f(x) dx \leq M.$$

By the intermediate value theorem of the same exercise, there exists $c \in [a, b]$ so that (1.21) is satisfied. ∎

Exercise 1.25 (Absolute Riemann integrable $\not\Rightarrow$ Riemann integrable) *It is tempting to try to generalize item 1 of Proposition 1.23 to obtain that if $|f(x)|$ is Riemann integrable on $[a, b]$, then so too is $f(x)$ Riemann integrable on $[a, b]$, and that the above bound applies. Prove by example that this is not true. Hint: Can $|f(x)|$ be continuous everywhere and $f(x)$ be continuous nowhere?*

When $f(x)$ is not bounded, or the interval is not compact, some of the above seemingly natural properties are no longer necessarily valid.

To exemplify, recall that an integral such as $\int_a^\infty f(x)dx$ is called an **improper integral** and is defined to exist when the limit of integrals over bounded intervals $[a, N]$ exists as $N \to \infty$. In this case we define:

$$(\mathcal{R}) \int_a^\infty f(x)dx \equiv \lim_{N \to \infty} \int_a^N f(x)dx. \tag{1.22}$$

Similarly, when the double limit exists,

$$(\mathcal{R}) \int_{-\infty}^\infty f(x)dx \equiv \lim_{M, N \to \infty} \int_{-M}^N f(x)dx. \tag{1.23}$$

Exercise 1.26 (Item 2, Proposition 1.23) *Show that if the improper integrals of $f(x)$ and $g(x)$ exist in either sense above, then so too does the improper integral of $k_1 f(x) + k_2 g(x)$, and item 2 remains true.*

Example 1.27 (Proposition 1.23 generalization failures) *Weakening assumptions in Proposition 1.23 can lead to failed conclusions.*

Item 1: By the Lebesgue existence theorem, any example with almost everywhere continuous f must be an example where f is unbounded, or the interval is unbounded. In such examples, the integral of $f(x)$ may exist only due to positive and negative cancellations on subintervals. The situation here is reminiscent of examples of conditionally convergent series that are not absolutely convergent.

For example, define $f(x)$ on $[0, \infty)$ by:

$$f(x) = (-1)^{n+1}/n, \text{ for } n - 1 \le x < n.$$

Then $f(x)$ is bounded and continuous except on the positive integers, a set of measure 0. Further,

$$\int_0^\infty f(x)dx \equiv \lim_{N \to \infty} \int_0^N f(x)dx = \sum_{n=1}^\infty (-1)^{n+1}/n.$$

This summation is well defined and finite, and in fact equal to $\ln 2$ as is verified by the Taylor series expansion for $\ln(1 + x)$.

However,

$$\int_0^\infty |f(x)| \, dx = \sum_{n=1}^\infty 1/n,$$

the harmonic series, which diverges.

Item 5: Using $f(x)$ above, it can be shown as an exercise that while $\int_0^\infty f(x)dx$ exists, the integral $\int_{A_N} f(x)dx$ is unbounded in N for either $A_N \equiv \bigcup_{i=0}^N [2i, 2i + 1]$ or $A_N \equiv \bigcup_{i=1}^N [2i - 1, 2i]$, diverging to ∞, respectively $-\infty$, as $N \to \infty$. Thus item 5 cannot be generalized to infinite unions of subintervals of the unbounded interval $[0, \infty)$.

Exercise 1.28 *Develop an almost everywhere continuous function $f(x)$ on the compact interval $[0,1]$ that is Riemann integrable, but where $|f(x)|$ is not. Hint: This function cannot be bounded. Consider a countable partition of $[0,1]$ and define $f(x) = a_n$ on each interval so that $\int_0^1 f(x)dx$ is the alternating harmonic series.*

Show how item 5 also fails for this example, in the sense that it cannot be generalized to infinite disjoint unions of subintervals.

It should be noted that the potential for failure is greater than that illustrated by this example and exercise. The insights come from the **Riemann series theorem** by **Bernhard Riemann** (1826–1866), which applies to every convergent series that is not absolutely convergent.

To set the stage, recall that when a series is absolutely convergent, the value of the sum is independent of the ordering of the series, and is thus independent of so-called series rearrangements. When convergent but not absolutely convergent, Riemann's theorem states that such a series can be rearranged to obtain any value in $\mathbb{R} \bigcup \{\pm\infty\}$.

The idea underlying the proof is that such a series is composed of two divergent series, one diverging to ∞, one to $-\infty$. This is seen above for the alternating harmonic series. Thus by rearranging, the series terms can be made to converge to any value. See **Reitano** (2010), for example.

Using the same idea for $f(x)$ in Example 1.28, one can show that for any $\alpha \in \mathbb{R} \bigcup \{\pm\infty\}$, there exists an ordered collection of intervals $\{[n_j, n_j + 1]\}_{j=1}^{\infty}$, so that with $A_N \equiv \bigcup_{j=1}^{N} [n_j, n_j + 1]$:

$$\int_{A_N} f(x)dx \to \alpha.$$

This was illustrated for $\alpha = \pm\infty$ as a failure of item 5 in Example 1.27.

1.1.6 Evaluating Integrals

There are a limited number of functions of interest for which the value of a Riemann integral can be evaluated directly or happily using only the definitions. There are many "tricks" employed in a typical calculus book for reducing, or otherwise converting such an integral for easier valuation. At the heart of most such methods is the so-called Fundamental Theorem of Calculus, often abbreviated FTC. This foundational result is sometimes stated with the respective qualifiers of Version I and Version II.

What we will call Version I of the fundamental theorem relates the integral of a derivative to the underlying function. Recall the notion of **continuously differentiable**.

Definition 1.29 (Continuously differentiable) *A function $f(x)$ is said to be **continuously differentiable on** $[a,b]$, or to be a **differentiable function with** $f'(x)$ **continuous on** $[a,b]$, if $f'(x)$ exists and is continuous on (a,b), if $f'(x)$ exists at $x = a, b$ as a one-sided derivative, and that so defined, $f'(x)$ is continuous on $[a,b]$.*

One analogously applies this terminology to partial derivatives, and to higher order derivatives.

Proposition 1.30 (FTC Version I: Integral of a Derivative) *Let $f(x)$ be a differentiable function with $f'(x)$ continuous on $[a,b]$. Then*

$$(\mathcal{R}) \int_a^b f'(x)dx = f(b) - f(a). \tag{1.24}$$

Proof. *This integral exists by continuity of $f'(x)$ and Proposition 1.15, and thus equals a limit of Riemann sums:*

$$\int_a^b f'(x)dx = \lim_{\mu \to 0} \sum_{i=1}^n f'(\widetilde{x}_i)\Delta x_i,$$

where $\Delta x_i \equiv x_i - x_{i-1}$ and tags $\widetilde{x}_i \in [x_{i-1}, x_i]$ can be arbitrarily chosen. By the mean value theorem choose $\widetilde{x}_i \in (x_{i-1}, x_i)$ so that $f'(\widetilde{x}_i)\Delta x_i = f'(x_i) - f'(x_{i-1})$, and the proof is complete since these sums "telescope" to the result in (1.24). ■

With a change of notation we obtain the version of this theorem most familiar to calculus students. In this context, $F(x)$ is called an **antiderivative** of the integrand $f(x)$, which means that $F'(x) = f(x)$. Antiderivatives are only unique up to an additive constant, since if $F'(x) = G'(x)$, then $F(x) = G(x) + c$.

Proposition 1.31 (FTC Version I: Antiderivative of an Integrand) *Let $f(x)$ be a continuous function on $[a, b]$, and $F(x)$ any function with $F'(x) - f(x)$. Then:*

$$(\mathcal{R}) \int_a^b f(x)dx = F(b) - F(a). \tag{1.25}$$

Thus to evaluate the integral of $f(x)$, it is enough to identify an antiderivative. But this identification can be difficult and sometimes impossible. The various tricks referenced above are typically implemented to convert the integrand to one for which this identification is simplified.

Rewriting the statement in (1.24) by substituting the notationally fixed b by the variable x, this result can be restated as follows.

Proposition 1.32 (FTC Version I: Integral of a Derivative) *Every continuously differentiable function $f(x)$ on $[a, b]$ can be expressed as the Riemann integral of its derivative:*

$$f(x) = f(a) + (\mathcal{R}) \int_a^x f'(y)dy. \tag{1.26}$$

Version II contemplates the derivative of a Riemann integral with variable upper limit of integration.

Proposition 1.33 (FTC Version II: Derivative of an Integral) *Let $f(x)$ be a continuous function on $[a, b]$. Define $F(x)$ on this interval with arbitrary value $F(a)$:*

$$F(x) = F(a) + (\mathcal{R}) \int_a^x f(y)dy. \tag{1.27}$$

Then $F(x)$ is differentiable on (a, b) and:

$$F'(x) = f(x). \tag{1.28}$$

If $f(x)$ is continuous and Riemann integrable on $(-\infty, b]$ and $F(x)$ defined in (1.27) with $a = -\infty$, then (1.28) remains valid.

More generally, if $f(x)$ is bounded and Riemann integrable on $[a, b]$, and hence continuous except possibly on a set of measure 0, then $F(x)$ in (1.27) is differentiable almost everywhere on (a, b), and (1.28) is satisfied at each continuity point of $f(x)$.

Proof. *This integral exists for any $x \in [a, b]$ by continuity of $f(x)$ and Proposition 1.15. If $a \leq x < x + h \leq b$, then by Proposition 1.23:*

$$\frac{F(x + h) - F(x)}{h} = \frac{1}{h} \int_x^{x+h} f(y) dy. \tag{1}$$

The mean value theorem of Corollary 1.24 obtains that there exists $c_h \in [x, x + h]$ so that:

$$\frac{F(x + h) - F(x)}{h} = f(c_h).$$

By continuity of $f(x)$, $f(c_h) \to f(x)$ as $h \to 0$. The same result is obtained if $a \leq x + h < x \leq b$, and thus $F(x)$ is differentiable with $F'(x) = f(x)$.

For the case $a = -\infty$, the same proof works since $f(x)$ is assumed integrable.

If $f(x)$ is bounded and Riemann integrable on $[a, b]$, then (1) remains true, and thus:

$$\frac{F(x + h) - F(x)}{h} - f(x) = \frac{1}{h} \int_x^{x+h} g(y) dy,$$

where $g(y) = f(y) - f(x)$.

Now f is continuous almost everywhere by Lebesgue's existence theorem, and is continuous at x if and only if g is continuous at x, and then $g(x) = 0$. The proof is completed by proving that if $g(y)$ is continuous at x with $g(x) = 0$, then as $h \to 0$:

$$\frac{1}{h} \int_x^{x+h} g(y) dy \to 0.$$

By continuity at x, given $\epsilon > 0$ there exists δ so that $|g(y) - g(x)| = |g(y)| < \epsilon$ for $|y - x| < \delta$. Then by Proposition 1.23, for all $\delta' \leq \delta$:

$$\left| \frac{1}{\delta'} \int_x^{x+\delta'} g(y) dy \right| \leq \frac{1}{\delta'} \int_x^{x+\delta'} |g(y)| \, dy < \epsilon.$$

As this is true for all ϵ, the proof is complete. ∎

Remark 1.34 (I vs. II) *The notation for Version II of the fundamental theorem is different from that of Version I, and this is deliberate because they address different questions.*

In Version I we are given $f(x)$ and $f'(x)$, and the fundamental theorem relates these by a Riemann integral as in (1.24) and (1.26). The same statement is made in (1.25), but changing notation to $F(x)$ and $F'(x) = f(x)$.

For Version II, we are given the integrand $f(x)$, which is integrated to produce a new function $F(x)$, for which we have no a priori reason to suspect is related by differentiation to $f(x)$. The conclusion in (1.28) is that it is so related, and indeed $F'(x) = f(x)$.

Remark 1.35 (On $(\mathcal{R}) \int_b^a f(x) dx$, $a < b$) *Given $a < b$, the Riemann integral $(\mathcal{R}) \int_b^a f(x) dx$ is not formally definable in terms of Riemann or Darboux sums, but can be defined consistently using properties above:*

$$\int_b^a f(x) dx \equiv - \int_a^b f(x) dx. \tag{1.29}$$

This follows formally from the fundamental theorem above in (1.25):

$$- \int_a^b f(x) dx = F(a) - F(b) = \int_b^a f(x) dx.$$

With this definition, recall item 4 of Proposition 1.23, that if $a < c < b$, then:

$$\int_a^b f(x)dx = \int_a^c f(x)dx + \int_c^b f(x)dx.$$

This then remains true for $c < a < b$, or $a < b < c$, given (1.29).

Besides providing the value of the integral $(\mathcal{R}) \int_a^b f(y)dy$ in (1.25) when the antiderivative $F(x)$ of $f(x)$ can be identified, Version I in (1.24) provides a justification for two of the most useful techniques for integrating complicated expressions.

1. **Integration by parts:** If it is assumed that $f(x)$ and $g(x)$ are continuously differentiable functions, and hence $(f(x)g(x))'$ is continuous, then by (1.24):

$$(\mathcal{R}) \int_a^b (f(x)g(x))'dx = f(b)g(b) - f(a)g(a).$$

But then, since $(f(x)g(x))' = f'(x)g(x) + f(x)g'(x)$, this formula can be expressed using Proposition 1.23:

$$(\mathcal{R}) \int_a^b f'(x)g(x)dx = f(b)g(b) - f(a)g(a) - (\mathcal{R}) \int_a^b f(x)g'(x)dx. \qquad (1.30)$$

This identity is known as **integration by parts** because we are "partially" integrating $f'(x)g(x)$ by just integrating $f'(x)$, but then we must differentiate $g(x)$. Differentiation often simplifies functions, so this can often be used effectively with some forethought.

A simple example and exercise.

Example 1.36 ($\int_1^e \ln x dx$ **)** *To evaluate this integral, let $g(x) = \ln x$ and $f'(x) = 1$. Then $g'(x) = 1/x$ and $f(x) = x + c$ for any constant c. For the definite integral here, this c term will cancel out and is typically set equal to 0. Then:*

$$\int_1^e \ln x dx = e \ln e - 1 \ln 1 - \int_1^e 1 dx = 1.$$

Exercise 1.37 *Evaluate* $\int_0^\infty x^3 e^{-x} dx$.

Integration by parts can be generalized to differentiable functions $f(x)$ and $g(x)$ but then we must assume that $f'(x)$ and $g'(x)$ are continuous almost everywhere to justify these integrals, recalling Proposition 1.22. This technique will also be found in other integration theories. For example, in this book we develop versions for both the Lebesgue and Riemann-Stieltjes integrals.

2. **Method of substitution:** Assume that $g(x)$ is continuous with antiderivative $G(x)$, and $h(x)$ is continuously differentiable. Then $(G(h(x)))'$ is continuous, and so by (1.24):

$$(\mathcal{R}) \int_a^b (G(h(x)))' dx = G(h(b)) - G(h(a)).$$

Since $(G(h(x)))' = g(h(x))h'(x)$, this can be written:

$$(\mathcal{R}) \int_a^b g(h(x))h'(x)dx = G(h(b)) - G(h(a)). \tag{1.31}$$

To apply this for a given integral $(\mathcal{R}) \int_a^b f(x)dx$, the trick is to identify $g(x)$ and $h(x)$ so that $f(x) = g(h(x))h'(x)$.

The name **method of substitution** comes from the notational manipulation one performs in the original integral, formally "substituting" $u = h(x)$, defining $du = h'(x)dx$, and noting that the u-limits of integration are $h(a)$ and $h(b)$:

$$(\mathcal{R}) \int_a^b g(h(x))h'(x)dx = (\mathcal{R}) \int_{h(a)}^{h(b)} g(u)du.$$

The goal of the substitution is to see if this obtains a simpler integral in u, whereby one can identify the antiderivative $G(u)$. Reassembling the pieces obtains (1.31).

This approach will see significant generalizations in the general integration theory of Book V, there under the terminology of "change of variables."

Example 1.38 ($\int_0^{2\pi} \sin^{10} x \cos x dx$) *Here we recognize that* $\cos x$ *is the derivative of* $\sin x$ *and so the substitution* $u = \sin x$ *obtains:*

$$\int_0^{\pi/2} \sin^{10} x \cos x dx = \int_0^1 u^{10} du = \frac{1}{11}.$$

Exercise 1.39 *Evaluate* $\int_0^\infty x^3 e^{-x^2} dx$. *Hint: Use both methods.*

The final result relating Riemann integrals and derivatives is called the **Leibniz integral rule,** and named after **Gottfried Wilhelm Leibniz** (1646–1716).

Proposition 1.40 (Leibniz integral rule) *Assume that* $a(x)$ *and* $b(x)$ *are continuously differentiable on* $[x_0, x_1]$, *and that* $f(x,t)$ *and the* x-*partial derivative* $f_x(x,t) \equiv \frac{\partial f(x,t)}{\partial x}$ *are continuous for* $a(x) \le t \le b(x)$ *for* $x_0 \le x \le x_1$.
    Then:

$$F(x) \equiv (\mathcal{R}) \int_{a(x)}^{b(x)} f(x,t)dt,$$

is differentiable on $x_0 \le x \le x_1$, *with:*

$$F'(x) = (\mathcal{R}) \int_{a(x)}^{b(x)} f_x(x,t)dt + f(x,b(x))b'(x) - f(x,a(x))a'(x). \tag{1.32}$$

Proof. *Assume that we have proved (1.32) for constant* $a(x) = u$ *and* $b(x) = v$, *and thus if* $F(x,u,v) \equiv \int_u^v f(x,t)dt$ *then:*

$$\frac{\partial F(x,u,v)}{\partial x} = \int_u^v f_x(x,t)dt. \tag{1}$$

For $f(x,t)$ continuous as assumed, $F(x,u,v)$ is also differentiable in u and v for $a(x) \leq u, v \leq b(x)$ by (1.28) and (1.29). Thus $F(x) \equiv F(x, a(x), b(x))$ is differentiable in x by the chain rule since $a(x)$ and $b(x)$ are differentiable. Using the chain rule, (1), and then two applications of (1.28):

$$F'(x) = \frac{\partial F(x,u,v)}{\partial x} + \frac{\partial F(x,u,v)}{\partial u}\frac{da}{dx} + \frac{\partial F(x,u,v)}{\partial v}\frac{db}{dx}$$

$$= \int_{a(x)}^{b(x)} f_x(x,t)dt + f(x, b(x))b'(x) - f(x, a(x))a'(x),$$

which is (1.32).

Simplifying notation, let $F(x) \equiv \int_a^b f(x,t)dt$, where we assume that $f(x,t)$ and $f_x(x,t)$ are continuous for $a \leq t \leq b$ and $x_0 \leq x \leq x_1$. Then (1) is equivalent to:

$$\lim_{h \to 0} \left[\frac{F(x+h) - F(x)}{h} - \int_a^b f_x(x,t)dt \right] = 0, \tag{2}$$

for $x \in [x_0, x_1]$. As noted above, for $x = x_0$ we restrict $h > 0$, and for $x = x_1$ we restrict $h < 0$.

Now by definition:

$$F(x+h) - F(x) = \int_a^b [f(x+h,t) - f(x,t)]\, dt,$$

while by continuity of $f(x,t)$ and the mean value theorem, there exists $0 < \theta < 1$ with $\theta \equiv \theta(x,h,t)$ so that:

$$f(x+h,t) - f(x,t) - hf_x(x+\theta h, t).$$

Rewriting (2), it must be proved that:

$$\lim_{h \to 0} \int_a^b [f_x(x+\theta h, t) - f_x(x,t)]\, dt = 0. \tag{3}$$

Since $f_x(x,t)$ is continuous in $[x_0, x_1] \times [a, b]$, it is uniformly continuous by Exercise 1.14. Thus given $\epsilon > 0$, there exists δ so that $|f_x(x+\theta h, t) - f_x(x,t)| < \epsilon$ if $|(x+\theta h, t) - (x,t)| = \theta |h| < \delta$. Taking $|h| < \delta$ accomplishes this, and thus by item 1 of Proposition 1.23, if $|h| < \delta$:

$$\left| \int_a^b [f_x(x+\theta h, t) - f_x(x,t)]\, dt \right| \leq \int_a^b |f_x(x+\theta h, t) - f_x(x,t)|\, dt < \epsilon(b-a).$$

As ϵ is arbitrary, this proves (3). ∎

1.2   The Integral and Function Limits

An important question that often arises in mathematics is related to a sequence of functions $\{f_n(x)\}_{n=1}^\infty$ that are known to converge in some sense to a function $f(x)$. If each function in the sequence is known to have a certain property, can it then be concluded that $f(x)$ will also have this property? In a typical application, the functions in the sequence are simple enough in some sense to directly establish that they have the identified property. The question of whether we can infer that this property is also then shared by $f(x)$ is important because it might be very difficult to establish this conclusion directly.

For example, in Proposition I.3.47, it was proved that Lebesgue measurability is preserved under **pointwise convergence** of Lebesgue measurable $f_n(x)$ to $f(x)$ on a Lebesgue measurable set E, and this generalizes to measurability defined with respect to any measure. Thus for example, if $f(x)$ is the pointwise limit of simple functions, which are measurable by definition, then such $f(x)$ must be measurable. On the other hand, it is well known that continuity is a property that does not transfer well under pointwise convergence. See Example 1.42.

Recall that by "pointwise" convergence is meant:

Definition 1.41 (Pointwise convergence) *A sequence of functions $\{f_n(x)\}_{n=1}^{\infty}$ converges pointwise to $f(x)$ on a set E if for any $x \in E$ and any $\epsilon > 0$, there is an $N = N(x, \epsilon)$ so that $|f_n(x) - f(x)| < \epsilon$ for all $n \geq N$.*

For pointwise convergence, it is possible that each function in the sequence is continuous, yet $f(x)$ is not.

Example 1.42 *Define the functions:*

$$f_n(x) = \begin{cases} 1, & x \leq 0, \\ 1 - nx, & 0 < x \leq \frac{1}{n}, \\ 0, & x > \frac{1}{n}, \end{cases} \qquad f(x) = \begin{cases} 1, & x \leq 0, \\ 0, & x > 0. \end{cases}$$

Although $f_n(x) \to f(x)$ for every x, the continuity of $f_n(x)$ is lost at $x = 0$ because the convergence is increasingly "slow" as x nears 0. In detail, for given $0 < x < 1$ and $\epsilon > 0$, $|f_n(x) - f(x)| < \epsilon$ for $n \geq N \simeq (1 - \epsilon)/x$, and thus N increases without bound as x nears 0.

This example also reveals a potential solution to the transferability problem for continuity. That is, if $f_n(x) \to f(x)$ uniformly, or with "speed" that is independent of x, continuity will be preserved.

Definition 1.43 (Uniform convergence) *A sequence of functions $\{f_n(x)\}_{n=1}^{\infty}$ converges uniformly to $f(x)$ on a set E if for any $\epsilon > 0$, there is an $N = N(\epsilon)$ so that $|f_n(x) - f(x)| < \epsilon$ for all $n \geq N$ and all $x \in E$.*

Exercise 1.44 (Uniform convergence $\Rightarrow$ continuity transfer) *Show that if $\{f_n(x)\}_{n=1}^{\infty}$ are continuous, and $f_n(x) \to f(x)$ uniformly on E, then $f(x)$ is continuous on E. Hint: By the triangle inequality, for any n fixed:*

$$|f(x) - f(y)| \leq |f(x) - f_n(x)| + |f_n(x) - f_n(y)| + |f_n(y) - f(y)|.$$

Remark 1.45 (Continuity transfer $\not\Rightarrow$ uniform convergence) *If $\{f_n(x)\}_{n=1}^{\infty}$ are continuous and $f_n(x) \to f(x)$ pointwise on a compact set E with $f(x)$ continuous, must this convergence be uniform?*

The answer is "no" and a simple counterexample on $[0, 1]$ is:

$$f_n(x) = \begin{cases} nx, & 0 \leq x \leq 1/n, \\ 2 - nx, & 1/n < x \leq 2/n, \\ 0, & 2/n < x \leq 1. \end{cases}$$

Then $\{f_n(x)\}_{n=1}^{\infty}$ are continuous and $f_n(x) \to f(x) \equiv 0$ pointwise on $[0, 1]$, but this convergence cannot be uniform since $f_n(1/n) - f(1/n) = 1$ for all n.

Beyond the property of continuity, we are also interested in whether the property of Riemann integrability is transferred from the functions $f_n(x)$ to the function $f(x)$, and if so transferred, whether the values of the integrals converge as well.

Example 1.46 *Although pointwise convergence did not preserve continuity in Example 1.42, it did preserve both Riemann integrability over bounded intervals, $[-1, 1]$ say, as well as the convergence of these integrals:*

$$\int_{-1}^{1} f_n(x)dx = \frac{1}{2n} \to 0 = \int_{-1}^{1} f(x)dx.$$

1.2.1 Positive Results on Limits

Returning to the general question, if $f_n(x)$ is Riemann integrable over $[a, b]$ for all n and $f_n(x) \to f(x)$ pointwise, will it be the case that $\int_a^b f(x)dx$ exists as a Riemann integral, and if so does:

$$\int_a^b f_n(x)dx \to \int_a^b f(x)dx?$$

If not, what kind(s) of convergence of Riemann integrable functions will assure Riemann integrability of $f(x)$ and the convergence of integral values, and over what intervals will these results apply?

One important positive result is given next.

Proposition 1.47 (Riemann integral convergence 1) *If $\{f_n(x)\}_{n=1}^{\infty}$ is a sequence of continuous functions on a closed and bounded interval $[a, b]$, and there is a function $f(x)$ so that $f_n(x) \to f(x)$ uniformly on this interval, then $f(x)$ is Riemann integrable on $[a, b]$, and:*

$$\int_a^b f_n(x)dx \to \int_a^b f(x)dx. \tag{1.33}$$

Proof. *That $\int_a^b f_n(x)dx$ exists for all n follows by Proposition 1.15 because these functions are continuous and the interval is closed and bounded. Also, uniform convergence assures the continuity of $f(x)$ by Exercise 1.44, and hence also proves the existence of $\int_a^b f(x)dx$. The only question remaining is that of convergence of the values of the integrals in (1.33). By linearity of the integral from Proposition 1.23, this convergence is equivalent to proving that:*

$$\int_a^b [f_n(x) - f(x)] \, dx \to 0.$$

By uniform convergence, given $\epsilon > 0$ there is an N so that $|f_n(x) - f(x)| < \epsilon$ for all x for $n \geq N(\epsilon)$. Hence for any partition, $a = x_0 < x_1 < \cdots < x_m = b$, and tags $\widetilde{x}_j \in [x_{j-1}, x_j]$, the associated Riemann sum of $f_n(x) - f(x)$ is bounded for $n \geq N(\epsilon)$:

$$\left| \sum_{j=1}^{m} [f_n(\widetilde{x}_j) - f(\widetilde{x}_j)] (x_j - x_{j-1}) \right| \leq \sum_{j=1}^{m} |f_n(\widetilde{x}_j) - f(\widetilde{x}_j)| (x_j - x_{j-1})$$
$$< \epsilon(b - a).$$

This is true for all partitions, and it follows that for $n \geq N(\epsilon)$:

$$\int_a^b [f_n(x) - f(x)] \, dx < \epsilon(b - a).$$

Consequently, $\int_a^b [f_n(x) - f(x)] \, dx \to 0$, and (1.33) is proved by adding finite $\int_a^b f(x)dx$ to this convergence result and applying item 2 of Proposition 1.23. ∎

This result can be generalized in that the assumption of continuity of $f_n(x)$ can be relaxed to the assumption that these functions are bounded and Riemann integrability. By the Lebesgue existence theorem of Proposition 1.22, this is equivalent to assuming that each of these functions is bounded and continuous except on a set of measure zero. In the continuous case above, boundedness did not have to be specified since every continuous function attains its maximum and minimum value on every closed and bounded interval $[a, b]$ by Exercise 1.13, or on any compact set, and is thus automatically bounded.

Note that we do not assume that the sequence $\{f_n(x)\}_{n=1}^{\infty}$ is **uniformly bounded** in the next result, and it may thus seem possible that these bounds could diverge to ∞. What prevents this from happening is the assumption of uniform convergence of $f_n(x) \to f(x)$, and this assures uniform boundedness.

Proposition 1.48 (*Riemann integral convergence 2*) *If $\{f_n(x)\}_{n=1}^{\infty}$ is a sequence of bounded, Riemann integrable functions on a closed and bounded interval $[a, b]$, and there is a function $f(x)$ so that $f_n(x) \to f(x)$ uniformly, then $f(x)$ is Riemann integrable and (1.33) holds.*

Proof. *To prove that such $f(x)$ is Riemann integrable it is enough to prove that $f(x)$ is bounded and continuous except on a set of measure zero, and then apply the Lebesgue existence theorem.*

To this end let E_n denote the set of discontinuity points of $f_n(x)$, where each has measure 0 by Lebesgue's theorem, and let $E = \bigcup E_n$. Then E has measure 0 since for each n there exist intervals $\{I_{n,m}\}_{m=1}^{\infty}$ with $E_n \subset \bigcup_{m=1}^{\infty} I_{n,m}$ and $\sum_{m=1}^{\infty} m(I_{n,m}) < \epsilon/2^n$. Thus $E \subset \bigcup_{n,m} I_{n,m}$, and by subadditivity of Lebesgue measure, $\sum_{n,m} m(I_{n,m}) < \epsilon$.

We next show that $f(x)$ is continuous outside E, but note that in general, $f(x)$ may also be continuous on many, even all of the points in E.

Given $\epsilon > 0$, it follows by uniform convergence that there is $N = N(\epsilon)$ so that $|f(y) - f_n(y)| < \epsilon$ for all $y \in [a, b]$ and all $n \geq N$. If $x \in [a, b] - E$, then since $f_N(x)$ is continuous at x, there is a δ_N so that $|f_N(x) - f_N(y)| < \epsilon$ if $|x - y| < \delta_N$. By the triangle inequality, if $|x - y| < \delta_N$:

$$
\begin{aligned}
|f(x) - f(y)| &\leq |f(x) - f_N(x)| + |f_N(x) - f_N(y)| + |f(y) - f_N(y)| \\
&< 3\epsilon.
\end{aligned}
$$

Thus $f(x)$ is continuous outside E, a set of measure 0.

Boundedness of $f(x)$ also follows from uniform convergence and the boundedness of $f_N(x)$:

$$
\begin{aligned}
|f(x)| &\leq |f(x) - f_N(x)| + |f_N(x)| \\
&< \epsilon + C_N.
\end{aligned}
$$

To prove convergence of integrals in 1.33, uniform continuity implies that for all $x \in [a, b]$ and $n \geq N$:

$$-\epsilon < f(x) - f_n(x) < \epsilon.$$

This implies by Proposition 1.23 that for $n \geq N$:

$$-\epsilon(b - a) < \int_a^b [f(x) - f_n(x)]\, dx < \epsilon(b - a).$$

As ϵ is arbitrary, (1.33) is proved. ∎

1.2.2 Generalization Counterexamples

Proposition 1.48 provides a positive conclusion, that uniform convergence of bounded Riemann integrable functions on $[a, b]$ implies Riemann integrability of the limit function and the convergence of integrals. However, while uniform convergence is sufficient, it is not a necessary condition for this conclusion given the earlier Example 1.46 of only pointwise convergence.

The following examples demonstrate the difficulty in generalizing any of the assumptions in this proposition. We will return to discuss these examples in the context of the Lebesgue integral in Section 2.6.2.

Example 1.49

1. **Pointwise convergence of continuous $f_n(x)$ on a bounded interval $[a, b]$ with $f(x)$ integrable.**

   a. **Integrals converge:** *Examples 1.42 and 1.46 provide continuous $f_n(x)$ converging pointwise to discontinuous $f(x)$, but where $f(x)$ is integrable and the sequence of integrals converge in the sense of (1.33).*

   b. **Integrals do not converge:** *Define almost everywhere continuous $f_n(x)$ on $[0,1]$ for $n \geq 1$:*

   $$f_n(x) = \begin{cases} 2^n, & 1/2^n \leq x \leq 1/2^{n-1}, \\ 0, & \text{elsewhere.} \end{cases}$$

   Then $f_n(x)$ converges pointwise on $[0,1]$, but not uniformly, to the continuous function $f(x) \equiv 0$. Also, for all n:

   $$\int_0^1 f_n(x)dx = 1 \neq \int_0^1 f(x)dx.$$

   Consequently, pointwise convergence of almost everywhere continuous functions on a compact interval to an integrable function does not assure convergence of integrals.

Exercise 1.50 *Generalize the result in 1.b. to make $f_n(x)$ continuous with $f_n(x) \to 0$ pointwise on $[0,1]$, but again where the integrals do not converge. Hint: Make $f_n(x)$ continuous on $[1/2^{n+1}, 1/2^{n-2}]$, with all integrals equal 1.*

2. **Uniform convergence of continuous $f_n(x)$ on an unbounded interval $[a, \infty)$ with $f(x)$ integrable.**

   a. **Integrals converge:** *Define $f_n(x)$ on $[1, \infty)$ for $n \geq 1$:*

   $$f_n(x) = x^{-(n+1)}/n.$$

   Then since $0 \leq f_n(x) \leq 1/n$ for all $x \geq 1$, this continuous function sequence converges uniformly to the continuous function $f(x) \equiv 0$. In addition, the improper integrals are well defined with $\int_1^\infty f_n(x)dx = 1/n^2$. Consequently,

   $$\int_1^\infty f_n(x)dx \to 0 = \int_1^\infty f(x)dx.$$

b. Integrals do not converge: *Modifying 2.a., define $f_n(x)$ on $[1, \infty)$ for $n \geq 1$:*

$$f_n(x) = \frac{1}{n} x^{-(n+1)/n}.$$

Then each $f_n(x)$ is again continuous and bounded, $f_n(x) \to f(x) \equiv 0$ uniformly on $[1, \infty)$, and the improper integrals are well defined with $\int_1^\infty f_n(x)dx = 1$. Consequently,

$$\int_1^\infty f_n(x)dx = 1 \neq \int_1^\infty f(x)dx.$$

3. Pointwise convergence of Riemann integrable $f_n(x)$ with $f(x)$ not integrable.

a. Convergence on a bounded interval: *For any ordering $\{r_j\}_{j=1}^\infty$ of the rational numbers in $[0, 1]$, define $f_n(x)$ on $[0, 1]$:*

$$f_n(x) = \begin{cases} 1, & x = r_j, \ 1 \leq j \leq n, \\ 1/n, & \text{elsewhere.} \end{cases}$$

Then $f_n(x)$ is bounded and continuous except at n points and hence is Riemann integrable with $\int_0^1 f_n(x)dx = 1/n$. However, $f_n(x) \to f(x)$ pointwise with:

$$f(x) = \begin{cases} 1, & x \text{ rational,} \\ 0, & x \text{ irrational,} \end{cases}$$

which is bounded but nowhere continuous and hence not Riemann integrable.

b. Convergence on an unbounded interval: *For any ordering $\{r_j\}_{j=1}^\infty$ of the rational numbers on $[1, \infty)$, define $f_n(x)$:*

$$f_n(x) = \begin{cases} 1, & x = r_j, \ 1 \leq j \leq n, \\ x^{-(n+1)}, & \text{elsewhere.} \end{cases}$$

Then $f_n(x)$ is bounded and continuous except at n points and hence is Riemann integrable with $\int_1^\infty f_n(x)dx = 1/n$. However, $f_n(x) \to f(x)$ pointwise with:

$$f(x) = \begin{cases} 1, & x \text{ rational,} \\ 0, & x \text{ irrational,} \end{cases}$$

which is bounded but nowhere continuous and hence not Riemann integrable.

1.3   The Integral in $\mathbb{R}^n$

In this section, we extend the development of the Riemann integral on $\mathbb{R}$ to $\mathbb{R}^n$, beginning with the Riemann and Darboux criteria for existence. We then investigate equivalence of these and the Cauchy criterion, prove existence of such integrals for continuous functions, and derive the Lebesgue existence theorem. Properties of integrals are then summarized, and finally we turn to the issue of valuation of integrals using iterated integrals.

1.3.1 The Riemann Approach

Given $f : \mathbb{R}^n \to \mathbb{R}$, the development of the Riemann integral of f over a rectangle $R \equiv \prod_{j=1}^{n}[a_j, b_j]$ proceeds as follows. Each interval $[a_j, b_j]$ is partitioned as in (1.1) using $\{x_{j,i}\}_{i=1}^{m_j}$:

$$a_j = x_{j,0} < x_{j,1} < \cdots < x_{j,m_j-1} < x_{j,m_j} = b_j,$$

with mesh size μ of these n partitions now defined by:

$$\mu \equiv \max_{j,i}\{x_{j,i} - x_{j,i-1}\}. \tag{1.34}$$

Thus the maximum is defined over $1 \leq j \leq n$ and $1 \leq i \leq m_j$.

These interval partitions lead to a partition of R in a natural way, and we denote such a partition by P or P_μ to identify the mesh size. Defining $A_{j,i} \equiv [x_{j,i-1}, x_{j,i}]$:

$$\bigcup_I \prod_{j=1}^{n} A_{j,i_j} = R, \tag{1.35}$$

where $I = \{(i_1, i_2, ..., i_n) | 1 \leq i_j \leq m_j\}$ is the index set which identifies the $\prod_{j=1}^{n} m_j$ subrectangles induced by this partition.

We call $A_J \equiv \prod_{j=1}^{n} A_{j,i_j}$ a **partition subrectangle of the partition** $P \equiv \{A_J\}$, where $J \in I$.

Notation 1.51 (On Lebesgue measure m^n) *In Chapter I.7, when Lebesgue measure on $\mathbb{R}^n$ was derived as an example of a product measure, the notational convention of m^n was introduced as a product of the Lebesgue measures m on $\mathbb{R}$. But in this book, we will drop this now unnecessary extra notation, and use m for both the Lebesgue measure on $\mathbb{R}$ and on $\mathbb{R}^n$. There should be no confusion given the context.*

The definition of Riemann integrability of a function $f(x)$ defined on a rectangle R is a notational generalization of Definition 1.2, again using approximating **Riemann sums**.

Definition 1.52 (Riemann integral of f over R) *The **Riemann integral of** $f(x)$ **over bounded** $R \equiv \prod_{j=1}^{n}[a_j, b_j]$ is defined by:*

$$(\mathcal{R}) \int_R f(x)dx = \lim_{\mu \to 0} \sum_{J \in I} f(\tilde{x}_J) |A_J|, \tag{1.36}$$

*when this limit exists, is finite, and has value that is independent of the **subinterval tags** $\tilde{x}_J \in A_J$, where $|A_J| = m(A_J)$, the Lebesgue (product) measure of A_J:*

$$|A_J| \equiv \prod_{j=1}^{n}(x_{j,i_j} - x_{j,i_j-1}). \tag{1.37}$$

That is, for any $\epsilon > 0$ there is a δ so that:

$$\left| \sum_{J \in I} f(\tilde{x}_J) |A_J| - (\mathcal{R}) \int_R f(x)dx \right| < \epsilon,$$

*for all **tagged partitions** $\{\tilde{x}_J, A_J\}_{J \in I}$ of mesh size $\mu \leq \delta$.*

*The function $f(x)$ is called the **integrand**; and R is the **domain of integration** of the Riemann integral.*

It may seem surprising to have a Lebesgue measure appear in the definition of a Riemann integral. Buy recall that Lebesgue measure equals the ordinary notion of "length" for intervals in $\mathbb{R}$, and analogously as a product measure, equals the ordinary "volume" of the rectangle A_J in $\mathbb{R}^n$.

1.3.2   The Darboux Approach

To develop the Darboux approach, we introduce a **step function** defined on a rectangle R.

Definition 1.53 (Step function) *A bounded function $\varphi(x)$ defined on R is a **step function** if there exists a partition P on R as in (1.35) so that:*

$$\varphi(x) = \sum_{J \in I} a_J \chi_{A_J}(x),$$

where each A_J is a partition subrectangle, $\{a_J\}_{J \in I} \subset \mathbb{R}$, and the summation is over all $J \equiv (i_1, i_2, ..., i_n) \in I$.

The next result derives the Riemann integral of a step function. It will not be lost on the reader that this result requires a lot of effort given the simplicity of such functions. But note that most of this work is to prove integrability. If we knew that such a function was integrable, by some existence theorem for example, then we could have derived the result in (1.38) using any convenient sequence of partitions with $\mu \to 0$. And for this, we could have simply used all refinements (Definition 1.59) of the partition underlying $\varphi(x)$ and a simple calculation, since then every Riemann sum obtains the final answer by finite additivity of Lebesgue measure.

Proposition 1.54 (Riemann integral of a step function) *Given a step function $\varphi(x)$, the Riemann integral over R exists and is given by:*

$$(\mathcal{R}) \int_R \varphi(x)dx \equiv \sum_{J \in I} a_J |A_J|. \tag{1.38}$$

Proof. *Given $\epsilon > 0$ we prove that there exists δ so that for any partition $P_{\mu'} = \{B_L\}_{L \in K}$ of R with mesh size $\mu' \leq \delta$, and tags $\{\widetilde{y}_L\}_{L \in K}$:*

$$\left| \sum_{L \in K} \varphi(\widetilde{y}_L)|B_L| - \sum_{J \in I} a_J |A_J| \right| < \epsilon. \tag{1}$$

For a general partition $P_{\mu'}$, denote interval partition points:

$$a_j = y_{j,0} < y_{j,1} < \cdots < y_{j,n_j-1} < y_{j,n_j} = b_j.$$

By definition $y_{j,0} = x_{j,0} = a_j$ and $y_{j,n_j} = x_{j,m_j} = b_j$ for all j, where $\{x_{j,i}\}$ denote the points of the partition P underlying $\varphi(x)$. Initially choose $\delta < \mu$, the mesh size of P, so that at most one $x_{j,i} \in [y_{j,l-1}, y_{j,l}]$ for any such interval.

Now consider:

$$\sum_{L \in K} \varphi(\widetilde{y}_L)|B_L|.$$

If given $B_L \subset A_J$ for some A_J then of necessity $\varphi(\widetilde{y}_L) = a_J$ for any tag $\widetilde{y}_L$. In the general case, there exists an index subcollection $\{J'\} \equiv I' \subset I$ so that $B_L \cap A_{J'} \neq \emptyset$ for all $J' \in I'$, and $B_L \cap A_J = \emptyset$ for all $J \in I - I'$. Thus for any tag, $\varphi(\widetilde{y}_L) \in \{a_{J'}\}_{J' \in I'}$.

Now $B_L = \bigcup_{J'} B_L \bigcap A_{J'}$, and thus by finite additivity:

$$\varphi(\widetilde{y}_L)|B_L| = \sum_{J' \in I'} a_{J'} |B_L \bigcap A_{J'}| + \sum_{J' \in I'} (\varphi(\widetilde{y}_L) - a_{J'})|B_L \bigcap A_{J'}|. \tag{2}$$

Noting that the case $B_L \subset A_J$ also satisfies (2) with $\varphi(\widetilde{y}_L) = a_{J'}$, this obtains by finite additivity:

$$\sum_{L \in K} \varphi(\widetilde{y}_L)|B_L|$$
$$= \sum_{L \in K} \left[\sum_{J' \in I'} a_{J'} |B_L \bigcap A_{J'}| + \sum_{J' \in I'} (\varphi(\widetilde{y}_L) - a_{J'})|B_L \bigcap A_{J'}| \right]$$
$$= \sum_{J \in I} a_J |A_J| + \sum_{L \in K} \sum_{J' \in I'} (\varphi(\widetilde{y}_L) - a_{J'})|B_L \bigcap A_{J'}|.$$

Now $\{a_J\}_{J \in I}$ is a bounded set so let $M \equiv \max\{|a_J - a_{J'}|\}_{J,J' \in I}$. Then all $|\varphi(\widetilde{y}_L) - a_{J'}| \leq M$ since each $\varphi(\widetilde{y}_L) = a_J$ for some J. Further, $\sum_{J' \in I'} |B_L \bigcap A_{J'}| \leq \delta^n$ by definition of mesh size. By construction, at most one $x_{j,i} \in [y_{j,l-1}, y_{j,l}]$ for any such interval, and thus the total nonzero count in the summation $\sum_{L \in K}$ is $\prod_{j=1}^n m_j$, reflecting the partitions in P.

Hence:

$$\left| \sum_{L \in K} \varphi(\widetilde{y}_L) |B_L| - \sum_{J \in I} a_J |A_J| \right| < M\delta^n \prod_{j=1}^n m_j,$$

and this can be made arbitrarily small by choosing δ small. ∎

Corollary 1.55 (Riemann integral of a step function) *Given a step function $\varphi(x)$ defined on a rectangle R, and rectangle $R' \subset R$, define:*

$$(\mathcal{R}) \int_{R'} \varphi(x)dx = (\mathcal{R}) \int_R \varphi(x)\chi_{R'}(x)dx. \tag{1.39}$$

Then $\varphi(x)\chi_{R'}(x)$ is a step function on R, and:

$$(\mathcal{R}) \int_{R'} \varphi(x)dx \equiv \sum_{J \in I} a_J \left| A_J \bigcap R' \right|. \tag{1.40}$$

In particular with $R = R'$:

$$(\mathcal{R}) \int_R \varphi(x)dx = \sum_{J \in I} \int_{A_J} \varphi(x)dx$$

Proof. *Left as an exercise. Hint: Characterize the partition underlying $\varphi(x)\chi_{R'}(x)$.* ∎

The definition of the Darboux integral of a function f over a rectangle R now generalizes Definition 1.7:

Definition 1.56 (Darboux integral of f over R) *Let f be a bounded function defined on $R \equiv \prod_{j=1}^n [a_j, b_j]$ with:*

$$\sup_{\varphi_\mu \leq f} \int_R \varphi_\mu(x)dx = I = \inf_{f \leq \psi_\mu} \int_R \psi_\mu(x)dx, \tag{1.41}$$

where $\varphi_\mu(x)$ and $\psi_\mu(x)$ denote step functions with $\varphi_\mu(x) \leq f(x) \leq \psi_\mu(x)$, defined relative to arbitrary partitions of R of mesh size μ, and with associated Riemann integrals as given by (1.38).

*Then define the **Darboux integral of $f(x)$ over $R \equiv \prod_{j=1}^n [a_j, b_j]$** by:*

$$(\mathcal{D}) \int_R f(x)dx = I. \tag{1.42}$$

If the extrema in (1.41) are unequal, the Darboux integral of $f(x)$ over R does not exist.

By boundedness of $f(x)$, the supremum and infimum in (1.41) are defined over all partitions of mesh size μ, and all step functions defined over such partitions. But given any partition P_μ of mesh size μ and arbitrary $\varphi_\mu(x) \leq f(x) \leq \psi_\mu(x)$ as above, one can without loss of generality replace φ_μ and ψ_μ with:

$$\varphi_\mu(x) \equiv \sum_{J \in I} m_J \chi_{A_J}(x), \quad \psi_\mu(x) \equiv \sum_{J \in I} M_J \chi_{A_J}(x), \tag{1.43}$$

whereas in (1.5):

$$m_J \equiv \inf\{f(x)|x \in A_J\}, \quad M_J \equiv \sup\{f(x)|x \in A_J\}.$$

It is common to see (1.41) stated explicitly with this convention.

Definition 1.57 (Upper/lower Darboux integrals) *The integrals of $\varphi_\mu(x)$ and $\psi_\mu(x)$ as defined in (1.43) are called the **lower Darboux sum**, respectively, **upper Darboux sum**, of $f(x)$ over R **with partition** P_μ. These are often denoted $L(f, P_\mu)$ and $U(f, P_\mu)$, and thus:*

$$L(f, P_\mu) \equiv \sum_{J \in I} m_J |A_J|, \quad U(f, P_\mu) \equiv \sum_{J \in I} M_J |A_J|. \tag{1.44}$$

*If $f(x)$ is a bounded function defined on $R \equiv \prod_{j=1}^{n} [a_j, b_j]$, the **lower Darboux integral** of f over R is defined:*

$$\underline{\int_R} f(x)dx \equiv \sup_{\varphi_\mu \leq f} \int_R \varphi_\mu(x)dx = \sup_{P_\mu} L(f, P_\mu). \tag{1.45}$$

*Similarly, the **upper Darboux integral of** f **over** R is defined:*

$$\overline{\int_R} f(x)dx \equiv \inf_{f \leq \psi_\mu} \int_R \psi_\mu(x)dx = \inf_{P_\mu} U(f, P_\mu). \tag{1.46}$$

A function $f(x)$ is defined to be Darboux integrable over R when the upper and lower Darboux integrals agree:

$$\sup_{P_\mu} L(f, P_\mu) = \inf_{P_\mu} U(f, P_\mu). \tag{1.47}$$

Remark 1.58 *The upper and lower Darboux integrals always exist and are finite for bounded $f(x)$. This follows because if $|f(x)| \leq M$, then certainly $-M \leq m_J, M_J \leq M$ for all J, so these integrals are bounded by $\pm M |R|$.*

1.3.3   Riemann and Darboux Equivalence

The goal of this section is to prove that the Darboux and Riemann notions of the integral are equivalent for bounded functions, and again we do this by proving equivalence with the Cauchy criterion. The first step is to investigate the effect of partition refinements on step function integrals, generalizing Exercise 1.1, and this requires a definition.

Definition 1.59 (Refinement of a partition) *Given a partition P_μ of bounded $R = \prod_{j=1}^{n} [a_j, b_j]$ into subrectangles $\{A_J\}_{J \in I}$, a partition $P_{\mu'}$ of R into subrectangles $\{B_K\}_{K \in L}$ is called a **refinement of** P_μ, if for any $K \in L$ there is a $J \in I$ so that $B_K \subset A_J$.*

Equivalently, if $\{B_K\}_{K \in L}$ is defined by the interval points $\left\{\{y_{j,i}\}_{i=0}^{n_j}\right\}_{j=1}^{n}$, and $\{A_J\}_{J \in I}$ defined by $\left\{\{x_{j,i}\}_{i=0}^{m_j}\right\}_{j=1}^{n}$, then $\{B_K\}_{K \in L}$ is a refinement of $\{A_J\}_{J \in I}$ if $\{x_{j,i}\}_{i=0}^{m_j} \subset \{y_{j,i}\}_{i=0}^{n_j}$ for each j.

The preliminary result needed below states that every lower Darboux sum is majorized by every upper Darboux sum, independent of the partitions used, and further, refinements improve estimates.

Proposition 1.60 (Refinements improve estimates) *Given any two partitions $A = \{A_J\}_{J \in I}$ and $B = \{B_K\}_{K \in L}$ of $R = \prod_{j=1}^{n} [a_j, b_j]$, if C is a refinement of both A and B, then:*

$$\int_R \varphi_A(x)dx \leq \int_R \varphi_C(x)dx \leq \int_R \psi_C(x)dx \leq \int_R \psi_B(x)dx. \tag{1.48}$$

Here $\varphi_.(x) \leq f(x) \leq \psi_.(x)$, and the subscripts denote the defining partitions for $\varphi(x)$ and $\psi(x)$ of (1.43).

Proof. *Assume that for each j that partition A is defined by $\{x_{j,i}\}_{i=0}^{m_j}$ and partition B is defined by $\{y_{j,i}\}_{i=0}^{n_j}$. Now define partition C_0 by $\{x_{j,i}\}_{i=0}^{m_j} \bigcup \{y_{j,i}\}_{i=0}^{n_j}$ for each j, counting repeated points once, and thus C_0 is the minimal common refinement of both A and B. Further, any C as above is a refinement of C_0.*

Denoting by φ_C and ψ_C, the lower and upper step functions defined with respect to such C, the middle inequality of (1.48) is true by definition,. To complete the proof, it is enough to prove the outer inequalities when C is defined with a single extra point in a single component j, and then apply induction.

For example, let $j = 1$. If $\{x_{1,i}\}_{i=0}^{m_1}$ defines partition A, let C be defined with $\{x_{1,i}\}_{i=0}^{m_1} \bigcup \{z\}$ with $x_{1,0} < z < x_{1,1}$, and assume that A and C have the same partition points $\{x_{j,i}\}_{i=0}^{m_j}$ for $j > 1$. Then the first and third inequalities in (1.48) follow from a simple observation. If A_J is any subrectangle of A with first interval component $[x_{1,0}, x_{1,1}]$, and m_J and M_J denote the associated bounds for $f(x)$ on A_J, then:

$$m_J = \min(m_J', m_J''), \quad \max(M_J', M_J'') = M_J.$$

Here m_J' and m_J'' denote the infima of $f(x)$ on the two rectangles A_J' and A_J'' in C that union to A_J, and M_J' and M_J'' are similarly defined in terms of suprema.

Thus with $A_J = A_J' \bigcup A_J''$, then by Proposition 1.54 and Corollary 1.55:

$$
\begin{aligned}
\int_{A_J} \varphi_A(x)dx &= m_J |A_J| \\
&\leq m_J' |A_J'| + m_J'' |A_J''| \\
&= \int_{A_J} \varphi_C(x)dx.
\end{aligned}
$$

The same inequality holds for the integrals over any A_J, and then by Corollary 1.55 is true for A, proving the left inequality in (1.48).

The right inequality is proved analogously. ∎

Exercise 1.61 (On $\mu \to 0$) *Prove that as a corollary to Proposition 1.60, that in the definition of Darboux integral and (1.41):*

$$\sup_{\varphi_\mu \leq f} \int_a^b \varphi_\mu(x)dx = \lim_{\mu \to 0} \sup_{\varphi_\mu \leq f} \int_a^b \varphi_\mu(x)dx, \tag{1.49}$$

where the supremum on the right is over all φ_μ with given μ. The same is true with infima of $\psi_\mu(x)$-integrals.

As for the Riemann integral on $\mathbb{R}$, the significance of the next result is that it not only proves equivalence of the Riemann and Darboux notions of integrability but also provides a simpler criterion in (1.50) for integrability, which is useful for the existence results that follow. It is called the **Cauchy criterion for existence of the Riemann integral,** named after **Augustin-Louis Cauchy** (1789–1857).

Proposition 1.62 (Cauchy criterion for the Riemann integral) *A bounded function $f(x)$ is Riemann integrable on a bounded rectangle $R = \prod_{j=1}^n [a_j, b_j]$ by Definition 1.52 if and only if it is Darboux integrable by Definition 1.56, and then the integrals agree.*

In either case, $f(x)$ is so integrable if and only if for any $\epsilon > 0$ there is a partition P_μ of R with subrectangles $\{A_J\}_{J \in I}$, so that:

$$0 \leq U(f, P_\mu) - L(f, P_\mu) < \epsilon, \tag{1.50}$$

where:

$$L(f, P_\mu) \equiv \sum_{J \in I} m_J |A_J|, \quad U(f, P_\mu) \equiv \sum_{J \in I} M_J |A_J|,$$
$$m_J \equiv \inf\{f(x)|x \in A_J\}, \quad M_J \equiv \sup\{f(x)|x \in A_J\}. \tag{1.51}$$

Further, (1.50) is then also satisfied for every refinement of P_μ.

Proof. *As noted above, we show that the requirement of either definition of integrability is equivalent to the above Cauchy criterion, and thus each is equivalent to the other.*

Given $\epsilon > 0$, assume that (1.50) is satisfied. It follows from (1.48) and the definition of infimum and supremum, that for such P_μ:

$$L(f, P_\mu) \leq \sup_{\varphi \leq f} \int_R \varphi(x)dx \leq \inf_{f \leq \psi} \int_R \psi(x)dx \leq U(f, P_\mu).$$

Thus:

$$0 \leq \inf_\psi \int_R \psi(x)dx - \sup_\varphi \int_R \varphi(x)dx < \epsilon, \tag{1}$$

and since this is true for all ϵ, it follows that $f(x)$ is Darboux integrable on R by Definition 1.56.

Now (1.49) implies that there exists μ' so that for every step function $\varphi'_{\mu_1}(x)$ in (1.43) with partition mesh size $\mu_1 \leq \mu'$:

$$\sup_{\varphi_\mu \leq f} \int_R \varphi(x)dx \leq \int_R \varphi'_{\mu_1}(x)dx + \epsilon/2.$$

Similarly, there exists μ'' so that for every step function $\psi''_{\mu_2}(x)$ in (1.43) with partition mesh size $\mu_2 \leq \mu''$:

$$\inf_{f \leq \psi_\mu} \int_R \psi_\mu(x)dx \geq \int_R \psi''_{\mu_2}(x)dx - \epsilon/2.$$

Then by (1.50) and (1), if $\mu_3 \leq \min\{\mu', \mu''\}$:

$$\int_R \psi''_{\mu_3}(x)dx - \int_R \varphi'_{\mu_3}(x)dx < 2\epsilon.$$

Thus given any such partition $\{A_J\}$ and associated step functions, and interval tags $\{\widetilde{x}_J\}$ with $\widetilde{x}_J \in A_J$:

$$\int_R \varphi''_{\mu_3}(x)dx \leq \sum_{J \in I} f(\widetilde{x}_J)|A_J| \leq \int_R \varphi''_{\mu_3}(x)dx + 2\epsilon. \tag{2}$$

As (2) is satisfied for every partition with mesh size $\mu_3 \leq \min\{\mu', \mu''\}$, it is thus true taking a limit as $\mu_3 \to 0$. Since ϵ is arbitrary, $f(x)$ is Riemann integrable on R by Definition 1.52.

Further, by (2) and (1), the Darboux and Riemann integrals agree.

Conversely, if $f(x)$ is Darboux integrable by Definition 1.56 and $\epsilon > 0$ is given, then again by definition of extrema there exists partitions of R of mesh sizes μ_1 and μ_2, and step functions $\varphi_{\mu_1} \leq f \leq \psi_{\mu_2}$, so that:

$$0 \leq \int_a^b \psi_{\mu_2}(x)dx - \int_a^b f(x)dx < \epsilon/2,$$

$$0 \leq \int_a^b f(x)dx - \int_a^b \varphi_{\mu_1}(x)dx < \epsilon/2.$$

Thus:

$$0 \le \int_a^b \psi_{\mu_2}(x)dx - \int_a^b \varphi_{\mu_1}(x)dx < \epsilon. \tag{3}$$

It then follows from (1.48) that (3) is satisfied with both $\psi(x)$ and $\varphi(x)$ defined relative to the common refinement of these partitions, and this is (1.50).

Similarly, if f is Riemann integrable by Definition 1.52 and $\epsilon > 0$ is given, then there is a δ so that for all tagged partitions $\{\widetilde{x}_J, A_J\}_{J \in I}$ of mesh size $\mu \le \delta$:

$$\left| \sum\nolimits_{J \in I} f(\widetilde{x}_J)|A_J| - (\mathcal{R}) \int_R f(x)dx \right| < \epsilon/3.$$

Thus for any such partition and any two sets of tags $\{\widetilde{x}_J\}_{J \in I}$ and $\{\widetilde{x}'_J\}_{J \in I}$:

$$\left| \sum\nolimits_{J \in I} f(\widetilde{x}_J)|A_J| - \sum\nolimits_{J \in I} f(\widetilde{x}'_J)|A_J| \right| < 2\epsilon/3.$$

By definition of M_J and m_J, we can choose such tags so that these Riemann sums are within $\epsilon/6$ of $\sum_{J \in I} M_J|A_J|$ and $\sum_{J \in I} m_J|A_J|$, and this obtains (1.50).

Finally, if (1.50) is satisfied for $\{A_J\}_{J \in I}$, it is also satisfied for every refinement by (1.48). ∎

1.3.4 On Existence of the Integral

The goal of this section is to verify integrability of continuous functions, and prove the generalization provided by the Lebesgue existence theorem. The first proof is greatly simplified by the Cauchy criterion, as was the case for Proposition 1.15.

Proposition 1.63 (Continuous ⇒ Riemann integrable) *If $f(x)$ is continuous on a bounded rectangle $R = \prod_{j=1}^n [a_j, b_j]$, then f is Riemann integrable on R.*
Proof. *To prove Riemann integrability, it is enough to prove (1.50) and apply Proposition 1.62.*

By Exercise 1.14, continuity of $f(x)$ on R implies uniform continuity on R. Given $\epsilon > 0$, let δ' be defined by uniform continuity with $\epsilon' = \epsilon/m(R)$, so:

$$|f(y) - f(x)| < \epsilon' \text{ whenever } |y - x| < \delta'.$$

Given a partition P_μ of R with subrectangles $\{A_J\}_{J \in I}$ and mesh size μ, if $x, y \in A_J$ then by (1.34):

$$|y - x|^2 = \sum\nolimits_{i=1}^n (y_i - x_i)^2 \le n\mu^2.$$

Thus $|y - x| < \delta'$ if $\mu < \delta'/\sqrt{n}$.

Defining $\delta = \delta'/\sqrt{n}$, this obtains that if $\mu \le \delta$ then $M_J - m_J < \epsilon'$ for all J. Thus:

$$\sum\nolimits_{J \in I} (M_J - m_J)|A_J| < \epsilon' m(R) = \epsilon.$$

∎

As was the case for the Riemann integral on $\mathbb{R}$, continuity is sufficient but not necessary for the existence of $(\mathcal{R}) \int_R f(x)dx$ for bounded $f(x)$. Indeed, as in the earlier development, there is a Lebesgue existence theorem that provides the same necessary and sufficient condition for a bounded function, that $f(x)$ be continuous almost everywhere, meaning except on a set of Lebesgue measure 0.

As for Definition 1.18, there is a simple and intuitive characterization for sets of Lebesgue measure zero, with the same justification as was given there. First, $m \equiv m^*$ on Lebesgue measurable sets by Proposition I.7.20 (recalling Section I.7.6.1), and all sets with $m^*(A) = 0$ are Lebesgue measurable. The Lebesgue outer measure m^* is defined in terms of rectangle covers by I.(6.3). While such covers are defined in terms of the algebra generated by right semi-closed rectangles $H_i = \prod_{j=1}^{n}(a_{i,j}, b_{i,j}]$, for Lebesgue measure, $m(H_i) = m(G_i) = \prod_{j=1}^{n}(b_{i,j} - a_{i,j})$, where $G_i = \prod_{j=1}^{n}(a_{i,j}, b_{i,j})$.

Definition 1.64 (Set of measure 0 in $\mathbb{R}^n$) *A set $E \subset \mathbb{R}^n$ has **measure 0,** or more formally **Lebesgue measure 0,** if for any $\epsilon > 0$ there is a finite or countable collection of open rectangles $\{G_i\}$ with $G_i = \prod_{j=1}^{n}(a_{i,j}, b_{i,j})$, so that $E \subset \bigcup_i G_i$ and $\sum_i |G_i| < \epsilon$, where $|G_i| \equiv m(G_i) \equiv \prod_{j=1}^{n}(b_{i,j} - a_{i,j})$ is the Lebesgue measure of G_i.*

Next we up-dimension the notion of the **oscillation of a function** at a point x from Definition 1.19.

Definition 1.65 (Oscillation of $f(x)$ at x) *Given an open rectangle $R^o \equiv \prod_{j=1}^{n}(a_j, b_j)$, the **oscillation of $f(x)$ on R^o**, denoted $\omega(R^o)$, is defined:*

$$\omega(R^o) = M_i - m_i,$$

where:

$$m_i \equiv \inf_{x \in R^o}\{f(x)\}, \quad M_i \equiv \sup_{x \in R^o}\{f(x)\}.$$

*The **oscillation of $f(x)$ at x**, denoted $\omega(x)$, is defined:*

$$\omega(x) = \inf\{\omega(R^o)|x \in R^o\}.$$

Exercise 1.66 *This exercise both generalizes Exercise 1.20 and Proposition 1.21, and proves that the "boundary" of a rectangle has measure 0.*

1. **Characterizing discontinuity**: *Repeat Exercise 1.20 and show that with:*

$$E_N = \{x|\omega(x) \geq 1/N\}, \qquad E \equiv \bigcup_{N \geq 1} E_N = \{x|\omega(x) > 0\},$$

   that E is the collection of discontinuities of $f(x)$. Then prove that E_N is closed for all N, and thus E is a Borel set, $E \in \mathcal{B}(\mathbb{R}^n)$, and specifically an $\mathcal{F}_\sigma$-set (Notation I.2.16).

2. **Oscillation and Cauchy criterion**: *Prove a generalization of Proposition 1.21: If $\omega(x) < k$ for all x in a bounded rectangle $R = \prod_{j=1}^{n}[a_j, b_j]$, then for some partition of R with subrectangles $\{A_J\}_{J \in I}$:*

$$\sum_{J \in I}(M_J - m_J)|A_J| < k|R|, \tag{1.52}$$

   where m_J and M_J are defined as in (1.51).

3. **Measure of ∂R**: *Given a bounded rectangle $R = \prod_{j=1}^{n}[a_j, b_j]$, prove that the boundary ∂R:*

$$\partial R \equiv \prod_{j=1}^{n}[a_j, b_j] - \prod_{j=1}^{n}(a_j, b_j),$$

   has Lebesgue measure 0. Hint: By Definition 1.64, you must prove that for any $\epsilon > 0$ there is a collection of rectangles $\{R_j\}_{j=1}^{m}$ so that $\partial R \subset \bigcup_{j=1}^{m} R_j$ and $\sum_{j=1}^{m} m(R_j) < \epsilon$. Show that this boundary is a union of 2^n "faces" defined by:

$$c_i \times \prod_{j \neq i}[a_j, b_j], \quad \text{where } c_i = a_i, \text{ or } c_i = b_i.$$

   Each such face is contained in $[c_i - \delta, c_i + \delta] \times \prod_{j \neq i}[a_j, b_j]$ for any δ. Consider $n = 2$ for a visual of this construction.

Remark 1.67 (Riemann integral of a step function) *That ∂R has Lebesgue measure zero was in fact demonstrated indirectly in the Proposition 1.54 derivation of the Riemann integral of a step function. The reader is encouraged to identify where in the proof the boundaries of the step function partition were covered by subrectangles of arbitrarily small measure.*

We now have the tools to prove Lebesgue's result.

Proposition 1.68 (Lebesgue's Existence Theorem) *If $f(x)$ is a bounded function on a bounded rectangle $R = \prod_{j=1}^{n}[a_j, b_j]$, then:*

$$(\mathcal{R}) \int_R f(x)dx$$

exists if and only if $f(x)$ is continuous almost everywhere, meaning except on a set of points $E \equiv \{x_\alpha\}$ of Lebesgue measure 0.

Proof. *Assume that this integral exists. We prove that E_N has measure 0 for all N, and thus as a countable union, so too does E by countable subadditivity of Lebesgue measure. Since $f(x)$ is bounded, Proposition 1.62 obtains for any $\epsilon > 0$ and integer N, there is a partition P_μ of R with subrectangles $\{A_J\}_{J \in I}$, so that:*

$$\sum_{J \in I}(M_J - m_J)|A_J| < \frac{\epsilon}{N}.$$

If any of the discontinuity points $x \in E_N$ of $f(x)$ are in the partition subrectangle boundaries, these points have measure 0 by item 3 of Exercise 1.66.

Focusing on the discontinuity points within the interiors of the partition subrectangles, denote by $\{A_{J_k}\}_{k=1}^{m}$ the subset of the subrectangles that have at least one such discontinuity point in its interior. Thus:

$$E_N \subset \bigcup_{k=1}^{m} A_{J_k} \bigcup\bigcup_{J \in I} \partial A_J. \tag{1}$$

For any A_{J_k}-subrectangle, $M_{J_k} - m_{J_k} \geq 1/N$ by definition of E_N, since $1/N$ is the infimum over all open rectangles that contain these discontinuities. Thus:

$$\frac{1}{N}\sum_{k=1}^{m}|A_{J_k}| \leq \sum_{J \in I}(M_J - m_J)|A_J| < \frac{\epsilon}{N}. \tag{2}$$

Since $m\left(\bigcup_{J \in I} \partial A_J\right) = 0$, we obtain from (1), (2), and finite subadditivity of m :

$$m(E_N) \leq \sum_{k=1}^{m}|A_{J_k}| < \epsilon.$$

Since ϵ is arbitrary, it follows that $m(E_N) = 0$.

Conversely, assume that $f(x)$ is bounded on $R = \prod_{j=1}^{n}[a_j, b_j]$ and that $m(E) = 0$. Then $m(E_N) = 0$ for any N, and so for any $\epsilon > 0$ there exists an at most countable collection of open rectangles $\{R_j\}$ with $E_N \subset \bigcup R_j$ and $\sum m(R_j) < \epsilon$. As E_N is closed by Exercise 1.66 and bounded, it is compact. By the Heine-Borel theorem of Proposition I.2.27, there exists a finite subcollection $\{R_k\}_{k=1}^{m}$ with the same properties, so $E_N \subset \bigcup_{k=1}^{m} R_k$ and $\sum_{k=1}^{m} m(R_k) < \epsilon$.

To create a partition for R, let $R_k \equiv \prod_{j=1}^{n}(a_{j,k}, b_{j,k})$. For each j, let $\{a_{j,k}, b_{j,k}\}_{k=1}^{m}$ induce a partition of $[a_j, b_j]$, and let $\{A_J\}_{J \in I}$ be the partition subrectangles of R created by these interval partitions. By construction, each R_k is either contained in a unique A_J, or a finite union of such A_J, so $\bigcup_{k=1}^{m} R_k \subset \bigcup_{J \in I'} A_J$ for a subcollection $I' \subset I$. Further by *construction,* $\bigcup_{k=1}^{m} \bar{R}_k = \bigcup_{J \in I'} A_J$, *where* $\bar{R}_k$ *is the closure of* R_k. *Thus* $m(R_k) = m(\bar{R}_k)$ *by Exercise 1.66, and finite subadditivity obtains:*

$$m\left(\bigcup_{J \in I'} A_J\right) \leq \sum_{k=1}^{m} m(\bar{R}_k) < \epsilon. \tag{3}$$

Let $\{A_J\}_{J \in I} = \{A_J\}_{J \in I'} \bigcup \{A_J\}_{J \in I''}$. *Then* $\bigcup_{k=1}^{m} R_k \bigcap \bigcup \{A_J\}_{J \in I''} = \emptyset$ *by construction, and thus* $\omega(x) < 1/N$ *on* $\{A_J\}_{J \in I''}$. *By item 2 of Exercise 1.66, there is a partition of each such* A_J *with* $J \in I''$ *for which (1.52) is satisfied with* $k = 1/N$. *With* $\{B_K\}_{K \in L}$ *the total collection of subrectangles for all* A_J *with* $J \in I''$:

$$\sum_{K \in L} (M_K - m_K) \, m(B_K) < \frac{1}{N} \sum_{K \in L} m(B_K). \tag{4}$$

Returning to the original partition, but with refined subrectangles $\{C_K\}_{K \in I' \cup L} \equiv \{A_J\}_{J \in I'} \bigcup \{B_K\}_{K \in L}$, *note that by boundedness,* $(M_K - m_K) \leq M - m$ *for all* $K \in I'$ *with* m, M *defined on* R. *It then follows from (3) and (4):*

$$\begin{aligned} \sum_{K \in I' \cup L} (M_K - m_K) \, m(C_K) \quad &< \quad \frac{1}{N} \sum_{K \in L} m(B_K) + (M - m)\,\epsilon \\ &\leq \quad \frac{m(R)}{N} + (M - m)\,\epsilon. \end{aligned}$$

As $\epsilon > 0$ *and* N *are arbitrary, this proves the existence of* $(\mathcal{R}) \int_a^b f(x) dx$ *by Proposition 1.62.* ∎

To generalize the definition of Riemann integral to **unbounded rectangles**, which are again called **improper integrals**, we generalize the approach of (1.22) and (1.23).

For example, if R is an unbounded rectangle, let $\{R_n\}_{n=1}^{\infty}$ be a nested collection of bounded rectangles, meaning $R_n \subset R_{n+1}$, with $\bigcup_{n=1}^{\infty} R_n = R$. Then when the limit exists, define:

$$(\mathcal{R}) \int_R f(x) dx = \lim_{n \to \infty} (\mathcal{R}) \int_{R_n} f(x) dx. \tag{1.53}$$

To generalize the above definition of Riemann integral over R to an integral over an arbitrary set $E \subset \mathbb{R}^n$ is in general quite difficult and requires restrictions on the nature of E. The most general approach requires E to be a **Jordan region,** named after Camille Jordan (1838–1922), who introduced the notion of the **Jordan measurability** of E.

Generalizing item 3 of Exercise 1.66, define the **boundary of a bounded set** E:

$$\partial E \equiv \{x \in \mathbb{R}^n | x \text{ is a limit point of } E \text{ and } \widetilde{E}\},$$

where $\widetilde{E}$ is the complement of E. Equivalently,

$$\partial E \equiv \{x \in \mathbb{R}^n | \text{for all } r > 0, \; B_r(x) \cap E \neq \emptyset \text{ and } B_r(x) \cap \widetilde{E} \neq \emptyset\},$$

where $B_r(x)$, the open "ball" about x of radius r, is defined as in Definition I.2.10 by $B_r(x) = \{y | \, |y - x| < r\}$.

Exercise 1.69 *Show that for a rectangle as in item 3 of Exercise 1.66, that the two definitions agree.*

Definition 1.70 (Jordan region) *A bounded set* $E \subset \mathbb{R}^n$ *is defined to be a **Jordan region** if for any* $\epsilon > 0$ *there exists a collection of open rectangles* $\{A_J\}_{J \in I}$ *so that* $\partial E \subset \bigcup_{J \in I} A_J$ *and* $\sum_{J \in I} m(A_J) < \epsilon$.

Intuitively, a Jordan region E has an arbitrarily "thin" boundary. Formally, $E \subset \mathbb{R}^n$ is defined to be a **Jordan region** if ∂E has Lebesgue measure 0, defined with respect to the Lebesgue product measure on $\mathbb{R}^n$ (Chapter I.7). A bounded rectangle $R \equiv \prod_{j=1}^{n}[a_j, b_j]$ is a Jordan region by item 3 of Exercise 1.66, and neither the collection of irrationals in R nor the collection of rationals in R is a Jordan region.

For such sets, $(\mathcal{R})\int_E f(x)dx$ can be defined in much the same way as was the integral over a rectangle R. First f is redefined to equal 0 outside E, then the upper and lower bounding step functions are defined with respect to rectangles for which $A_J \cap E \neq \emptyset$.

As we have no further application for this development, we leave it here, but with an illustrative exercise.

Exercise 1.71 *Prove that as a consequence of the approach of the above paragraph, that for a bounded, Riemann integrable function defined on R:*

$$(\mathcal{R})\int_R f(x)dx = (\mathcal{R})\int_{R'} f(x)dx, \tag{1.54}$$

where $R' \equiv \prod_{j=1}^{n}(a_j, b_j)$. Hint: Given a partition of mesh size μ, compare the values of the bounding Darboux sums over R and R'. Here, all of the subrectangles of R satisfy $A_J \cap R' \neq \emptyset$. But the collection of subrectangles $\{A'_J\}$ with $A_J \cap R' \neq \emptyset$ but $A_J \not\subseteq R'$ have arbitrarily small measure as $\mu \to 0$.

1.3.5 Properties of the Integral

Properties of the Riemann integral are the same as those seen in Proposition 1.23, with appropriate changes in notation. We document this result here for completeness and leave the proofs as an exercise. As noted above, existence of integrals below is assured by the Lebesgue existence theorem, so one only has to show these results using any of the criteria identified in Proposition 1.62.

Proposition 1.72 (Properties of the Riemann integral) *If $f(x)$ and $g(x)$ are bounded and Riemann integrable on a bounded rectangle $R = \prod_{j=1}^{n}[a_j, b_j]$, then suppressing the $(\mathcal{R})$-notation:*

1. **Triangle inequality:** *The integral $\int_R |f(x)|\,dx$ exists, and if $|f(x)| \leq c$ on R, then:*

$$\left|\int_R f(x)dx\right| \leq \int_R |f(x)|\,dx \leq cm(R),$$

where $|R| \equiv m(R)$, the Lebesgue measure of R, as given in (1.37).

Further, if $c_1 \leq f(x) \leq c_2$:

$$c_1|R| \leq \int_R f(x)dx \leq c_2|R|.$$

2. *For any constants k_1, k_2, $k_1 f(x) + k_2 g(x)$ is Riemann integrable and:*

$$\int_R (k_1 f(x) + k_2 g(x))\,dx = k_1 \int_R f(x)dx + k_2 \int_R g(x)dx.$$

3. *If R' is a bounded rectangle with $R' \subset R$, then $\int_{R'} f(x)dx$ exists where we define as in (1.39):*

$$\int_{R'} f(x)dx = \int_R \chi_{R'}(x)f(x)dx. \tag{1}$$

4. If $R = R' \bigcup R''$, a disjoint union of bounded rectangles, then $\int_{R'} f(x)dx$ and $\int_{R''} f(x)dx$ exist by item 3, and:

$$\int_R f(x)dx = \int_{R'} f(x)dx + \int_{R''} f(x)dx.$$

5. For a finite disjoint union of bounded rectangles $E \equiv \bigcup_{i=1}^n R_j \subset R$, the integral $\int_E f(x)dx$ exists as defined in (1), and:

$$\int_E f(x)dx = \sum_{i=1}^n \int_{R_i} f(x)dx.$$

Proof. *Left as an exercise. Hint: Revise the proofs of Proposition 1.23 using (1.36) and Proposition 1.62. Also note that for any rectangle R, the boundary of R has Lebesgue measure 0 by item 3 of Exercise 1.66.* ■

Item 1 of Proposition 1.72 has a simple yet useful corollary. Continuity is essential for this result as in the 1-dimensional case, and the reader is invited to provide examples of Riemann integrable functions for which this result fails.

Corollary 1.73 (Mean value theorem for integrals) *If $f(x)$ is continuous on $R = \prod_{j=1}^n [a_j, b_j]$, then there exists $c \in R$ so that:*

$$\int_R f(x)dx = f(c)|R|. \tag{1.55}$$

Proof. *Continuity of $f(x)$ obtains by Exercise 1.13 that $m \leq f(x) \leq M$ for all $x \in R$, and that these extremes are achieved with points in R. Then by item 1 of Proposition 1.72:*

$$m \leq \frac{1}{|R|} \int_R f(x)dx \leq M.$$

By the intermediate value theorem of the same exercise, there exists $c \in R$ so that (1.21) is satisfied. ■

Exercise 1.74 (Absolute Riemann integrable $\nRightarrow$ Riemann integrable) *Recalling Exercise 1.25, it is tempting to think that if $|f(x)|$ is Riemann integrable on R, then so too is $f(x)$ Riemann integrable on R, and that the above bound in item 1 of Proposition 1.72 applies. Prove by example that this is not true.*

1.3.6 Evaluating Integrals

Looking at the various criteria for the existence and value of the Riemann integral of $f(x)$ over a bounded rectangle $R \subset \mathbb{R}^n$ provided by Proposition 1.62, it is likely clear that outside of very easy examples, this calculation will be very difficult. An example of a very easy valuation is for a constant function, $f(x) = c$, since then Proposition 1.72 obtains:

$$\int_R f(x)dx = c|R|,$$

where $|R| = m(R)$ is easy to calculate by (1.37).

Example 1.75 ($f(x) = x_1$) *The next easiest example is* $f(x) = x_j$, *for which we assume* $j = 1$ *for notational simplicity, and* $R \equiv \prod_{j=1}^{n}[a_j, b_j]$. *As a continuous function, existence of the integral is assured by Proposition 1.63, and so we can seek to evaluate this integral with Riemann sums.*

Each interval $[a_j, b_j]$ *can be partitioned as in (1.1) using* $\{x_{j,i}\}_{i=1}^{m_j}$:

$$a_j = x_{j,0} < x_{j,1} < \cdots < x_{j,m_j-1} < x_{j,m_j} = b_j,$$

but a moment of thought justifies that since $f(x)$ *only depends on* x_1, *only* $[a_1, b_1]$ *need be so partitioned.*

This follows because in the Riemann sums of (1.36), $f(\widetilde{x}_J)$ *only depends on the first component of* $\widetilde{x}_J$, *recalling that* $J \equiv (i_1, i_2, ..., i_n)$. *Thus we can use the same* $\widetilde{x}_J$ *for every subrectangle* $A_J \equiv \prod_{j=1}^{n} A_{j,i_j}$ *with given* $A_{1,i_1} \equiv [x_{1,i_1-1}, x_{1,i_1}]$. *The union of the subrectangles with given* A_{1,i_1} *is then:*

$$\bigcup_{J \in I'} A_J = [x_{1,i_1-1}, x_{1,i_1}] \times \prod_{j=2}^{n}[a_j, b_j],$$

where I' *is the subset of* I *with fixed* i_1. *This obtains:*

$$\sum_{J \in I'} m(A_J) = (x_{1,i_1} - x_{1,i_1-1}) \prod_{j=2}^{n}(b_j - a_j).$$

Thus for any partition of R:

$$\sum_{J \in I} f(\widetilde{x}_J) m(A_J) = \prod_{j=2}^{n}(b_j - a_j) \sum_{i=1}^{m_1} \widetilde{x}_{1,i}(x_{1,i} - x_{1,i-1}),$$

where $\widetilde{x}_{1,i} \in A_{1,i}$. *This summation is recognized as the Riemann sum for the integral of* $g(x) = x$ *over* $[a_1, b_1]$, *and thus:*

$$(\mathcal{R})\int_R x_1 dx = \left(\frac{b_1 + a_1}{2}\right)|R|.$$

The same approach works equally well for $f(x) = x_j$.

Thus by Proposition 1.72:

$$(\mathcal{R})\int_R \sum_{j=1}^{n} e_j x_j dx = |R| \sum_{j=1}^{n} e_j \left(\frac{b_j + a_j}{2}\right).$$

Note that this is an example of the mean value theorem of Corollary 1.73, where $c = (b + a)/2$.

The reader is invited to attempt to integrate $f(x) = \prod_{j=1}^{n} x_j$ directly to motivate the need for a better approach. The following result will provide an algorithm for this and more general examples.

Proposition 1.76 (Riemann integral iteration, continuous $f(x)$**)** *Let* $f(x)$ *be continuous on a bounded rectangle* $R = \prod_{j=1}^{n}[a_j, b_j]$. *Then with* $R' \equiv \prod_{j=2}^{n}[a_j, b_j]$ *and* $x' \equiv (x_2, ..., x_n)$, *the function:*

$$g(x_1) \equiv \int_{R'} f(x_1, x_2, ..., x_n) dx',$$

is continuous on $[a_1, b_1]$, *and:*

$$(\mathcal{R})\int_R f(x)dx = (\mathcal{R})\int_{a_1}^{b_1} g(x_1)dx_1. \tag{1.56}$$

Proof. *First, $g(x_1)$ is well defined pointwise by Proposition 1.63, since $f(x') \equiv f(\cdot, x_2, ..., x_n)$ is continuous and bounded on R'. It then follows from (1.36) that:*

$$g(x_1) = \lim_{\mu' \to 0} \sum_{J \in I} f(x_1, \widetilde{x}'_J) |A'_J|, \tag{1}$$

where $\widetilde{x}'_J \in A'_J$, and $\{A'_J\}$ is the collection of partition subrectangles of mesh size μ' with $\bigcup_{J \in I} A'_J = R'$.

For continuity of $g(x_1)$:

$$\begin{aligned}
|g(x_1) - g(y_1)| \leq &\left| g(x_1) - \sum_{J \in I} f(x_1, \widetilde{x}'_J) |A'_J| \right| \\
&+ \left| g(y_1) - \sum_{J \in I} f(y_1, \widetilde{x}'_J) |A'_J| \right| \\
&+ \sum_{J \in I} |f(x_1, \widetilde{x}'_J) - f(y_1, \widetilde{x}'_J)| \, |A'_J|.
\end{aligned} \tag{2}$$

The first two terms in (2) can be made arbitrarily small for μ' small by (1).

Now f is uniformly continuous on R by Exercise 1.14, so given $\epsilon > 0$ there exists δ such that $|f(x) - f(y)| < \epsilon$ if $|x - y| < \delta$. For any partition of R' with mesh size $\mu' < \delta/\sqrt{n}$, it can be verified that if $\widetilde{x}'_J \in A'_J$ and $|x_1 - y_1| < \delta/\sqrt{n}$, then:

$$|(x_1, \widetilde{x}'_J) - (y_1, \widetilde{x}'_J)| < \delta. \tag{3}$$

Thus by (3), the third term in (2) is bounded by $\epsilon m(R')$. This proves that $g(x_1)$ is continuous on $[a_1, b_1]$, and thus Riemann integrable by Proposition 1.63.

We prove (1.56) using the upper and lower Darboux integrals. Let $\{A'_J\}_{J \in I}$ again denote the collection of partition subrectangles of partition $P_{\mu'}$ of mesh size μ' with $\bigcup_{J \in I} A'_J = R'$, and similarly $\{I_k\}_{k=1}^N$ a partition P_μ of $[a_1, b_1]$ of the mesh size μ. Then $\{I_k \times A'_J\}_{k,J}$ creates a partition $P_{\mu \times \mu'}$ of R of mesh size bounded by $\sqrt{2} \max\{\mu, \mu'\}$. In the notation of Proposition 6.62, modified for clarity, let $x \equiv (x_1, x')$ and define:

$$\varphi(f, P_{\mu \times \mu'}) \equiv \sum_{k,J} m_{k,J} \chi_{I_k \times A'_J}(x_1, x'), \quad m_{k,J} \equiv \inf\{f(x_1, x') | (x_1, x') \in I_k \times A'_J\}.$$

We similarly define $\psi(f, P_{\bar{\mu} \times \bar{\mu}'})$ in terms of $M_{J,k} \equiv \sup\{f(x_1, x') | (x_1, x') \in I_k \times A'_J\}$, and partitions $P_{\bar{\mu}'}$ of R' and $P_{\bar{\mu}}$ of $[a_1, b_1]$.

Since $m(I_k \times A'_J) = m(I_k) \, m(A'_J) \equiv |I_k| |A'_J|$ for all measurable rectangles (Proposition I.7.20):

$$\int_R \varphi(f, P_{\mu \times \mu'}) dx = \sum_{k=1}^N \left(\sum_{J \in I} m_{k,J} |A'_J| \right) |I_k|.$$

Letting $m_J(x_1) \equiv \inf\{f(x_1, x') | x' \in A'_J\}$, then $m_{k,J} \leq m_J(x_1)$ for any $x_1 \in I_k$ and so:

$$\begin{aligned}
\int_R \varphi(f, P_{\mu \times \mu'}) dx &\leq \sum_{k=1}^N \left(\sum_{J \in I} m_J(x_1) |A'_J| \right) |I_k| \\
&= \sum_{k=1}^N \left(\int_{R'} \varphi(f, P_{\mu'}) dx' \right) |I_k|.
\end{aligned} \tag{4}$$

Here $\varphi(f, P_{\mu'}) = \varphi(f, P_{\mu'})(x_1)$ is the subordinate step function for $f(x_1, x')$ over R' as a function of $x_1 \in I_k$.

Now $f(x_1, x')$ is Riemann integrable over R' for each x_1 as noted above, with integral $g(x_1)$ by definition. Further, $g(x_1)$ equals the supremum of sums over all partitions $P_{\mu'}$ or R' with subrectangles $\{A'_J\}_{J \in I}$:

$$g(x_1) = \sup_{\varphi(x') \leq f(x_1, x')} \int_{R'} \varphi(x') dx'.$$

This implies that $\int_{R'} \varphi(f, P_{\mu'})dx' \leq g(x_1)$ *for any* $x_1 \in I_k$, *and thus on this interval:*

$$\int_{R'} \varphi(f, P_{\mu'})dx' \leq \inf_{x_1 \in I_k} g(x_1).$$

Combining this with (4) *obtains:*

$$\int_R \varphi(f, P_{\mu \times \mu'})dx \leq \sum_{k=1}^{N} \inf_{x_1 \in I_k} g(x_1) |I_k| = \int_{a_1}^{b_1} \varphi(g, P_{\mu})dx_1, \tag{5}$$

where $\varphi(g, P_\mu)$ *is the subordinate step function for* $g(x_1)$ *induced by* P_μ.

A nearly identical set of steps with $\psi(f, P_{\bar{\mu} \times \bar{\mu}'})$ *defined above, and* $\psi(g, P_{\bar{\mu}})$ *the superior step function for* $g(x_1)$ *induced by* $P_{\bar{\mu}}$:

$$\int_R \psi(f, P_{\bar{\mu} \times \bar{\mu}'}) \geq \sum_{k=1}^{N} \sup_{x_1 \in I_k} g(x_1) |I_k| = \int_{a_1}^{b_1} \psi(g, P_{\bar{\mu}})dx_1. \tag{6}$$

Since by (1.6):

$$\int_{a_1}^{b_1} \varphi(g, P_{\mu})dx_1 \leq \int_{a_1}^{b_1} \psi(g, P_{\bar{\mu}})dx_1,$$

it follows from (5) *and* (6):

$$\int_R \varphi(f, P_{\mu \times \mu'})dx \leq \int_{a_1}^{b_1} \varphi(g, P_{\mu})dx_1 \leq \int_{a_1}^{b_1} \psi(g, P_{\bar{\mu}})dx_1 \leq \int_R \psi(f, P_{\bar{\mu} \times \bar{\mu}'}). \tag{7}$$

As (7) *is valid for any pairs of partitions* $(P_{\mu'}, P_\mu)$ *and* $(P_{\bar{\mu}'}, P_{\bar{\mu}})$ *of* $(R', [a_1, b_1])$, *we can take suprema on the left two integrals and infima on the right two. The outer integrals then equal* $(\mathcal{R}) \int_R f(x)dx$, *and thus the extrema of the inner integrals are equal. It follows by Definition 1.56 that these must equal* $(\mathcal{R}) \int_{a_1}^{b_1} g(x_1)dx_1$ *and the proof is complete.* ∎

Of course, there was nothing special about x_1 in the above result, and we could have afforded this special treatment to any component variable. Further, the inner integral $\int_{R'} f(x_1, x_2, ..., x_n)dx'$ that defined $g(x_1)$ is yet another Riemann integral, now on $R' \subset \mathbb{R}^{n-1}$, to which we can apply Proposition 1.76 again.

Iterating, we can thus evaluate a Riemann integral in terms of what are called **iterated integrals.**

Corollary 1.77 (Riemann integral by iterated integrals) *If* $f(x)$ *is continuous on a bounded rectangle* $R = \prod_{j=1}^{n}[a_j, b_j]$, *then:*

$$(\mathcal{R}) \int_R f(x)dx = (\mathcal{R}) \int_{a_n}^{b_n} \cdots \int_{a_1}^{b_1} f(x_1, x_2, ..., x_n)dx_1 \ldots dx_n. \tag{1.57}$$

Further, these iterated integrals can be implemented in any order.

Proof. *Proposition 1.76 can be applied with any component variate* x_j *first, and then the integral defined by* $g(x_j)$ *is a Riemann integral on* $'R' \subset \mathbb{R}^{n-1}$ *to which this result can be applied again, and so forth.* ∎

Example 1.78

1. $f(x) = \prod_{j=1}^{n} x_j$:

   To evaluate $(\mathcal{R}) \int_R f(x)dx$ with $R = \prod_{j=1}^{n}[a_j, b_j]$, we apply (1.57):

$$\int_R \prod_{j=1}^{n} x_j dx = \int_{a_n}^{b_n} \cdots \int_{a_1}^{b_1} \prod_{j=1}^{n} x_j dx_1 \ldots dx_n$$

$$= 2^{-1}\left(b_1^2 - a_1^2\right) \int_{a_n}^{b_n} \cdots \int_{a_2}^{b_2} \prod_{j=2}^{n} x_j dx_2 \ldots dx_n$$

$$\vdots$$

$$= |R| \prod_{j=1}^{n} \left(\frac{b_j + a_j}{2}\right)$$

   Note that this is another example of the mean value theorem of Corollary 1.73, where $c = (b + a)/2$.

2. $f(x) = \frac{\partial^n g}{\partial x_1 \ldots \partial x_n}$:

   Assume that $g(x_1, x_2, ..., x_n)$ is differentiable and $f(x)$ as defined is continuous. Then applying (1.57) with $R = \prod_{j=1}^{n}[a_j, b_j]$, and then (1.24):

$$\int_R \frac{\partial^n g}{\partial x_1 \ldots \partial x_n} dx = \int_{a_n}^{b_n} \cdots \int_{a_1}^{b_1} \frac{\partial^n g}{\partial x_1 \ldots \partial x_n} dx_1 \ldots dx_n$$

$$= \int_{a_n}^{b_n} \cdots \int_{a_2}^{b_2} \frac{\partial^{n-1} g(b_1, x_2, ..., x_n)}{\partial x_2 \ldots \partial x_n} dx_2 \ldots dx_n$$

$$- \int_{a_n}^{b_n} \cdots \int_{a_2}^{b_2} \frac{\partial^{n-1} g(a_1, x_2, ..., x_n)}{\partial x_2 \ldots \partial x_n} dx_2 \ldots dx_n.$$

   Simplifying notation:

$$I_n = I_{n-1}(b_1) - I_{n-1}(a_1),$$

   where $I_{n-1}(e)$ implies the integral of $\partial^{n-1} g / \partial x_2 \ldots \partial x_n$ over $\prod_{j=2}^{n}[a_j, b_j]$ with $x_1 = e$ fixed.

   Continuing in this fashion, we eventually obtain a summation of 2^n terms, since the number of terms doubles at each step. Further, each term in the sum equals $g(c_1, ..., c_n)$, where each $c_j \in \{a_j, b_j\}$. In other words, each $(c_1, ..., c_n)$ is a vertex of R. In addition, the coefficient of $g(c_1, ..., c_n)$ must be ± 1, and will be negative exactly when the number of $c_j = a_j$ is odd.

   Thus:

$$\int_R \frac{\partial^n g}{\partial x_1 \ldots \partial x_n} dx = \sum_c sgn(c_1, ..., c_n) g(c_1, ..., c_n),$$

   where $sgn(c_1, ..., c_n) = -1$ if the number of $c_j = a_j$ is odd, and is $+1$ otherwise.

   The observant reader may recognize this as item 2 of Example I.8.6 in the development of Borel measures on $\mathbb{R}^n$.

Remark 1.79 (Generalizations of Proposition 1.76) *Proposition 1.76 and Corollary 1.77 can be generalized beyond the standard iterated integral approach, and the assumption of continuity of f can be relaxed somewhat.*

1. Other Decompositions: *Let $x = (x_1, x_2) \in \mathbb{R}^n$ with $x_1 \in \mathbb{R}^p$ and $x_2 \in \mathbb{R}^{n-p}$ for $1 \leq p \leq n-1$. Then with the same proof and a small change of notation one obtains:*

Proposition 1.80 $(x = (x_1, x_2) \in \mathbb{R}^n$ with $x_1 \in \mathbb{R}^p$ and $x_2 \in \mathbb{R}^{n-p})$ *Let f be continuous on a bounded rectangle $R = \prod_{j=1}^n [a_j, b_j]$. With $R_2 \equiv \prod_{j=p+1}^n [a_j, b_j]$ and $x = (x_1, x_2)$, the function:*

$$g(x_1) \equiv \int_{R_2} f(x_1, x_2) dx_2,$$

is continuous on $R_1 \equiv \prod_{j=1}^p [a_j, b_j]$, and:

$$(\mathcal{R}) \int_R f(x) dx = (\mathcal{R}) \int_{R_1} g(x_1) dx_1.$$

Proof. *Left as an exercise, largely following the proof above.* ∎

2. On Continuity: *By Proposition 1.68, a bounded function need not be continuous on $R = \prod_{j=1}^n [a_j, b_j]$ to be Riemann integrable, only continuous almost everywhere. To generalize the above to such functions, one immediately encounters the question: If $f(x_1, x_2, ..., x_n)$ is only continuous almost everywhere on R, must $f(\cdot, x_2, ..., x_n)$ be continuous almost everywhere on $R' = \prod_{j=2}^n [a_j, b_j]$? The answer is "no" because the notion of a set having Lebesgue measure 0 is fundamentally related to dimension.*

For example, assume that f is continuous on R except on $S \equiv \{x \in R | x_1 = c \equiv (a_1 + b_1)/2\}$. As a set $S \subset \mathbb{R}^n$, $m(S) = 0$ since:

$$S \subset [c - \epsilon, c + \epsilon] \times \prod_{j=2}^n [a_j, b_j].$$

In $\mathbb{R}^n$ the set on the right has Lebesgue measure $2\epsilon \prod_{j=2}^n (b_j - a_j)$, which can be made as small as desired.

But now $f(\cdot, x_2, ..., x_n)$ is continuous on R' if $x_1 \neq c$, and continuous nowhere if $x_1 = c$.

Thus to generalize beyond continuity, we will not be able to assume that the restricted function $f(\cdot, x_2, ..., x_n)$ is continuous almost everywhere for each x_1, and thus Riemann integrable.

The following result can be proved as above, and generalized to the case in item 1. However, it does not lend itself to iteration as in the continuous case in Proposition 1.76. For its statement, recall Definition 1.57, and the observation of Remark 1.58 that upper and lower Darboux integrals always exist.

Proposition 1.81 (Riemann integral iteration, integrable $f(x)$) *If $f(x)$ is Riemann integrable on a bounded rectangle $R = \prod_{j=1}^n [a_j, b_j]$, then with $R' \equiv \prod_{j=2}^n [a_j, b_j]$ and $x' = (x_2, ..., x_n)$, define the partial upper and lower Darboux integrals:*

$$g^+(x_1) \equiv \overline{\int_{R'}} f(x_1, x_2, ..., x_n) dx',$$

$$g^-(x_1) \equiv \underline{\int_{R'}} f(x_1, x_2, ..., x_n) dx'.$$

Then $g^+(x_1)$ and $g^-(x_1)$ are Riemann integrable over $[a_1, b_1]$, and:

$$(\mathcal{R}) \int_R f(x) dx = (\mathcal{R}) \int_{a_1}^{b_1} g^+(x_1) dx_1 = (\mathcal{R}) \int_{a_1}^{b_1} g^-(x_1) dx_1.$$

Proof. *Looking to the above proof, the demonstration leading to* (4) *there remains unchanged, so we again obtain:*

$$\int_R \varphi(f, P_{\mu \times \mu'}) dx \leq \sum_{k=1}^{N} \left(\sum_{J \in I} m_J(x_1) m(A_J') \right) m(I_k)$$

$$= \sum_{k=1}^{N} \left(\int_{R'} \varphi(f, P_{\mu'}) dx' \right) m(I_k). \tag{4}$$

Here $\varphi(f, P_{\mu'}) = \varphi(f, P_{\mu'})(x_1)$ *is the subordinate step function for* $f(x_1, x')$ *over* R' *as a function of* $x_1 \in I_k$.

Now $g^-(x_1)$ *equals the supremum of sums over all partitions* $P_{\mu'}$ *of* R' *with subrectangles* $\{A_J'\}_{J \in I}$:

$$g^-(x_1) = \sup_{\varphi(x') \leq f(x_1, x')} \int_{R'} \varphi(x') dx'.$$

This implies that $\int_{R'} \varphi(f, P_{\mu'}) dx' \leq g^-(x_1)$ *for any* $x_1 \in I_k$, *and thus on this interval:*

$$\int_{R'} \varphi(f, P_{\mu'}) dx' \leq \inf_{x_1 \in I_k} g^-(x_1).$$

Combining this with (4) *obtains:*

$$\int_R \varphi(f, P_{\mu \times \mu'}) dx \leq \int_{a_1}^{b_1} \varphi(g^-, P_\mu) dx_1 \leq \int_{a_1}^{b_1} \psi(g^-, P_\mu) dx_1. \tag{5}$$

Here, $\varphi(g^-, P_\mu)$ *is the subordinate step function for* $g^-(x_1)$ *using* $\inf_{x_1 \in I_k} g^-(x_1)$ *for each interval* I_k, *and* $\psi(g^-, P_\mu)$ *the superior step function for* $g^-(x_1)$ *using* $\sup_{x_1 \in I_k} g^-(x_1)$.

A nearly identical set of steps with $\psi(f, P_{\mu \times \mu'})$ *defined above obtains:*

$$\int_R \psi(f, P_{\mu \times \mu'}) \geq \int_{a_1}^{b_1} \psi(g^+, P_\mu) dx_1. \tag{6}$$

Now $g^- \leq g^+$ *obtains:*

$$\int_{a_1}^{b_1} \varphi(g^-, P_\mu) dx_1 \leq \int_{a_1}^{b_1} \psi(g^+, P_{\bar{\mu}}) dx_1,$$

and it follows from (5) *and* (6):

$$\int_R \varphi(f, P_{\mu \times \mu'}) dx \leq \int_{a_1}^{b_1} \varphi(g^-, P_\mu) dx_1 \leq \int_{a_1}^{b_1} \psi(g^-, P_\mu) dx_1 \leq \int_R \psi(f, P_{\mu \times \mu'}). \tag{7}$$

As f *is Riemann integrable over* R, *this obtains that* $g^-(x_1)$ *is Riemann integrable over* $[a_1, b_1]$ *with:*

$$(\mathcal{R}) \int_R f(x) dx = (\mathcal{R}) \int_{a_1}^{b_1} g^-(x_1) dx_1.$$

A similar analysis obtains the Riemann integrability of $g^+(x_1)$ *and is left as an exercise.* ■

3. **Fubini's and Tonelli's theorems:** *The above results will be generalized substantially in Book V, and be applicable to general Lebesgue-Stieltjes integrals on product spaces. Such integrals include the Lebesgue integrals of the next chapter as a special case. But*

as it is no easier to derive these results in the special Lebesgue case, we will defer the general development to Book V.

Fubini's Theorem *is named after* **Guido Fubini** *(1879–1943). There are several versions of this result and we present here the one applicable in the Lebesgue integral context of the next chapter. In the notation of item 1 above, let $x = (x_1, x_2) \in \mathbb{R}^n$ with $x_1 \in \mathbb{R}^p$ and $x_2 \in \mathbb{R}^{n-p}$ for $1 \le p \le n - 1$. Assume that $f(x_1, x_2)$ is Lebesgue integrable over a product set $R_1 \times R_2$. Then Fubini's theorem states that $f(x_1, \cdot)$ is Lebesgue measurable and integrable over R_1 for* **almost all** *x_2, that $f(\cdot, x_2)$ is Lebesgue measurable and integrable over R_2 for* **almost all** *x_1. Further, the Lebesgue integral of these iterated integrals exists almost everywhere, are integrable, and:*

$$\int_{R_1 \times R_2} f(x)dx = \int_{R_2} \left[\int_{R_1} f(x_1, x_2)dx_1 \right] dx_2 \qquad (1.58)$$
$$= \int_{R_1} \left[\int_{R_2} f(x_1, x_2)dx_2 \right] dx_1.$$

Tonelli's Theorem *is named after* **Leonida Tonelli** *(1885–1946) and addresses the same question as does Fubini's theorem. Here we assume that $f(x_1, x_2)$ is nonnegative and Lebesgue measurable on a product set $R_1 \times R_2$. Then the various measurability results of Fubini related to $f(x_1, \cdot)$ and $f(\cdot, x_2)$ and their respective integrals hold, as does (1.58), though all integrals may be infinite.*

To apply Fubini's theorem requires that one first establish the integrability of $f(x)$ over $R_1 \times R_2$, which could be difficult. But then it assures existence of iterated integrals and the identity in (1.58). While requiring the extra assumption of nonnegativity, Tonelli's result can be used even if the integrability of $f(x)$ is unknown. Specifically, if one can evaluate either of the iterated integrals and confirm it is finite, then all the integrals in (1.58) agree and are thus finite. If any iterated integral is infinite, then all integrals again agree, and thus all are infinite.

For Lebesgue measurable $f(x)$, Tonelli's result can be used to prove integrability of $|f(x)|$. This then implies integrability of $f(x)$, and then the iterated results of Fubini apply. While Riemann integrability of $|f(x)|$ does not imply such integrability of $f(x)$ as noted in Exercises 1.25 and 1.74, this implication is valid for Lebesgue integrals below, and the Lebesgue-Stieltjes integrals of Book V.

2

The Lebesgue Integral

The approach to integration developed in this chapter was informally introduced in Chapter I.1, and originally developed by **Henri Lebesgue** (1875–1941) in 1904. Here we develop some of the existence theory for the Lebesgue integral using a standard sequential approach. Specifically, we first define the Lebesgue integral of simple functions, then prove in steps how this definition obtains the integrals of bounded measurable functions, then positive measurable functions, then finally general measurable functions. The reader will note that this development provides an example of the generalization of the Darboux framework discussed in Remark 1.9.

One of the more striking things about the following development of Lebesgue integration theory is that it is effectively dimensionless. That is, until we get to the differentiation theory of the next chapter that is restricted to $n = 1$, the integration theory itself can be developed with the same steps without committing, other than notationally, whether the sets and functions are defined on $\mathbb{R}$ or defined on $\mathbb{R}^n$. For background on Lebesgue measure spaces, the reader is referred to Chapter I.1 for the Lebesgue measure space $(\mathbb{R}, \mathcal{M}_L(\mathbb{R}), m)$, and to Chapter I.7 and especially Section I.7.6 for $(\mathbb{R}^n, \mathcal{M}_L(\mathbb{R}^n), m)$.

The development that follows will periodically refer to Book I results that were often explicitly developed within the 1-dimensional Lebesgue measure space. However as was often noted there, these results remain true in n-dimensional Lebesgue measure spaces as well as often in more general measure spaces. The justification for such extensions is that the proofs of the referenced results will be seen to reflect only standard properties of sigma algebras and measurable functions, and thus by simply reimagining the notation to refer to the more general situation, one obtains the more general proof.

One limitation in such generalizations is that some of the Book I results explicitly depended on the completeness of the 1-dimensional Lebesgue measure space. Thus while these results generalize well to n-dimensional Lebesgue measure space or other measure spaces that are also complete, such results will typically not be valid in more general settings without completeness.

The development below can largely be transplanted to take place on a general measure space. As this would introduce a level of abstraction not needed at this point, this general development is deferred to Book V.

2.1 Integrating Simple Functions

Lebesgue integration theory begins with the explicit definition of the Lebesgue integral for the simplest measurable functions imaginable, which are thus unsurprisingly called "simple functions." Once this definition is proved to be consistent and some properties of this integral developed, the remainder of this chapter proves that this definition extends in a well-defined way to more general classes of functions.

DOI: 10.1201/9781003264590-2

Notation 2.1 (On Lebesgue measure m^n; $\mathbb{R}^n$) *As noted in Notation 1.51, the notational convention of m^n was introduced in Chapter I.7 for Lebesgue measure on $\mathbb{R}^n$, where it was derived as an example of a product measure based on the Lebesgue measure m on $\mathbb{R}$. In this book we will drop this now unnecessary extra notation and use m for both the Lebesgue measure on $\mathbb{R}$ and on $\mathbb{R}^n$. There should be no confusion given the context.*

Also, while we generally denote the real Euclidean space by $\mathbb{R}^n$ as is conventional, it is also the case that in some proofs the variable n is called upon for higher service. Then this space will be denoted $\mathbb{R}^p$ or similar, to avoid confusion.

Definition 2.2 (Simple function) *A **simple function** $\varphi(x)$ defined on $(\mathbb{R}^n, \mathcal{M}_L(\mathbb{R}^n), m)$ is a bounded function given by:*

$$\varphi(x) = \sum\nolimits_{i=1}^{m} a_i \chi_{A_i}(x), \qquad (2.1)$$

where:

1. *$a_i \neq 0$ for all i.*

2. *$\{A_i\}_{i=1}^m \subset \mathcal{M}_L \equiv \mathcal{M}_L(\mathbb{R}^n)$ are **disjoint Lebesgue measurable sets** with $m(\bigcup_{i=1}^m A_i) < \infty$.*

3. *$\chi_{A_i}(x)$ is the **characteristic function** or **indicator function** for A_i, defined in (1.8).*

Exercise 2.3 (On disjointness of $\{A_i\}_{i=1}^m$) *It is not necessary to assume that $\{A_i\}_{i=1}^m$ are disjoint, but it is convenient for simplifying proofs.*

Prove that if $\varphi(x)$ is defined as in (2.1) with general $\{A_i\}_{i=1}^m$, that it can be equivalently defined with disjoint $\{B_j\}_{j=1}^M$ with $m(\bigcup_{i=1}^m A_i) = m(\bigcup_{j=1}^M B_j)$, and appropriate $\{b_j\}_{j=1}^M$, noting that some b_j may equal 0 and will be discarded. Hint: Use induction starting with $\{A_1, A_2\}$. Then with $C \equiv A_1 \bigcap A_2$:

$$A_1 \bigcup A_2 = C \bigcup (A_1 - C) \bigcup (A_2 - C).$$

Now if this result is true for $\{A_i\}_{i=1}^m$, apply the above to each A_i and A_{m+1} to prove true for $\{A_i\}_{i=1}^{m+1}$. Now identify $\{b_j\}_{j=1}^M$.

Next, recalling the discussion of Remark 1.9, we **define** the **Lebesgue integral of a simple function:**

Definition 2.4 (Lebesgue integral of a simple function) *The Lebesgue integral of the simple function in (2.1) is defined by:*

$$(\mathcal{L}) \int \varphi(x) dx = \sum\nolimits_{i=1}^{m} a_i m(A_i). \qquad (2.2)$$

Notation 2.5 (dx vs. dm) *In much of this chapter, Lebesgue integrals will be denoted as in 2.2, with a dx differential and an $(\mathcal{L})$-modifier preceding the integral sign. This is a standard notation, especially in a development that seeks to explore relationships between this integral and the Riemann counterpart.*

However, one will also commonly see the following notational convention:

$$(\mathcal{L}) \int f(x) dx \equiv \int f(x) dm,$$

in which the Lebesgue measure m underlying this integral is highlighted. This latter notational convention will be reflected somewhat in Chapter 4 in which Riemann-Stieltjes integrals are developed, and extensively in Book V where the notation $d\mu$ will replace dm.

Remark 2.6 (On (2.2)) *The definition of* $(\mathcal{L}) \int \varphi(x)dx$ *for simple* $\varphi(x)$ *reflects two basic "axioms" of integration theory:*

1. *For measurable* A:

$$(\mathcal{L}) \int \chi_A(x)dx = m(A).$$

2. *Lebesgue integrals are* **linear** *as operators on the integrand. That is, given integrable* $f(x)$ *and* $g(x)$, *and* $a, b \in \mathbb{R}$:

$$\int (af(x) + bg(x))\, dx = a \int f(x)dx + b \int g(x)dx.$$

Both of these requirements will be seen in the general development of Book V as:

$$\int \chi_A(x)d\mu = \mu(A),$$
$$\int (af(x) + bg(x))\, d\mu = a \int f(x)d\mu + b \int g(x)d\mu.$$

We will see that building this linearity property into simple functions is enough to make integrals linear in general.

As was the case for step functions in Example 1.10, it must be verified that the above definition is well defined. However, even from (2.2), it follows that simple function representations can be allowed to differ on sets of Lebesgue measure 0 without changing the value of this integral. So in the following proposition, we require only (2.4) for any two such representations, rather than that $\bigcup_i A_i = \bigcup_j A'_j$.

Recalling Definition I.4.1, (2.4) can also be stated:

$$m \left(\bigcup_j A'_j \triangle \bigcup_i A_i \right) = 0,$$

using the **symmetric difference** operator $\triangle$:

$$A \triangle B \equiv (A - B) \bigcup (B - A). \tag{2.3}$$

To accommodate this slightly more general setting, we prove that given simple functions with $\varphi(x) = \psi(x)$ a.e., meaning **almost everywhere** or outside a set of Lebesgue measure zero, then:

$$(\mathcal{L}) \int \varphi(x)dx = (\mathcal{L}) \int \psi(x)dx.$$

Proposition 2.7 (Well definedness of (2.2)) *Assume that the simple function* $\psi(x)$ *is given:*

$$\psi(x) = \sum_{j=1}^{m'} a'_j \chi_{A'_j}(x),$$

with disjoint Lebesgue measurable $\{A'_j\}_{j=1}^{m'} \subset \mathcal{M}_L(\mathbb{R}^n)$, $a'_j \neq 0$ *for all* j, *and* $\varphi(x) = \psi(x)$ *a.e. with* $\varphi(x)$ *as given in (2.1).*

Then:

$$m \left(\bigcup_i A_i - \bigcup_j A'_j \right) = m \left(\bigcup_j A'_j - \bigcup_i A_i \right) = 0, \tag{2.4}$$

and:

$$\sum_{j=1}^{m'} a'_j m(A'_j) = \sum_{i=1}^{m} a_i m(A_i).$$

Hence, the Lebesgue integral of a simple function is well defined by (2.2), and this integral is unchanged by redefinitions on sets of Lebesgue measure zero.

Proof. *Since $a_i \neq 0$ and $a'_j \neq 0$ for all i, j, it follows that $\varphi(x) = 0$ exactly on $\left(\bigcup_i A_i\right)^c$, and $\psi(x) = 0$ exactly on $\left(\bigcup_j A'_j\right)^c$, where B^c denotes the complement of B. Thus since:*

$$\bigcup_i A_i - \bigcup_j A'_j = \{x | \varphi(x) > 0 \text{ and } \psi(x) = 0\},$$

this set has measure 0 since $\varphi(x) = \psi(x)$ a.e. Repeating the argument derives (2.4).

It follows from this that $m\left(\bigcup_j A'_j\right) = m\left(\bigcup_i A_i\right)$ since:

$$\bigcup_i A_i = \left(\bigcup_i A_i \bigcap \bigcup_j A'_j\right) \bigcup \left(\bigcup_i A_i - \bigcup_j A'_j\right),$$
$$\bigcup_j A'_j = \left(\bigcup_i A_i \bigcap \bigcup_j A'_j\right) \bigcup \left(\bigcup_j A'_j - \bigcup_i A_i\right),$$

and thus these sets agree outside sets of measure 0 by (2.4).

Define Lebesgue measurable $B_{ij} = A_i \bigcap A'_j$, and note that:

$$\bigcup_j B_{ij} = A_i \bigcap \bigcup_j A'_j,$$

and thus by finite additivity of m and (2.4):

$$\sum_{j=1}^{m'} m(B_{ij}) = m(A_i). \tag{1}$$

The same derivation obtains:

$$\sum_{i=1}^{m} m(B_{ij}) = m(A'_j). \tag{2}$$

Now on any B_{ij} with $m(B_{ij}) > 0$, it follows from $\varphi(x) = \psi(x)$ a.e. that $a_i = a'_j$ on B_{ij}, so define $b_{ij} = a_i = a'_j$. If $m(B_{ij}) = 0$ then choose b_{ij} arbitrarily, but finite. Defining:

$$\phi(x) = \sum_{j=1}^{m'} \sum_{i=1}^{m} b_{ij} \chi_{B_{ij}}(x),$$

it follows from (1) and (2) that:

$$\sum_{j=1}^{m'} a'_j m(A'_j) = \sum_{j=1}^{m'} \sum_{i=1}^{m} b_{ij} m(B_{ij}) = \sum_{i=1}^{m} a_i m(A_i).$$

■

We now record two basic properties of the Lebesgue integral of simple functions.

Proposition 2.8 (Basic properties) *Let $\varphi(x)$ and $\psi(x)$ be simple functions. Then suppressing the $(\mathcal{L})$-notation:*

1. *If $\varphi(x) \leq \psi(x)$ except on a set of Lebesgue measure 0, then:*

$$\int \varphi(x)dx \leq \int \psi(x)dx. \tag{2.5}$$

2. *For any real constants a and b:*

$$\int [a\varphi(x) + b\psi(x)]dx = a \int \varphi(x)dx + b \int \psi(x)dx, \tag{2.6}$$

and this generalizes to any finite summation by induction.

Proof. *Left as an exercise.* ∎

We next define the Lebesgue integral of function over a Lebesgue measurable set E. This definition will apply generally below, so it is stated herein that general context despite the open question currently of the existence of such integrals beyond simple functions.

Definition 2.9 (On $\int_E$) *If $E \in \mathcal{M}_L(\mathbb{R}^n)$ is a Lebesgue measurable set and $f(x)$ a Lebesgue measurable function, then when the integral on the right exists, define:*

$$(\mathcal{L}) \int_E f(x)dx \equiv (\mathcal{L}) \int \chi_E(x)f(x)dx, \tag{2.7}$$

where $\chi_E(x)$ denotes the characteristic function of E defined in (1.8).

When $n = 1$ and $E = [a, b]$, it is customary to use the notation of Riemann integration:

$$(\mathcal{L}) \int \chi_{[a,b]}(x)f(x)dx \equiv (\mathcal{L}) \int_a^b f(x)dx.$$

Remark 2.10 ($\int_E$ for simple functions) *The existence of the integral on the right in (2.7) is assumed in this definition. But note that if $\varphi(x)$ is the simple function in (2.1), then the integral on the right does indeed exist because $\chi_E(x)\varphi(x)$ is then also a simple function.*

Specifically:

$$\chi_E(x)\varphi(x) = \sum_{i=1}^m a_i \chi_{A_i \cap E}(x),$$

and thus by (2.2):

$$(\mathcal{L}) \int_E \varphi(x)dx = \sum_{i=1}^m a_i m\left(A_i \bigcap E\right). \tag{2.8}$$

This implies that if $m(E) = 0$, then for all simple functions $\varphi(x)$:

$$(\mathcal{L}) \int_E \varphi(x)dx = 0.$$

Example 2.11 *Define the simple function:*

$$\varphi(x) = \sum_{j=1}^m \frac{j}{m} \chi_{I_j}(x),$$

where $I_j = [(j-1)/m, j/m]$. Then:

$$(\mathcal{L}) \int_0^1 \varphi(x)dx = \sum_{j=1}^m \frac{j}{m} m(I_j) = (m+1)/2m.$$

Let $E \subset [0,1]$ denote the collection of irrationals, which is Lebesgue measurable with $m(E) = 1$. This follows by finite additivity since $\tilde{E} \equiv [0,1] - E$ are the rationals in $[0,1]$, a set of measure 0. Then by (2.6):

$$\begin{aligned}
(\mathcal{L}) \int_E \varphi(x)dx &\equiv (\mathcal{L}) \int_0^1 \sum_{j=1}^m \frac{j}{m} \chi_{I_j \cap E}(x)dx \\
&= \sum_{j=1}^m \frac{j}{m} m\left(I_j \bigcap E\right) \\
&= (m+1)/2m.
\end{aligned}$$

One the other hand:

$$(\mathcal{L}) \int_{\tilde{E}} \varphi(x)dx = 0,$$

since $m(I_j \cap \tilde{E}) = 0$ for all j.

Exercise 2.12 *Generalizing Example 2.11, show that for a simple function $\varphi(x)$ and any $E \in \mathcal{M}_L$:*

$$\int \varphi(x)dx = \int_E \varphi(x)dx + \int_{\tilde{E}} \varphi(x)dx,$$

where we suppress the notational $(\mathcal{L})$.

2.2   Integrating General Functions

As was the case for Riemann integrals, Lebesgue integrals will ultimately be defined with unbounded measurable functions as integrands and/or on unbounded measurable domains. However, in contrast to Riemann integration, the integral $(\mathcal{L}) \int_E f(x)dx$ will only be **defined** to exist if $(\mathcal{L}) \int_E |f(x)| \, dx$ exists. In other words, $f(x)$ will be defined to be Lebesgue integrable only when $f(x)$ is absolutely Lebesgue integrable.

A powerful motivation for defining a general measurable function to be Lebesgue integrable only in such cases is the desire to be able to decompose the domains of integrals in an arbitrary way, and preserve both integrability and additivity.

Generalizing Exercise 2.12, let $\{E_j\}$ be disjoint Lebesgue measurable sets with $E' \equiv \bigcup_j E_j \subset E$. Then if the integral of $f(x)$ exists over E, we would like to be sure that $\int_{E'} f(x)dx$ is well defined, and further:

$$\int_{E'} f(x)dx = \sum_j \int_{E_j} f(x)dx.$$

As noted in Example 1.27 and Exercise 1.28, such an identity need not be satisfied for Riemann integrals when $f(x)$ is not absolutely Riemann integrable.

The approach followed for the development of the general Lebesgue theory starts with the definition of the integral of **simple functions** accomplished above. The next steps below are then:

1. Generalize to the Lebesgue integral of **bounded measurable functions** that equal 0 outside sets of finite measure. These integrals will be defined in terms of the Lebesgue integrals of dominant and subordinate simple functions, using the Darboux framework discussed in Chapter 1.

2. Develop a general theory for **nonnegative Lebesgue measurable functions,** $f(x) \geq 0$, using subordinate bounded functions that equal 0 outside sets of finite measure, and their integrals from step 1.

3. Express a **general measurable function** as the sum of a "positive" and "negative" part and extend the results in step 2 to this situation. A consequence of this approach is that for a general measurable function $f(x)$ to be Lebesgue integrable, both the positive and the negative parts of the function must be integrable. As will be seen, this assures the integrability of $|f(x)|$.

2.3   Bounded Measurable Functions

In this section, we extend the definition of Lebesgue integral from simple functions to bounded functions that equal 0 outside a set of finite measure. Since simple functions are bounded and 0 outside such sets, this first extension is relatively straightforward.

2.3.1 Definition of the Integral

For $E \in \mathcal{M}_L(\mathbb{R}^n)$ with $m(E) < \infty$, and a function $f(x)$ that is bounded on E, meaning $|f(x)| \leq M$ for $x \in E$, we seek to define $(\mathcal{L}) \int_E f(x) dx$. By Definition 2.9:

$$(\mathcal{L}) \int_E f(x) dx \equiv (\mathcal{L}) \int \chi_E(x) f(x) dx, \tag{2.9}$$

and thus our goal is to determine when $\chi_E(x) f(x)$ is Lebesgue integrable.

The approach of this section will be to bound $\chi_E(x) f(x)$ by subordinate and dominant simple functions:

$$\varphi(x) \leq \chi_E(x) f(x) \leq \psi(x),$$

and investigate "limits" of the associated simple function integrals. Since $\chi_E(x) f(x) = 0$ for $x \notin E$, there is no loss of generality by considering simple functions with:

$$\chi_E(x) \varphi(x) \leq \chi_E(x) f(x) \leq \chi_E(x) \psi(x). \tag{1}$$

Then, as in the Darboux framework, we investigate the infimum of such $(\mathcal{L}) \int_E \psi(x) dx$, and the supremum of such $(\mathcal{L}) \int_E \varphi(x) dx$, and determine when these agree. For this, note that these simple function integrals are well-defined by Definition 2.4 as applied in (2.8) of Remark 2.10.

In cases where the infimum and supremum agree, it follows by definition that there exists sequences $\{\varphi_n(x)\}$ and $\{\psi_n(x)\}$ so that for all n:

$$\chi_E(x) \varphi_n(x) \leq \chi_E(x) f(x) \leq \chi_E(x) \psi_n(x),$$

and dropping the $(\mathcal{L})$:

$$\lim_{n \to \infty} \int_E \psi_n(x) dx = \lim_{n \to \infty} \int_E \varphi_n(x) dx.$$

The integral $(\mathcal{L}) \int_E f(x) dx$ will then be defined by this common limit.

Remark 2.13 (On measurability of $f(x)$) *The following result states that given measurable E with $m(E) < \infty$, and $f(x)$ bounded on E, meaning $\chi_E(x) f(x)$ is bounded, the Lebesgue integral $(\mathcal{L}) \int_E f(x) dx$ exists in the above sense if and only if $\chi_E(x) f(x)$ is Lebesgue measurable. By Corollary I.3.27, such measurability means that for all $A \in \mathcal{B}(\mathbb{R}^n)$:*

$$f^{-1}(A) \bigcap E \in \mathcal{M}_L(\mathbb{R}^n).$$

Thus $(\mathcal{L}) \int_E f(x) dx$ exists for given E if and only if $f(x)$ is Lebesgue measurable when **restricted** *to E.*

More generally, this result implies that for bounded $f(x)$, the integral $(\mathcal{L}) \int_E f(x) dx$ exists for **all measurable** *E with $m(E) < \infty$ if and only if $f(x)$ is Lebesgue measurable.*

One direction is easy. If $f(x)$ is Lebesgue measurable then so too is $\chi_E(x) f(x)$ for any measurable E by Proposition I.3.30, and thus $(\mathcal{L}) \int_E f(x) dx$ exists for all measurable E with $m(E) < \infty$.

The other direction perhaps seems obvious but it is not. If $\chi_E(x) f(x)$ is Lebesgue measurable for all Lebesgue measurable E with $m(E) < \infty$, must $f(x)$ be Lebesgue measurable?

Exercise 2.14 *Prove this last statement is answered in the affirmative. Hint: This result is not true on a general measure space, and to prove it on $(\mathbb{R}^n, \mathcal{M}_L(\mathbb{R}^n), m)$ will use that this Lebesgue measure space is σ-finite (Definition I.5.34).*

Proposition 2.15 (Bounded f and $m(E) < \infty$: Integrable $\Leftrightarrow$ measurable) *Let $f(x)$ be bounded on a measurable set $E \in \mathcal{M}_L(\mathbb{R}^p)$ with $m(E) < \infty$, meaning that $\chi_E(x)f(x)$ is bounded. Then with simple functions $\varphi(x)$, $\psi(x)$ as in (1) above:*

$$\inf_{\psi \geq f} \int_E \psi(x)dx = \sup_{\varphi \leq f} \int_E \varphi(x)dx, \qquad (2.10)$$

if and only if $\chi_E(x)f(x)$ is Lebesgue measurable.

Thus given bounded $f(x)$, (2.10) is true for all $E \in \mathcal{M}_L(\mathbb{R}^p)$ with $m(E) < \infty$ if and only if $f(x)$ is Lebesgue measurable.

Proof. *If $\chi_E(x)f(x)$ is Lebesgue measurable and bounded, $|\chi_E(x)f(x)| \leq M$, then for given n define the level sets E_j for $-n \leq j \leq n$ by:*

$$E_j = \left\{ x \,|\, (j-1)\frac{M}{n} < f(x) \leq j\frac{M}{n} \right\} \bigcap E.$$

Note that $\{E_j\}_{j=-n}^{n}$ are measurable, disjoint, and satisfy $\bigcup_{j=-n}^{n} E_j = E$.
   Define simple functions:

$$\psi_n(x) = \frac{M}{n} \sum_{j=-n}^{n} j\chi_{E_j}(x),$$

$$\varphi_n(x) = \frac{M}{n} \sum_{j=-n}^{n} (j-1)\chi_{E_j}(x),$$

and note that $\varphi_n(x) \leq \chi_E(x)f(x) \leq \psi_n(x)$. In addition, for any n:

$$\inf_{\psi \geq f} \int_E \psi(x)dx \leq \int_E \psi_n(x)dx = \frac{M}{n} \sum_{j=-n}^{n} jm(E_j),$$

$$\sup_{\varphi \leq f} \int_E \varphi(x)dx \geq \int_E \varphi_n(x)dx = \frac{M}{n} \sum_{j=-n}^{n} (j-1)m(E_j).$$

   Thus by finite additivity:

$$\begin{aligned}
0 &\leq \inf_{\psi \geq f} \int_E \psi(x)dx - \sup_{\varphi \leq f} \int_E \varphi(x)dx \\
&\leq \frac{M}{n} \sum_{j=-n}^{n} m(E_j) \\
&= \frac{M}{n} m(E).
\end{aligned}$$

Letting $n \to \infty$, (2.10) is verified since $m(E) < \infty$.
   Conversely, the identity in (2.10) implies the existence of simple function sequences $\{\varphi_n(x)\}$ and $\{\psi_n(x)\}$ with $\chi_E(x)\varphi_n(x) \leq \chi_E(x)f(x) \leq \chi_E(x)\psi_n(x)$ and:

$$\lim_{n \to \infty} \int_E \psi_n(x)dx = \lim_{n \to \infty} \int_E \varphi_n(x)dx.$$

Choosing subsequences and renumbering obtains for all n:

$$0 < \int_E \psi_n(x)dx - \int_E \varphi_n(x)dx < 1/n. \qquad (2.11)$$

By Proposition I.3.47, which applies in general measure spaces, $\psi(x) \equiv \inf \chi_E(x)\psi_n(x)$ and $\varphi(x) \equiv \sup \chi_E(x)\varphi_n(x)$ are Lebesgue measurable, and:

$$\varphi(x) \leq \chi_E(x)f(x) \leq \psi(x).$$

We will prove that $\varphi(x) = \psi(x)$ a.e., meaning outside a set of Lebesgue measure 0, and thus $\chi_E(x)f(x) = \varphi(x) = \psi(x)$ a.e. Lebesgue measurability of $\chi_E(x)f(x)$ then follows by Proposition I.3.16 and Exercise I.3.15, since $(\mathbb{R}^p, \mathcal{M}_L(\mathbb{R}^p), m)$ is a complete measure space for all p.

To this end, $\varphi(x) \le \psi(x)$ by definition, so consider:

$$\{x|\varphi(x) < \psi(x)\} = \bigcup_k E_k, \tag{1}$$

where here $E_k \equiv \{x|\varphi(x) < \psi(x) - 1/k\}$. If $\varphi(x) < \psi(x) - 1/k$, then the bounds $\chi_E(x)\varphi_n(x) \le \varphi(x)$ and $\psi(x) \le \chi_E(x)\psi_n(x)$ imply that for all n:

$$E_k \subset \{x|\chi_E(x)\varphi_n(x) < \chi_E(x)\psi_n(x) - 1/k\}.$$

Then (2.11), (2.6), and $0 < \chi_E(x)\psi_n(x) - \chi_E(x)\varphi_n(x)$ obtain:

$$1/n > \int_E (\psi_n(x) - \varphi_n(x))\,dx \ge \int_{E_k} (\psi_n(x) - \varphi_n(x))\,dx \ge \frac{m(E_k)}{k}.$$

Thus for all n:

$$m(E_k) \le \frac{k}{n},$$

and $m(E_k) = 0$ for all k.

By (1) and countable subadditivity, $m\{x|\varphi(x) < \psi(x)\} = 0$. Thus $\varphi(x) = \psi(x)$ almost everywhere, and $\chi_E(x)f(x)$ is Lebesgue measurable.

The final statement is proved in Remark 2.13 and Exercise 2.14. ∎

Proposition 2.15 proves that for a bounded, measurable function $f(x)$ defined on a set E of finite measure, the Lebesgue integral can be defined in terms of either the infimum of simple functions $\psi(x)$ with $f(x) \le \psi(x)$ on E, or the supremum of simple functions $\varphi(x)$ with $\varphi(x) \le f(x)$ on E. In the following, we will choose whichever approach is more convenient for the given application.

For the following definition note that as a consequence of Remark 2.10, that if $m(E) = 0$ then:

$$(\mathcal{L}) \int_E f(x)dx = 0,$$

for all bounded Lebesgue measurable functions $f(x)$.

Definition 2.16 (Lebesgue integral of bounded $f(x)$, $m(E) < \infty$) *If $f(x)$ is a bounded, Lebesgue measurable function defined on a Lebesgue measurable set $E \in \mathcal{M}_L(\mathbb{R}^n)$ with $m(E) < \infty$, the Lebesgue integral of $f(x)$ over E is defined by either:*

$$(\mathcal{L}) \int_E f(x)dx = \inf_{\psi \ge f} \int_E \psi(x)dx, \tag{2.12}$$

or

$$(\mathcal{L}) \int_E f(x)dx = \sup_{\varphi \le f} \int_E \varphi(x)dx, \tag{2.13}$$

where ψ and φ are simple functions with integrals defined by (2.8).

2.3.2 Riemann Implies Lebesgue

In this section, we show that any **bounded function** that is Riemann integrable on $R \equiv \prod_{j=1}^{n}[a_j, b_j]$ is in fact Lebesgue integrable, and thus Lebesgue measurable by Proposition 2.15, and the values of these integrals agree. Consequently, at least for bounded functions on closed rectangles, Lebesgue integration is a generalization of Riemann integration.

Before proceeding, the reader might well think that this result is trivial based on the Chapter 1 development and Proposition 2.15.

Remark 2.17 (Is Proposition 2.18 needed?) *The conclusion of the next result, that Riemann integrability implies Lebesgue integrability, is sometimes thought to be obvious, but due to a somewhat hasty application of prior results.*

1. From the theory of Riemann integration, the **Lebesgue existence theorem for the Riemann integral** of Proposition 1.22 states:

   Proposition (Riemann Integrable): *If $f(x)$ is a bounded function on the rectangle $R \equiv \prod_{j=1}^{n}[a_j, b_j]$, then $(\mathcal{R})\int_R f(x)dx$ exists if and only if $f(x)$ is continuous except on a set of Lebesgue measure 0.*

2. *From Proposition 2.15:*

   Proposition (Lebesgue Integrable): *If $f(x)$ is a bounded function on Lebesgue measurable set E with $m(E) < \infty$, then $(\mathcal{L})\int_E f(x)dx$ exists if and only if $f(x)$ is Lebesgue measurable (on E).*

3. *Combining results on Lebesgue measurable functions from Book I:*

   Continuous vs. Measurable:

   a. *If $f(x)$ is continuous on a Lebesgue measurable set E, then it is Lebesgue measurable on E (Proposition I.3.11 and Remark I.3.10).*

   b. *If $f(x)$ is Lebesgue measurable on a Lebesgue measurable set E, and $g(x) = f(x)$ a.e., meaning except on a set of measure 0, then $g(x)$ is Lebesgue measurable (Proposition I.3.16).*

Hasty Conclusion: *If $f(x)$ is bounded and Riemann integrable on R, then it is continuous except on a set of measure zero. But if $f(x)$ is continuous except on a set of measure zero, then $f(x) = g(x)$ almost everywhere for $g(x)$ continuous, and hence $f(x)$ is Lebesgue measurable, and thus Lebesgue integrable on R by Proposition 2.15.*

Discussion: *The logic in this conclusion is faulty. The problem arises in the hasty identification of two similar-sounding notions discussed in Remark I.3.19. The first notion is that underlying the Riemann result, that $f(x)$ is continuous except on a set of measure 0. The second notion is that underlying the Lebesgue measurability result, that there is a continuous function $g(x)$ so that $f(x) = g(x)$ except on a set of measure 0. As exemplified in Book I, neither notion implies nor is implied by the other notion. Thus the Lebesgue measurability of $f(x)$ does not follow from Riemann integrability and this argument.*

Conclusion: *The result that bounded Riemann integrable implies Lebesgue measurable and thus Lebesgue integrable is not derivable from early results and requires a separate direct proof. The key to this proof is to show that Riemann integrable implies Lebesgue integrable, and then the Lebesgue measurability result follows from Proposition 2.15.*

We now turn to this needed proof. For a generalization of this result, see Proposition 2.56.

Proposition 2.18 (Riemann $\Rightarrow$ Lebesgue integrable on $R = \prod_{j=1}^{n}[a_j, b_j]$) *Let $f(x)$ be a bounded function defined on $R \equiv \prod_{j=1}^{n}[a_j; b_j]$, which is Riemann integrable. Then $f(x)$ is Lebesgue integrable on R and hence Lebesgue measurable on R, and:*

$$(\mathcal{L}) \int_R f(x)dx = (\mathcal{R}) \int_R f(x)dx. \tag{2.14}$$

Proof. *When it exists, the Riemann integral can be expressed in terms of equality of the infimum of the Riemann integrals of step functions $\widetilde{\psi} \geq f$, and the supremum of the Riemann integrals of step functions $\widetilde{\varphi} \leq f$, by Exercise 1.5 and Proposition 1.11. Since every step function is a simple function, this implies that with φ and ψ denoting simple functions:*

$$(\mathcal{R}) \int_R f(x)dx = \sup_{\widetilde{\varphi} \leq f} \int_R \widetilde{\varphi}(x)dx \leq \sup_{\varphi \leq f} \int_R \varphi(x)dx,$$

but also:

$$\inf_{\psi \geq f} \int_R \psi(x)dx \leq \inf_{\widetilde{\psi} \geq f} \int_R \widetilde{\psi}(x)dx = (\mathcal{R}) \int_R f(x)dx.$$

Now by definition:

$$\sup_{\varphi \leq f} \int_R \varphi(x)dx \leq \inf_{\psi \geq f} \int_R \psi(x)dx,$$

and so it follows that:

$$\inf_{\psi \geq f} \int_R \psi(x)dx = \sup_{\varphi \leq f} \int_R \varphi(x)dx = (\mathcal{R}) \int_R f(x)dx.$$

Thus $f(x)$ is Lebesgue integrable and (2.14) is satisfied.
Further by Proposition 2.15, $f(x)$ is Lebesgue measurable on R. ∎

2.3.3   Properties of the Integral

Continuing with the development of the Lebesgue integral of a bounded Lebesgue measurable function on a Lebesgue measurable set E with $m(E) < \infty$, we identify a number of results that are reminiscent of results from the Riemann integration context.

But note that items 2 and 3 below open a new door in integration theory, since altering a Riemann integrable function on a set of measure 0 can make this function no longer integrable. That said, the counterparts to 2 and 3 remain valid in the Riemann context if we also **assume** that both $f(x)$ and $g(x)$ are Riemann integrable.

We state this result assuming that the given functions are bounded and Lebesgue measurable. This simplifies notation, and is also the case of greatest applicability. But as noted in Proposition 2.15, these results are also valid for functions that are bounded and measurable only when restricted to the given domains of integration, by considering $f(x)\chi_E(x)$ and $g(x)\chi_E(x)$.

Proposition 2.19 (Properties of the integral) *If $f(x)$ and $g(x)$ are bounded, Lebesgue measurable functions, and $E \in \mathcal{M}_L(\mathbb{R}^n)$ with $m(E) < \infty$, then suppressing the $(\mathcal{L})$ notation:*

1. *For any $a, b \in \mathbb{R}$:*

$$\int_E [af(x) + bg(x)]dx = a \int_E f(x)dx + b \int_E g(x)dx.$$

2. *If $f(x) = g(x)$ a.e., then:*

$$\int_E f(x)dx = \int_E g(x)dx.$$

3. *If $f(x) \leq g(x)$ a.e., then:*

$$\int_E f(x)dx \leq \int_E g(x)dx.$$

4. *The **triangle inequality**:*

$$\left| \int_E f(x)dx \right| \leq \int_E |f(x)|\, dx. \tag{2.15}$$

5. *If $C_1 \leq f(x) \leq C_2$ a.e., then:*

$$C_1 m(E) \leq \int_E f(x)dx \leq C_2 m(E).$$

6. *If E_1 and E_2 are disjoint Lebesgue measurable sets of finite measure, and $E \equiv E_1 \bigcup E_2$, then:*

$$\int_E f(x)dx = \int_{E_1} f(x)dx + \int_{E_2} f(x)dx,$$

and this is then true for any finite disjoint union.

7. *If for all measurable $E' \subset E$:*

$$\int_{E'} f(x)dx = 0,$$

then $f(x) = 0$ a.e. on E.

Proof. *We prove these results in turn. Recall that the integral of $f(x)$ over a finite measurable set E is equal to the integral of $f(x)\chi_E(x)$ as noted in (2.7). To simplify notation, it is often convenient to suppress the $\chi_E(x)$ function, but with the understanding that statements such as $\varphi(x) \leq f(x)$ mean that such inequalities are valid on E, or equivalently, are valid when multiplied by $\chi_E(x)$.*

1. *Perhaps surprisingly, this first demonstration is the most subtle. First, $af(x) + bg(x)$ is bounded. With M_f, respectively M_g, denoting the bounds for $|f(x)|$, respectively $|g(x)|$:*

$$|af(x) + bg(x)| \leq |a|\, M_f + |b|\, M_g.$$

Lebesgue measurability follows from Proposition I.3.30, so the Lebesgue integral of $af(x) + bg(x)$ is well-defined by Proposition 2.15.

Given simple functions $\psi_{af}(x) \geq af(x)$ and $\psi_{bg}(x) \geq bg(x)$ obtains $\psi_{af}(x) + \psi_{bg}(x) \geq af(x) + bg(x)$, and so by Definition 2.4 and (2.6):

$$\begin{aligned}
\int_E [af(x) + bg(x)]dx &= \inf_{\psi \geq af+bg} \int_E \psi(x)dx \\
&\leq \int_E [\psi_{af}(x) + \psi_{bg}(x)]dx \\
&= \int_E \psi_{af}(x)dx + \int_E \psi_{bg}(x)dx,
\end{aligned}$$

This inequality remains valid after taking the infimum of the right-hand side, producing:

$$\int_E [af(x) + bg(x)]dx \leq \int_E af(x)dx + \int_E bg(x)dx.$$

On the other hand, given simple functions $\varphi_{af}(x) \leq af(x)$ and $\varphi_{bg}(x) \leq bg(x)$ obtains $\varphi_{af}(x) + \varphi_{bg}(x) \leq af(x) + bg(x)$, and the same logic using suprema proves that:

$$\int_E [af(x) + bg(x)]dx \geq \int_E af(x)dx + \int_E bg(x)dx,$$

and hence:

$$\int_E [af(x) + bg(x)]dx = \int_E af(x)dx + \int_E bg(x)dx.$$

For the last step of factoring out the constants a and b from these integrals, we illustrate a and consider the cases $a > 0$ and $a < 0$ separately, as the case $a = 0$ needs no discussion.

First note that we can identify bounding simple functions of $af(x)$ and $f(x)$ by $\psi_{af}(x) = a\psi_f(x)$, meaning if $\psi_f(x)$ bounds $f(x)$ then $a\psi_f(x)$ bounds $af(x)$, and conversely. If $a > 0$ then $\psi_{af}(x) \geq af(x)$ if and only if $\psi_f(x) \geq f(x)$, and so by Proposition 2.8:

$$\inf_{\psi_{af} \geq af} \int_E \psi_{af}(x)dx = a \inf_{\psi_f \geq f} \int_E \psi_f(x)dx.$$

When $a < 0$, this identification leads to the conclusion that $\psi_{af}(x) \geq af(x)$ if and only if $\psi_f(x) \leq f(x)$, and so:

$$\inf_{\psi_{af} \geq af} \int_E \psi_{af}(x)dx = a \sup_{\psi_f \leq f} \int_E \psi_f(x)dx.$$

Thus in either case, $\int_E af(x)dx = a \int_E f(x)dx$ by Definition 2.16.

2. *By assumption, $f - g = 0$ a.e., and so given simple functions $\varphi(x)$ and $\psi(x)$ with:*

$$\varphi(x) \leq f(x) - g(x) \leq \psi(x),$$

it follows that $\varphi(x) \leq 0$ a.e. and $\psi(x) \geq 0$ a.e. Thus by Definition 2.4, $\int_E \psi(x)dx \geq 0$ and $\int_E \varphi(x)dx \leq 0$. But since $f - g$ is Lebesgue integrable by part 1, (2.10) assures that the infimum of the $\psi(x)$ integrals equals the supremum of the $\varphi(x)$ integrals. This obtains that $\int_E [f(x) - g(x)]dx = 0$, and the result now follows from part 1.

3. *Here $g - f \geq 0$ a.e., and so for any simple function $\psi(x)$ with $g(x) - f(x) \leq \psi(x)$ it follows that $\psi(x) \geq 0$ a.e. and $\int_E \psi(x)dx \geq 0$ by Definition 2.4. Hence, the infimum of such integrals is also nonnegative. This infimum is $\int_E [g(x) - f(x)]dx$ by the integrability of $f - g$, and the final step is part 1.*

4. *Since* $f(x) \leq |f(x)|$ *everywhere, which is stronger than a.e., we conclude that* $\int_E f(x)dx \leq \int_E |f(x)| \, dx$ *by part 3. Similarly,* $-f(x) \leq |f(x)|$ *and by part 3,* $-\int_E f(x)dx \leq \int_E |f(x)| \, dx$. *Combining obtains (2.15).*

5. *The bounding simple functions defined by* $\varphi_0(x) = C_1\chi_E(x)$ *and* $\psi_0(x) = C_2\chi_E(x)$ *satisfy* $\varphi_0(x) \leq f(x) \leq \psi_0(x)$, *and thus by definition of the integral:*

$$\int_E f(x)dx \leq \int_E \psi_0(x)dx, \qquad \int_E f(x)dx \geq \int_E \varphi_0(x)dx.$$

The result follows since $\int_E C\chi_E(x)dx = Cm(E)$ *by (2.2).*

6. *By disjointness,* $\chi_{E_1 \bigcup E_2}(x) = \chi_{E_1}(x) + \chi_{E_2}(x)$, *and so the result follows by part 1 and (2.7).*

7. *Given* $\epsilon > 0$, *let* $E' = \{x \in E | f(x) \geq \epsilon\}$, *and note that* E' *is Lebesgue measurable. Then by item 5:*

$$\epsilon m(E') \leq \int_{E'} f(x)dx = 0,$$

and so $m(E') = 0$ *for all* $\epsilon > 0$. *Since* $\{f(x) > 0\} = \bigcap_{n=1}^\infty \{f(x) \geq 1/n\}$, *it follows that* $m\{f(x) > 0\} = 0$. *The same proof obtains* $m\{f(x) < 0\} = 0$.

∎

2.3.4   Bounded Convergence Theorem

The final result of this section investigates the Lebesgue integrability of a function $f(x)$ that is almost everywhere the pointwise limit of a sequence of bounded Lebesgue measurable functions $\{f_m(x)\}_{m=1}^\infty$, on a Lebesgue measurable set $E \in \mathcal{M}_L(\mathbb{R}^n)$ with $m(E) < \infty$. Recall Definition 1.41 for pointwise convergence on a set E. Pointwise convergence **almost everywhere on** E simply means this definition applies on E outside a set of measure 0. In the case where the integral of $f(x)$ exists, we seek the relationship between this integral's value and the values of the associated integral sequence.

Proposition 2.22 states affirmative results both on integrability of $f(x)$ and on the value of this integral. This is the first of several propositions collectively referred to as results on **integration to the limit.** It will be generalized below as the definition of the Lebesgue integral is generalized beyond bounded functions.

For this result we must assume more than the boundedness of each $f_m(x)$, and instead assume that the collection $\{f_m(x)\}_{m=1}^\infty$ is **uniformly bounded.** For the Riemann result of Proposition 1.47, uniform boundedness did not need to be explicitly assumed because it was assured by the assumption of uniform convergence. Here we only assume that convergence is pointwise almost everywhere.

Definition 2.20 (Uniformly bounded function sequence) *A sequence of functions* $\{f_m(x)\}_{m=1}^\infty$ *is* **uniformly bounded on a set** $E \in \mathcal{M}_L(\mathbb{R}^n)$ *if there exists* $M < \infty$ *so that* $|f_m(x)| \leq M$ *for all* m *and for all* $x \in E$.

To motivate the need for uniform boundedness:

Exercise 2.21 (Not uniformly bounded) *Produce an example of a pointwise convergent sequence of bounded, Lebesgue measurable functions on finite measurable* E, *for which the conclusion of the convergence of integrals in (2.17) below is false. Hint: Of necessity, this sequence cannot be uniformly bounded. With* $E = [0, 1]$, *can* $f \equiv 0$ *and each* f_m *be bounded and have integral 1?*

The proof below requires a technical result from Corollary I.4.9 of the **Severini-Egorov theorem** of Proposition I.4.8, named after **Dmitri Fyodorovich Egorov** (1869–1931) and **Carlo Severini** (1872–1951). This result states that if a Lebesgue measurable function sequence converges pointwise almost everywhere to a measurable function on a bounded set E, then it converges "almost" uniformly. While this result was framed in the context of Lebesgue measurable functions on $\mathbb{R}$, the proof will be seen to be perfectly general and applicable on $\mathbb{R}^n$.

We recall this result:

Severini-Egorov theorem: *Let $\{f_m(x)\}_{m=1}^{\infty}$ be a sequence of real-valued Lebesgue measurable functions defined on a Lebesgue measurable set $E \in \mathcal{M}_L(\mathbb{R}^n)$ with $m(E) < \infty$, and let $f(x)$ be a real-valued measurable function so that $f_m(x) \to f(x)$ almost everywhere for $x \in E$. Then given $\delta > 0$, there is a measurable set $A \subset E$ with $m(A) < \delta$, so that $f_m(x) \to f(x)$ uniformly on $E - A$. That is, for any $\epsilon > 0$ there exists N, so that:*

$$|f_m(x) - f(x)| < \epsilon,$$

for all $x \in E - A$ and all $m \geq N$.

We now state and prove an important first result on the Lebesgue integrability of the pointwise limit of a sequence of measurable functions, known as the **bounded convergence theorem**. This result, and all integration to the limit results, state that one can reverse two limiting processes. One limit relates to the convergence of the function sequence $\{f_m(x)\}_{m=1}^{\infty}$, while the other limit is in the value of an integral that is defined in terms of a supremum of the integrals of certain subordinate integrable functions.

More explicitly, (2.17) can be stated:

$$\int_E \lim_{m \to \infty} f_m(x)dx = \lim_{m \to \infty} \int_E f_m(x)dx. \tag{2.16}$$

Proposition 2.22 (Bounded convergence theorem) *Let $\{f_m(x)\}_{m=1}^{\infty}$ be a uniformly bounded sequence of Lebesgue measurable functions with bound M, defined on a Lebesgue measurable set $E \in \mathcal{M}_L(\mathbb{R}^n)$ with $m(E) < \infty$. If $f_m(x) \to f(x)$ almost everywhere for $x \in E$, then $f(x)$ is Lebesgue integrable on E, and:*

$$\int_E f(x)dx = \lim_{m \to \infty} \int_E f_m(x)dx. \tag{2.17}$$

Proof. *The limit function $f(x)$ is Lebesgue measurable by Corollary I.3.48, and $|f(x)| \leq M$ by the assumption on this sequence. Hence, $f(x)$ and all $f_m(x)$ are Lebesgue integrable by Proposition 2.15.*

By Corollary I.4.9, for any $\epsilon > 0$ and $\delta > 0$, there is a measurable set $A \subset E$ with $m(A) < \delta$, and an N, so that $|f_m(x) - f(x)| < \epsilon$ for all $x \in E - A$ and all $m \geq N$. Applying properties of Proposition 2.19:

$$\left| \int_E f_m(x)dx - \int_E f(x)dx \right|$$

$$\leq \int_E |f_m(x) - f(x)| \, dx$$

$$= \int_A |f_m(x) - f(x)| \, dx + \int_{E-A} |f_m(x) - f(x)| \, dx.$$

Now $|f_m(x) - f(x)| < \epsilon$ on $E - A$, and $|f_m(x) - f(x)| \leq 2M$ on A, so for $m \geq N$:

$$\left| \int_E f_m(x)dx - \int_E f(x)dx \right| \leq 2M\delta + \epsilon m(E - A)$$

$$\leq 2M\delta + \epsilon m(E).$$

Since ϵ and δ can be chosen arbitrarily, we obtain that as $m \to \infty$,

$$\left| \int_E f_m(x)dx - \int_E f(x)dx \right| \to 0,$$

which is (2.17). ∎

2.3.5   Evaluating Integrals

We begin with a remark that may initially surprise the reader.

Remark 2.23 (On evaluating Lebesgue integrals) *In practice, one rarely needs to actually evaluate the Lebesgue integral of a given measurable function. When this is required, it is usually the case that it will be possible to justify that the value of this integral equals that of the corresponding Riemann integral, and then to proceed in this more familiar setting.*

For example, by Proposition 2.18 and then Lebesgue's existence theorem for the Riemann integral, the Lebesgue integral of any bounded function over a bounded rectangle equals the associated Riemann integral when this function is continuous almost everywhere. Thus to encounter a bounded function that is Lebesgue integrable and not Riemann integrable, one needs to go outside the continuous almost everywhere class of functions. The Dirichlet function of Example 1.3 is such a function, but this is also an example for which the Lebesgue integral is easy to evaluate.

Given this, the reader may well be wondering, what is the purpose of this section and the corresponding sections below on evaluating integrals? More generally, what is the purpose of this chapter on Lebesgue integration?

As will be seen in later books, the role of an integration theory is fundamental in probability theory and thus also in finance. While these disciplines require an integration theory, it is also demonstrable that for all its beauty and applicability, the Riemann framework does not generalize to integrals in general measure and probability spaces.

For example, the Riemann theory requires the notion of continuity. For $f(x)$ to be continuous or continuous almost everywhere on a probability space would require that this space also have a topology (Definition I.2.15) with which to define this notion (Proposition I.3.12 and following comments). This would be an unnecessary burden in applications, since probability spaces in general have no other reason to need a notion of an "open" set and a topology.

Even more problematic is that the Riemann theory is more or less restricted to integration over rectangles and other fairly regular domains like the Jordan regions of Definition 1.70. On probability spaces, we will need to integrate over a much wider range of domains, and it has already been seen that the Lebesgue theory accommodates arbitrary measurable domains very well.

Thus for future applications, we will need a more robust, measure-theoretic framework for integration on probability spaces, and also in general measure spaces. The Lebesgue theory provides an all important first step in a framework that generalizes well to measure spaces as will be seen in Book V. Further, the Lebesgue theory provides this framework in the intuitively accessible setting of $\mathbb{R}$ and $\mathbb{R}^n$, and obtains integrals that, despite the dramatic shift in approach, often reproduce valuations from the familiar Riemann framework.

As for the importance of this and other sections on evaluating integrals, the approaches introduced will mostly be useful in derivations and in various proofs, both for theoretical investigations and in applications in probability theory and finance. The ultimate purpose of these results is to provide a simpler, more direct way to express the value of a Lebesgue integral than what is afforded by the definition.

These results will be seen to generalize in Book V, making even the most abstract integrals somewhat more accessible.

The definition of the Lebesgue integral $\int_E f(x)dx$ in (2.12) and (2.13) reflects uncountably many simple functions in the infimum and supremum calculations. The bounded convergence theorem gives a more practical and useful way to evaluate the Lebesgue integral of a bounded function using sequences of simple functions.

This result is stated using a convergent simple function sequence, rather than a general function sequence, because the integrals of simple functions require no effort as given in Definition 2.4.

Corollary 2.24 (Evaluating $\int_E f(x)dx$) *Let $f(x)$ be a bounded function defined on a Lebesgue measurable set $E \in \mathcal{M}_L(\mathbb{R}^n)$ with $m(E) < \infty$. If there exists a sequence $\{\varphi_m(x)\}_{m=1}^{\infty}$ of uniformly bounded simple functions defined on E with $\varphi_m(x) \to f(x)$ almost everywhere for $x \in E$, then $f(x)$ is Lebesgue integrable on E and:*

$$\int_E f(x)dx = \lim_{m \to \infty} \int_E \varphi_m(x)dx.$$

Proof. *Immediate from the above result.* ∎

Example 2.25 *Let $f(x)$ be measurable and bounded with $|f(x)| \leq M$ on a Lebesgue measurable set $E \in \mathcal{M}_L(\mathbb{R}^p)$ with $m(E) < \infty$. Recalling the construction in the proof of Proposition 2.15, for given n define the level sets E_j for $-n \leq j \leq n$ by:*

$$E_j = \left\{ x | (j-1)\frac{M}{n} < f(x) \leq j\frac{M}{n} \right\} \cap E.$$

The sets $\{E_j\}_{j=-n}^{n}$ are measurable, disjoint, and satisfy $\bigcup_{j=-n}^{n} E_j = E$.
The simple functions:

$$\psi_n(x) = \frac{M}{n} \sum_{j=-n}^{n} j \chi_{E_j}(x),$$

$$\varphi_n(x) = \frac{M}{n} \sum_{j=-n}^{n} (j-1) \chi_{E_j}(x),$$

satisfy:

$$\varphi_n(x) \leq f(x) \leq \psi_n(x),$$

and converge pointwise:

$$\varphi_n(x) \to f(x), \qquad \psi_n(x) \to f(x).$$

Thus using the $\psi_n(x)$ sequence:

$$\int_E f(x)dx = \lim_{n \to \infty} M \sum_{j=-n}^{n} \frac{j}{n} m(E_j), \tag{2.18}$$

with a similar expression using the $\varphi_m(x)$ sequence:

$$\int_E f(x)dx = \lim_{n \to \infty} M \sum_{j=-n}^{n} \frac{j-1}{n} m(E_j).$$

Because:

$$\int_E \psi_n(x)dx - \int_E \varphi_n(x)dx \leq \frac{M}{n} m(E),$$

the error in an approximation with either simple function is proportional to $\frac{1}{n}$.

2.4   Nonnegative Measurable Functions

In this section, we continue the process of generalizing the definition of Lebesgue integral beyond bounded $f(x)$ and sets $E \in \mathcal{M}_L(\mathbb{R}^n)$ with $m(E) < \infty$. To do so we will first take one step backward, now requiring $f(x)$ to be nonnegative, but also take two steps forward, by eliminating both the restriction on $m(E)$ and the boundedness requirement for $f(x)$.

The final section will remove the nonnegativity assumption on $f(x)$.

2.4.1   Definition of the Integral

We begin with a natural definition of the Lebesgue integral of nonnegative $f(x)$ in light of the previous section. Note that in general, the supremum in (2.19) need not be finite, nor does this definition require it.

Here we suppress the characteristic function $\chi_E(x)$ that was more prominently displayed notationally in the prior chapter. But in (2.19), we are defining $(\mathcal{L}) \int \chi_E(x) f(x) dx$ as in Definition 2.9, and are doing this in terms of the supremum of integrals of bounded functions $\chi_E(x) h(x)$ with $\chi_E(x) h(x) \leq \chi_E(x) f(x)$.

For the following definition, note that as a consequence of Remark 2.10 and the corresponding result for bounded measurable functions, that if $m(E) = 0$ then:

$$(\mathcal{L}) \int_E f(x) dx = 0,$$

for all nonnegative Lebesgue measurable functions $f(x)$.

Definition 2.26 (Lebesgue integral of nonnegative $f(x)$) *If $f(x)$ is a nonnegative Lebesgue measurable function defined on a Lebesgue measurable set $E \in \mathcal{M}_L(\mathbb{R}^n)$, define:*

$$(\mathcal{L}) \int_E f(x) dx = \sup_{h \leq f} \int_E h(x) dx, \qquad (2.19)$$

where $h(x)$ is bounded and Lebesgue measurable, with $m\{x | h(x) \neq 0\} < \infty$.

*The function $f(x)$ will be said to be **Lebesgue integrable** if this supremum is finite.*

*When the supremum is infinite, so $(\mathcal{L}) \int_E f(x) dx = \infty$, $f(x)$ will be deemed **not Lebesgue integrable**.*

Remark 2.27 (On Definition 2.26) *A few clarifying comments on this definition follow:*

1. *Note that we do not assume that $m(E) < \infty$. However, $\int_E h(x) dx$ exists for all such bounded Lebesgue measurable functions because we assume that $E' \equiv \{x | h(x) \neq 0\}$ has finite measure.*

   In detail, since $h(x) = h(x) \chi_{E'}(x)$, Definition 2.9 obtains:

$$\int_E h(x) dx = \int \chi_E(x) \chi_{E'}(x) h(x) dx = \int \chi_{E \cap E'}(x) h(x) dx = \int_{E \cap E'} h(x) dx.$$

2. *While $h(x)$ is not specified to be nonnegative in Definition 2.26, given any such $h(x) \leq f(x)$ we can define $g(x) = \max\{0, h(x)\}$, and then $g(x)$ is nonnegative and measurable, $g(x) \leq f(x)$ and:*

$$\int_E h(x)dx \leq \int_E g(x)dx \leq \sup_{h \leq f} \int_E h(x)dx = \int_E f(x)dx.$$

So for the supremum in (2.19), there is no loss of generality in restricting to only nonnegative, bounded $h(x)$.

3. *If $f(x)$ is Lebesgue integrable under the above definition and $E' = \{x | f(x) = \infty\}$, it must be the case that $m(E') = 0$.*

If $m(E') > 0$, then because $f \geq 0$ on E' we can define the sequence $h_n(x) = n$:

$$(\mathcal{L})\int_E f(x)dx \geq (\mathcal{L})\int_{E'} f(x)dx \geq (\mathcal{L})\int_{E'} h_n(x)dx = nm(E').$$

Hence, if $m(E') > 0$, such a function $f(x)$ can never be Lebesgue integrable. But $m(E') = 0$ does not assure Lebesgue integrability as item 1 of Example 2.28 demonstrates.

4. *If $f(x)$ is nonnegative and bounded, and E has finite measure, the supremum in (2.19) exists and is finite, and equals $(\mathcal{L})\displaystyle\int_E f(x)dx$ as defined in the prior section.*

To see this, recall the function series $\{\varphi_n(x)\}_{n=1}^{\infty}$ constructed in Proposition 2.15, which are then nonnegative in this context. For any bounded, Lebesgue measure function $h(x)$ with $h(x) \leq f(x)$, then since $\varphi_n(x) \to f(x)$ as an increasing sequence:

$$\int_E h(x)dx \leq \sup_n \int_E \varphi_n(x)dx.$$

The same is true of the supremum of such $h(x)$-integrals:

$$\sup_{h \leq f} \int_E h(x)dx \leq \sup_n \int_E \varphi_n(x)dx.$$

On the other hand:

$$\sup_n \int_E \varphi_n(x)dx \leq \sup_{h \leq f} \int_E h(x)dx,$$

since $\{\varphi_n(x)\}_{n=1}^{\infty}$ is contained in the allowable set of $h(x)$ functions.

Thus the supremum in (2.19) agrees with the supremum using $\{\varphi_n(x)\}_{n=1}^{\infty}$, and so the above integral equals that of the previous section.

Example 2.28

1. *With $f(x) = 1/x$ on $[1, \infty)$, let $h_n(x) = f(x)\chi_{[1,n]}(x)$.*

Then the Lebesgue integral of $h_n(x)$ exists by Proposition 2.15, and equals the Riemann integral by Proposition 2.18. Hence, the supremum in (2.19) is unbounded since $\int h_n(x)dx = \ln n$. The same conclusion follows for this function defined on $(0, 1]$ using the characteristic function of $[1/n, 1]$. Thus $f(x) = 1/x$ is not Lebesgue integrable on either $(0, 1]$ or $[1, \infty)$, but is Lebesgue integrable by the same argument on all compact $[a, b] \subset (0, \infty)$.

2. *Let $f(x) = 1/x^2$ be defined on $[1, \infty)$.*

Then $f(x)$ is Lebesgue integrable over $[1, n]$ and this integral equals that of the associated Riemann integral as in item 1:

$$(\mathcal{L}) \int_1^n f(x)dx = 1 - 1/n.$$

If nonnegative $h(x) \leq f(x)$ on $[1, n]$, then $\int_1^n h(x)dx \leq 1 - 1/n$, so the functions $h_n(x) = f(x)\chi_{[1,n]}(x)$ provide the suprema over each subinterval. The supremum over all such intervals is finite, and hence $f(x)$ is Lebesgue integrable with $(\mathcal{L}) \int_1^\infty f(x)dx = 1$.

2.4.2   Properties of the Integral

In this section, we prove some useful properties of the above integral, assigning some of the details of the proofs as exercises. Before beginning, we derive an alternative characterization of $\int_E f(x)dx$ for a nonnegative, Lebesgue measurable function.

Definition 2.26 characterizes this Lebesgue integral in terms of the supremum of the integrals of uncountably many subordinate, bounded, Lebesgue measurable functions, each 0 outside a set of finite measure. The next result states that this integral can be defined in terms of a countable collection of easy to define subordinate functions. See also Remark 2.30. As will be seen, this characterization simplifies certain proofs of the properties of this integral.

Recall that the characteristic function $\chi_A(x)$ in (2.20) is defined in (1.8), where here $A \equiv \{x \in \mathbb{R}^n \mid |x| \leq N\}$ and $|x|$ as given in (1.18).

Proposition 2.29 (Alternative definition of $\int_E f(x)dx$) *Let $f(x)$ be a nonnegative Lebesgue measurable function defined on a Lebesgue measurable set $E \in \mathcal{M}_L(\mathbb{R}^n)$. Then:*

$$(\mathcal{L}) \int_E f(x)dx = \lim_{N \to \infty} \int_E \min\{N, f(x)\}\chi_{\{|x| \leq N\}}(x)dx. \tag{2.20}$$

Thus $f(x)$ is Lebesgue integrable, respectively not so integrable, if and only if the limit in (2.20) is finite, respectively infinite.

Proof. *Note that $h_N(x) \equiv \min\{N, f(x)\}\chi_{|x| \leq N}(x)$ satisfies the requirements of Definition 2.26. First, $h_N(x)$ is Lebesgue measurable by Proposition I.3.47, noting that $\chi_{\{|x| \leq N\}}(x)$ is Lebesgue measurable since $\{|x| \leq N\}$ is a Borel set and thus Lebesgue measurable. Further, $h_N(x)$ is bounded by N, and is 0 outside $\{|x| \leq N\}$, a set of finite Lebesgue measure.*

Thus:

$$\sup_{h \leq f} \int_E h(x)dx \geq \sup_N \int_E h_N(x)dx = \lim_{N \to \infty} \int_E h_N(x)dx, \tag{1}$$

noting that the last equality follows from item 3 of Proposition 2.19 since $h_N(x) \leq h_{N'}(x)$ if $N \leq N'$.

Conversely, given $h(x)$ as in Definition 2.26, assume that $|h(x)| \leq M$. Then since $m\{h(x) \neq 0\} < \infty$, it follows that either $\{h(x) \neq 0\} \subset \{|x| \leq N\}$ for some N, or, this set is unbounded but for any $\epsilon > 0$ there exists N, so that:

$$m\left(\{h(x) \neq 0\} \bigcap \{|x| > N\}\right) < \epsilon.$$

In the first case, $h(x) \leq h_{N'}(x)$ with $N' \equiv \max\{N, M\}$, and so by Proposition 2.19:

$$\int_E h(x)dx \leq \int_E h_{N'}(x)dx. \tag{2}$$

In the second case, let $\epsilon > 0$ and the associated N be given, and define $A = \{|x| \leq N'\}$ with N' defined as above, and $\tilde{A} = \{|x| > N'\}$. Then by item 6 of Proposition 2.19, $\chi_A(x) + \chi_{\tilde{A}}(x) \equiv 1$ and $m\{h(x) \neq 0\} < \infty$:

$$\int_E h(x)dx = \int_E h(x)\chi_A(x)dx + \int_E h(x)\chi_{\tilde{A}}(x)dx$$
$$\leq \int_E h_{N'}(x)dx + \epsilon M.$$

The first integral bound is item 3 of Proposition 2.19 since $h(x)\chi_A(x) \leq h_{N'}(x)$, and the second is item 5 of this result. This implies:

$$\int_E h(x)dx \leq \lim_{N \to \infty} \int_E h_{N'}(x)dx. \tag{3}$$

Combining (2) and (3) obtains that:

$$\sup_{h \leq f} \int_E h(x)dx \leq \lim_{N \to \infty} \int_E h_{N'}(x)dx,$$

and this with (1) completes the proof. ∎

Remark 2.30 (On Proposition 2.29) *There was nothing unique about the characterization in (2.20). We leave it as an exercise to show that if $\{a_N\}_{N=1}^\infty$ and $\{b_N\}_{N=1}^\infty$ are nonnegative, increasing, and unbounded sequences, then:*

$$(\mathcal{L})\int_E f(x)dx = \lim_{N \to \infty} \int_E \max\{a_N, f(x)\}\chi_{\{|x| \leq b_N\}}(x)dx.$$

With the characterization of $\int_E f(x)dx$ for nonnegative $f(x)$ as given in (2.20), the proof of the next result will follow from Proposition 2.19. We provide details for a couple of parts and leave the remaining results as exercises. The reader is also encouraged to supply alternative proofs based on the Definition 2.26 characterization of this integral. We provide two such proofs below, for parts 1 and 2.

Proposition 2.31 (Properties of the integral) *If $f(x)$ and $g(x)$ are nonnegative Lebesgue measurable functions defined on a Lebesgue measurable set $E \in \mathcal{M}_L(\mathbb{R}^n)$, then suppressing the $(\mathcal{L})$ notation:*

1. *For any $a > 0$:*

$$\int_E af(x)dx = a\int_E f(x)dx.$$

2.

$$\int_E [f(x) + g(x)]dx = \int_E f(x)dx + \int_E g(x)dx$$

3. *If $f(x) = g(x)$ a.e., then:*

$$\int_E f(x)dx = \int_E g(x)dx.$$

4. *If $f(x) \leq g(x)$ a.e., then:*

$$\int_E f(x)dx \leq \int_E g(x)dx.$$

5. *If $C_1 \leq f(x) \leq C_2$ a.e. and $m(E) < \infty$, then:*

$$C_1 m(E) \leq \int_E f(x)dx \leq C_2 m(E).$$

6. *If $E' \subset E$ is measurable, then:*

$$\int_{E'} f(x)dx \leq \int_E f(x)dx.$$

7. *If $E = E_1 \bigcup E_2$, a union of disjoint Lebesgue measurable sets, then:*

$$\int_E f(x)dx = \int_{E_1} f(x)dx + \int_{E_2} f(x)dx,$$

and this is then true for any finite disjoint union.

8. *If for measurable E with $m(E) > 0$:*

$$\int_E f(x)dx = 0,$$

then $f(x) = 0$ a.e. on E.

Proof. *To simplify the proofs of items 1 and 2 using Definition 2.26, any function $h(x)$ used below as given in that definition will have the properties assumed there, adapted to Remark 2.27. That is, such $h(x)$ will be bounded, nonnegative, and Lebesgue measurable, with $m(E') \equiv m\{x|h(x) > 0\} < \infty$. Further, we will apply Proposition 2.19 to $\int_E h(x)dx$ even though it need not be the case that $m(E) < \infty$. The justification is the same as in item 1 of Remark 2.27, that this integral is equivalent to an integral over $E \bigcap E'$, and $m(E \bigcap E') < \infty$.*

1. *For this result, the proofs using either characterization are comparable.*

    a. **Using Definition 2.26:** *Since $a > 0$, it follows that $h \leq af$ if and only if $h' \equiv h/a \leq f$. By Proposition 2.19:*

$$\int_E h(x)dx = a \int_E h'(x)dx.$$

The result follows by taking suprema, noting that these integrals will be finite or infinite together.

    b. **Using Proposition 2.29:** *By (2.20):*

$$
\begin{aligned}
\int_E af(x)dx &= \lim_{N\to\infty} \int_E \min\{N, af(x)\}\chi_{\{|x|\leq N\}}(x)dx \\
&= a \lim_{N\to\infty} \int_E \min\{N/a, f(x)\}\chi_{\{|x|\leq N\}}(x)dx \\
&= a \int_E f(x)dx,
\end{aligned}
$$

where this last step follows from Remark 2.30.

2. This result will be proved using Definition 2.26, though the associated proof with Proposition 2.29 is much simpler once an inequality is verified.

    a. *Using Definition 2.26:* As for 1.a, if $h_f \leq f$ and $h_g \leq g$, then:

$$\int_E (h_f(x) + h_g(x))\, dx = \int_E h_f(x)dx + \int_E h_g(x)dx.$$

Taking suprema, then since $\{h_f(x) + h_g(x)\} \subset \{h(x)|h \leq f + g\}$, it follows that:

$$\int_E [f(x) + g(x)]dx \geq \int_E f(x)dx + \int_E g(x)dx. \tag{1}$$

On the other hand, given $h \leq f + g$ with $h \leq N$, there is no loss of generality in assuming that $f \leq h \leq f + g$, since we will be taking a supremum of such $h(x)$. Letting $h'_f(x) = \max\{N, f(x)\}\chi_{E'}(x)$ and $h'_g(x) = \max\{N, (h(x) - f(x))\}\chi_{E'}(x)$, were E' is defined relative to $h(x)$, it follows from Proposition I.3.47 that these functions are Lebesgue measurable. By construction, these functions are bounded and 0 outside E'. Also $h'_f \leq f$ and $h'_g \leq g$, and:

$$\int_E h(x)dx = \int_E h_f(x)dx + \int_E h_g(x)dx.$$

By taking a supremum over such h, and noting that these constructed subordinate functions to $f(x)$ and $g(x)$ are among all those possible, obtains:

$$\int_E [f(x) + g(x)]dx \leq \int_E f(x)dx + \int_E g(x)dx. \tag{2}$$

The proof is completed by (1) and (2).

    b. *Using Proposition 2.29:* The proof will follow from Remark 2.30 and Proposition 2.19 by verifying and using:

$$\min\{N, f(x)\} + \min\{N, g(x)\} \leq \min\{2N, f(x) + g(x)\}$$
$$\leq \min\{2N, f(x)\} + \min\{2N, g(x)\}.$$

3. – 7. These results follow from Proposition 2.19 using the alternative characterization in (2.20).

8. Assume that $f(x) \neq 0$ a.e. on E. Then $m(E') > 0$ for $E' = \{f > 0\}$, and there exists so that $m(E'_n) > 0$ for $E'_n = \{f > 1/n\}$. Otherwise, since $\bigcup E'_n = E'$ and $E'_n \subset E'_{n+1}$, continuity from below (Proposition I.2.44) of m would obtain that $m(E') = 0$, a contradiction. Given such n, measurability of E'_n, and item 4 obtain that:

$$\int_{E'_n} f(x)dx > m(E'_n)/n > 0.$$

However, by item 6:

$$\int_{E'_n} f(x)dx \leq \int_E f(x)dx = 0,$$

a contradiction.

■

Part 4 of Proposition 2.31 provides an interesting corollary that in essence states that every nonnegative Lebesgue integrable function can be used as a **test function** to identify other Lebesgue integrable functions.

Corollary 2.32 (Integrability comparison test) *Let $g(x)$ be a nonnegative Lebesgue measurable function, and $E \in \mathcal{M}_L(\mathbb{R}^n)$.*

1. *If $g(x)$ is Lebesgue integrable on E :*

    a. *Every nonnegative Lebesgue measurable function $f(x)$ with $f(x) \leq g(x)$ a.e. on E, is Lebesgue integrable on E.*

    b. *Every Lebesgue measurable function $f(x)$ with $|f(x)| \leq g(x)$ a.e. on E, is absolutely Lebesgue integrable on E, meaning $|f(x)|$ is Lebesgue integrable.*

2. *If $g(x)$ is not Lebesgue integrable on E:*

    a. *Every nonnegative Lebesgue measurable function $f(x)$ with $f(x) \geq g(x)$ a.e. on E, is not Lebesgue integrable on E.*

    b. *Every Lebesgue measurable function $f(x)$ with $|f(x)| \geq g(x)$ a.e. on E, is not absolutely Lebesgue integrable on E, meaning $|f(x)|$ is not Lebesgue integrable.*

Proof. *Immediate from part 4, noting that $|f(x)| = \max\{f(x), -f(x)\}$ is Lebesgue measurable by Proposition I.3.47.* ∎

Remark 2.33 *Items 1.b and 2.b address the integrability of $|f(x)|$ and leave open the question on the integrability of $f(x)$. It will be seen in Section 2.5 that $f(x)$ is Lebesgue integrable if and only if $|f(x)|$ is Lebesgue integrable, and thus this corollary also provides a test for the integrability of such $f(x)$.*

2.4.3   Fatou's Lemma

In this and the next section, we develop two key results of widespread applicability when integrating nonnegative function sequences. Like the bounded convergence theorem, these are results on integration to the limit. These results also apply to function sequences $\{|f_m(x)|\}_{m=1}^{\infty}$ for general measurable $\{f_m(x)\}_{m=1}^{\infty}$, as was the case in Remark 2.33, and will be generalized to other measure spaces in Book V.

The first result is called **Fatou's lemma,** and is the more general result in its application but provides "only" a conclusion of an upper bound. It is named after **Pierre Fatou** (1878–1929) and is known as a "lemma" to distinguish it from Fatou's theorem, which is a result in complex analysis.

The primary application of this lemma is in cases where the nonnegative functions $f_m(x)$ in the sequence are Lebesgue integrable, and the sequence of integrals $\left\{ \int_E f_m(x)dx \right\}_{m=1}^{\infty}$ has a finite limit inferior. Recalling Definition I.3.42, any function sequence or numerical sequence has a well-defined limit inferior, but this result need not be finite.

In some applications, this integral sequence may actually have a finite limit. In such cases, Fatou's lemma assures the integrability of any function $f(x)$ with $f(x) = \liminf f_m(x)$ a.e. The same will be true for $f(x) = \lim f_m(x)$ a.e. if this limit function exists. While this lemma does not in general provide the value of the integral of $f(x)$, we are provided an upper bound.

The result in (2.22) is sometimes stated as in (2.16):

$$\int_E [\liminf f_m(x)]dx \leq \liminf \int_E f_m(x)dx. \tag{2.21}$$

This statement emphasizes that we are interchanging two limiting processes: the limit inferior, and the value of an integral that is defined in terms of a supremum of the integrals of certain subordinate functions.

Proposition 2.34 (Fatou's lemma) *If* $\{f_m(x)\}_{m=1}^{\infty}$ *is a sequence of nonnegative Lebesgue measurable functions, and* $f(x) = \liminf f_m(x)$ *a.e. on a Lebesgue measurable set* $E \in \mathcal{M}_L(\mathbb{R}^n)$, *then:*

$$\int_E f(x)dx \leq \liminf \int_E f_m(x)dx. \tag{2.22}$$

Proof. *We prove this result for* $f(x) \equiv \liminf f_m(x)$ *and note that if* $\tilde{f}(x) = f(x)$ *a.e., then* (2.22) *will be satisfied with* $\tilde{f}(x)$ *in place of* $f(x)$. *This follows since* $\tilde{f}(x) = \liminf \tilde{f}_m(x)$ *with:*

$$\tilde{f}_m(x) \equiv \begin{cases} f_m(x), & \text{if } \tilde{f}(x) = f(x), \\ \tilde{f}(x), & \text{otherwise,} \end{cases}$$

and thus $\int_E f_m(x)dx = \int_E \tilde{f}_m(x)dx$ *for all* m *by item 3 of Proposition 2.31.*

Now $f(x)$ *is nonnegative by definition, and measurable by Proposition I.3.47. Define for* $m = 1, 2, ...,$

$$g_m(x) = \inf_{j \geq m} f_j(x).$$

Then $g_m(x)$ *is nonnegative and again measurable by Proposition I.3.47, and* $g_m(x) \leq f_m(x)$ *for all* m. *So by Proposition 2.31:*

$$\int_E g_m(x)dx \leq \int_E f_m(x)dx,$$

though the upper integral need not be finite. Also, $g_m(x) \leq g_{m+1}(x)$ *for all* m, *and by definition of limit inferior:*

$$f(x) \equiv \liminf f_m(x) = \lim g_m(x).$$

Given bounded and nonnegative measurable $h(x) \leq f(x)$ *with* $m(E') < \infty$ *for* $E' \equiv \{x \in E | h(x) \neq 0\}$, *define* $h_m(x) = \min\{g_m(x), h(x)\}$. *Then* $h_m(x)$ *is measurable, equal to* 0 *outside* E', *and* $h_m(x) \leq f_m(x)$. *Also,* $h_m(x) \to h(x)$ *on* E' *since* $g_m(x) \to f(x)$ *and* $h(x) \leq f(x)$. *Hence, since* $E' \subset E$ *and* $h_m(x) \leq h(x) \leq M$ *assures uniform boundedness, the bounded convergence theorem applies to* $\{h_m(x)\}$ *to obtain:*

$$\int_E h(x)dx = \lim_{m \to \infty} \int_E h_m(x)dx. \tag{1}$$

But $h_m(x) \leq f_m(x)$ *for each* m *and thus:*

$$\int_E h_m(x)dx \leq \int_E f_m(x)dx.$$

While the integral sequence on the left has a limit by the bounded convergence theorem, we cannot assert that the sequence on the right has a limit since $\{f_m(x)\}$ *need not be uniformly bounded. However, it must be the case that the limit of the sequence on the left cannot exceed the smallest accumulation point of the sequence on the right as this would violate the term-by-term inequalities. Hence:*

$$\lim_{m \to \infty} \int_E h_m(x)dx \leq \liminf \int_E f_m(x)dx. \tag{2}$$

Combining (1) *and* (2) *obtains that for all measurable* $h(x) \leq f(x)$ *with* $m(\{x \in E | h(x) \neq 0\}) < \infty$:

$$\int_E h(x)dx \leq \liminf \int_E f_m(x)dx.$$

Taking a supremum over all such $h(x)$ *completes the proof.* ∎

Remark 2.35 (Special cases for Fatou's lemma) *In cases where the function sequence* $\{f_m(x)\}_{m=1}^{\infty}$ *converges pointwise or pointwise a.e. to* $f(x)$, *and/or the integral sequence* $\int_E f_m(x)dx$ *has a limit, Fatou's lemma can be stated in terms of these limits instead of limits inferior. But this result is all the more applicable because it does not require the existence of such limits.*

1. *Fatou's lemma provides the exact value of* $\int_E f(x)dx$ *in cases where* $\liminf \int_E f_m(x)dx = 0$. *By nonnegativity of* $\{f_m(x)\}_{m=1}^{\infty}$, *it must be the case that* $\int_E f(x)dx \geq 0$, *and thus for such cases:*

$$\int_E f(x)dx = 0.$$

2. *If* $\{f_m(x)\}_{m=1}^{\infty}$ *is a sequence of Lebesgue measurable functions, and* $f(x) = \liminf f_m(x)$ *a.e., then* $\{|f_m(x)|\}_{m=1}^{\infty}$ *satisfies the assumption of this result. Then* $|f(x)| = \liminf |f_m(x)|$ *a.e., and thus:*

$$\int_E |f(x)|\, dx \leq \liminf \int_E |f_m(x)|\, dx.$$

Example 2.36 (Strict inequality in (2.22)) *With* $E = [0, \infty)$, *define* $f_m(x) = 1$ *on* $[m-1, m]$ *and* 0 *elsewhere. Then* $f_m(x) \to f(x)$ *for all* x *where* $f(x) \equiv 0$, *but* $\int_E f_m(x)dx = 1$ *for all* m, *and hence:*

$$0 = \int_E f(x)dx < \liminf \int_E f_m(x)dx = 1.$$

This example is generalizable to $\mathbb{R}^n$ *as an exercise.*

2.4.4   Lebesgue's Monotone Convergence Theorem

The second integration to the limit result applicable to nonnegative function sequences is **Lebesgue's monotone convergence theorem** and named after **Henri Lebesgue** (1875–1941). It is also called **Beppo Levi's theorem** and named after **Beppo Levi** (1875–1961). For this result, by adding the assumption that the function sequence is increasing, Fatou's inequality is strengthened to a conclusion of equality.

Similar to Fatou's lemma, this theorem does not assume that the functions in the sequence are integrable, and thus when not integrable, neither is the limit function. But given a monotonically increasing sequence of nonnegative **integrable** functions, which then obtains an increasing integral sequence by item 4 of Proposition 2.31, this theorem provides two important applications:

1. If the integral sequence is bounded and hence has a finite limit, then the limit function is integrable with integral equal to this limit.

2. If the integral sequence is unbounded, then the limit function is not integrable.

The conclusion below remains valid under the assumption that there is an N so that $\{f_m(x)\}_{m=1}^{\infty}$ is an increasing sequence of nonnegative Lebesgue measurable functions for $m \geq N$. In other words, because this is a statement about the limit function, the behavior of any finite collection of functions in the sequence does not alter the conclusion.

As was the case for Fatou's lemma, the result in (2.24) is sometimes expressed as:

$$\int_E \lim_{m \to \infty} f_m(x)dx = \lim_{m \to \infty} \int_E f_m(x)dx, \tag{2.23}$$

to emphasize the interchanging of two limiting processes.

Proposition 2.37 (Lebesgue's monotone convergence theorem) *If $\{f_m(x)\}_{m=1}^{\infty}$ is an increasing sequence of nonnegative Lebesgue measurable functions that converge almost everywhere to a function $f(x)$ on a Lebesgue measurable set $E \in \mathcal{M}_L(\mathbb{R}^n)$, then:*

$$\int_E f(x)dx = \lim_{m \to \infty} \int_E f_m(x)dx. \tag{2.24}$$

Thus $f(x)$ is Lebesgue integrable if and only if $\left\{ \int_E f_m(x)dx \right\}_{m=1}^{\infty}$ is bounded.

Proof. *As a limit of nonnegative Lebesgue measurable functions almost everywhere, $f(x)$ is also nonnegative, and measurable by Propositions I.3.47 and I.3.16.*

By Fatou's lemma:

$$\int_E f(x)dx \leq \liminf \int_E f_m(x)dx. \tag{1}$$

Since $\{f_m(x)\}_{m=1}^{\infty}$ is an increasing sequence, this implies by Proposition 2.31 that for all m:

$$\int_E f_m(x)dx < \int_E f(x)dx,$$

and so:

$$\limsup \int_E f_m(x)dx \leq \int_E f(x)dx. \tag{2}$$

As the limit superior cannot be smaller than the limit inferior, the results in (1) and (2) imply that these limits are equal, and (2.24) follows.

Since $\left\{ \int_E f_m(x)dx \right\}_{m=1}^{\infty}$ is an increasing sequence by item 4 of Proposition 2.31, the last conclusion follows by definition of Lebesgue integrable. ∎

Example 2.38 (Decreasing sequence $\{f_m(x)\}_{m=1}^{\infty}$) *Perhaps surprisingly, Lebesgue's monotone convergence theorem does not apply if $\{f_m(x)\}_{m=1}^{\infty}$ is a monotonically **decreasing** sequence of nonnegative Lebesgue measurable functions. Of course, if $m(E) < \infty$ and the functions of the sequence are uniformly bounded, (2.24) will be satisfied by the bounded convergence theorem of Proposition 2.22, so any counterexample of this must violate that theorem's assumptions.*

As a simple example on $\mathbb{R}$, define a decreasing sequence with $f_m(x) = \chi_{[m,\infty)}(x)$. Then $\int_{\mathbb{R}} f_m(x)dx = \infty$ for all m, but $f(x) \equiv 0$. Generalizing to $\mathbb{R}^n$ is an exercise.

But note that if we assume that any one of the functions of this monotonically decreasing sequence is integrable, then an affirmative result is assured as in Corollary 2.32.

We record next two important corollaries to the Lebesgue monotone convergence theorem. The first applies to the integral of a function series and provides a condition that allows the reversal of the two limiting processes: summation and integration. The second allows the decomposition of an integral into a countable number of disjoint domains, generalizing property 7 of Proposition 2.31.

Corollary 2.39 (Lebesgue's monotone convergence theorem) *If $\{f_m(x)\}_{m=1}^{\infty}$ is a sequence of nonnegative Lebesgue measurable functions and $f(x) = \sum_{m=1}^{\infty} f_m(x)$ a.e. on Lebesgue measurable $E \in \mathcal{M}_L(\mathbb{R}^n)$, then:*

$$\int_E f(x)dx = \sum_{m=1}^{\infty} \int_E f_m(x)dx. \qquad (2.25)$$

Proof. *Defining $g_m(x) = \sum_{k=1}^{m} f_k(x)$, it follows that $g_m(x)$ is nonnegative and Lebesgue measurable, and the sequence $\{g_m(x)\}_{m=1}^{\infty}$ is increasing. Further, $f(x) = \lim_{m\to\infty} g_m(x)$ a.e., so by Lebesgue's monotone convergence theorem:*

$$\int_E f(x)dx = \lim_{m\to\infty} \int_E g_m(x)dx.$$

But by Proposition 2.31:

$$\int_E g_m(x)dx = \sum_{k=1}^{m} \int_E f_k(x)dx,$$

and the result follows. ■

Corollary 2.40 (Lebesgue's monotone convergence theorem) *If $f(x)$ is a nonnegative Lebesgue integrable function on Lebesgue measurable set $E \in \mathcal{M}_L(\mathbb{R}^n)$, and $E = \bigcup_{j=1}^{\infty} E_j$ is a disjoint union of measurable sets, then:*

$$\int_E f(x)dx = \sum_{j=1}^{\infty} \int_{E_j} f(x)dx. \qquad (2.26)$$

Proof. *Left as an exercise. Hint: Definition 2.9.* ■

2.4.5   Evaluating Integrals

We recall the introductory comments in Remark 2.23, and continue with results on evaluation. The definition of the integral $\int_E f(x)dx$ in (2.19) reflects uncountably many bounded measurable functions in the supremum calculation, each of which is required to equal 0 outside a set of finite measure. This definition was simplified and recharacterized in Proposition 2.29 with a special sequence of such functions $\{f_m(x)\}_{m=1}^{\infty}$ with $f_m(x) \to f(x)$.

More generally, Lebesgue's monotone convergence theorem gives a simplified way to specify the Lebesgue integral of any nonnegative function.

Corollary 2.41 (Evaluating $\int_E f(x)dx$) *If $\{\varphi_m(x)\}_{m=1}^{\infty}$ is an increasing sequence of nonnegative simple functions defined on a Lebesgue measurable set $E \in \mathcal{M}_L(\mathbb{R}^n)$ with $\varphi_m(x) \to f(x)$ for almost all $x \in E$, then:*

$$\int_E f(x)dx = \lim_{m\to\infty} \int_E \varphi_m(x)dx. \qquad (2.27)$$

In other words, the Lebesgue integral can be evaluated using any such sequence of simple functions, the integrals of which are given in (2.2).

Proof. *Immediate from Proposition 2.37 since simple functions are 0 outside a set of finite measure by Definition 2.2, and are Lebesgue measurable by Proposition I.3.24.* ∎

Example 2.42 ($\int_E f(x)dx$ **with simple functions)** *Given nonnegative measurable $f(x)$ on measurable $E \in \mathcal{M}_L(\mathbb{R}^n)$, the proof of Proposition I.3.53 provides a construction of increasing simple functions $\{\varphi_m(x)\}_{m=1}^\infty$ for which $\varphi_m(x) \to f(x)$ for all $x \in E$. This construction did not require $f(x)$ to be bounded, nor E to be finitely measurable.*

For each m define $M \equiv m2^m + 1$ Lebesgue measurable sets, $\{A_j^{(m)}\}_{j=1}^M$ by:

$$A_j^{(m)} = \begin{cases} \{x \in E | (j-1)2^{-m} \le f(x) < j2^{-m}\}, & 1 \le j \le M-1, \\ \{x \in E | m \le f(x)\}, & j = M, \end{cases}$$

and define:

$$\varphi_m(x) = \sum_{j=1}^M (j-1)2^{-m}\chi_{A_j^{(m)}}(x).$$

Then $\{\varphi_m(x)\}_{m=1}^\infty$ is an increasing sequence of nonnegative simple functions, with $\varphi_m(x) \to f(x)$ for all $x \in E$.

If $m\{x \in E | f(x) = \infty\} > 0$, then $f(x)$ is not integrable as in item 3 of Remark 2.27. Letting E' denote this exceptional set:

$$\int_E \varphi_m(x)dx \ge m \int_{E'} dx \to \infty.$$

Otherwise, the conclusion from (2.27) is that:

$$\int_E f(x)dx = \lim_{m \to \infty} \sum_{j=1}^M (j-1)2^{-m}m^n\left(A_j^{(m)}\right), \tag{2.28}$$

where this limit may be finite or not.

2.5 General Measurable Functions

In this final section in the development of the Lebesgue integral, we extend the results from nonnegative measurable functions to general measurable functions.

2.5.1 Definition of the Integral

The final step in the development of the Lebesgue integral is to extend from nonnegative to general measurable functions, and this is relatively easy to do given the tools developed in the prior section. That this is so will be clear once we formalize a decomposition of $f(x)$.

Definition 2.43 (Positive/negative parts of $f(x)$) *Given $f(x)$, the **positive part of** $f(x)$, denoted $f^+(x)$, is defined by:*

$$f^+(x) = \max\{f(x), 0\}, \tag{2.29}$$

*and the **negative part of** $f(x)$, denoted $f^-(x)$, is defined by:*

$$f^-(x) = \max\{-f(x), 0\}. \tag{2.30}$$

Both the positive and negative parts of a function are nonnegative, and if f is measurable so too are these component functions by Proposition I.3.47. Thus we can apply the previous section's results to either part.

Exercise 2.44 (On $g(x) = f(x)$ **a.e.)** *If* $f(x)$ *is Lebesgue measurable and* $g(x) = f(x)$ *a.e., then* $g(x)$ *is Lebesgue measurable by Proposition I.3.16. Prove that if* $g(x) = f(x)$ *a.e., then* $g^+(x) = f^+(x)$ *a.e. and* $g^-(x) = f^-(x)$ *a.e.*

The original function and its absolute value are then recovered by:

$$f(x) = f^+(x) - f^-(x), \qquad |f(x)| = f^+(x) + f^-(x). \tag{2.31}$$

These identities will provide a basis for the definitions of $\int_E f(x)dx$, as well as an alternative approach to $\int_E |f(x)|\, dx$ to that already obtainable by the previous section as applied to nonnegative $|f(x)|$. See Remark 2.47.

For the following definition, note that as a consequence of Remark 2.10 and the corresponding result for nonnegative functions, that if $m(E) = 0$ then:

$$(\mathcal{L}) \int_E f(x)dx = 0,$$

for all Lebesgue measurable functions $f(x)$.

Definition 2.45 (Lebesgue integral of measurable $f(x)$**)** *A Lebesgue measurable function* $f(x)$ *is said to be **Lebesgue integrable** over a Lebesgue measurable set* $E \in \mathcal{M}_L(\mathbb{R}^n)$ *if both* $f^+(x)$ *and* $f^-(x)$ *are integrable over* E, *and in this case we define:*

$$(\mathcal{L}) \int_E f(x)dx = (\mathcal{L}) \int_E f^+(x)dx - (\mathcal{L}) \int_E f^-(x)dx. \tag{2.32}$$

When $f(x)$ *is Lebesgue integrable over* E, $|f(x)|$ *is also Lebesgue integrable over* E *and we define:*

$$(\mathcal{L}) \int_E |f(x)|\, dx = (\mathcal{L}) \int_E f^+(x)dx + (\mathcal{L}) \int_E f^-(x)dx. \tag{2.33}$$

If one of the functions $f^+(x)$ *and* $f^-(x)$ *is Lebesgue integrable and one is not, then* $f(x)$ *is said to be **not Lebesgue integrable** although it is then common to define* $(\mathcal{L}) \int_E f(x)dx = \infty$ *or* $(\mathcal{L}) \int_E f(x)dx = -\infty$ *as appropriate.*

If neither function is Lebesgue integrable, then $f(x)$ *is said to be **not Lebesgue integrable** and* $(\mathcal{L}) \int_E f(x)dx$, *which is formally* $\infty - \infty$, *is undefined, while* $(\mathcal{L}) \int_E |f(x)|\, dx = \infty$.

Remark 2.46 (On Definition 2.45) *In contrast to Riemann integration,* $f(x)$ *is integrable by the definition above if and only if* $|f(x)|$ *is integrable.*

Also, generalizing the case for nonnegative Lebesgue integrable functions, if general $f(x)$ *is Lebesgue integrable, then of necessity:*

$$m(\{x|f(x) = \pm\infty\}) = 0.$$

This follows since if $m(E \bigcup E') > 0$, *then at least one of* $E = \{x|f(x) = \infty\}$ *or* $E' = \{x|f(x) = -\infty\}$ *has positive measure. Hence, at least one of* $f^+(x)$ *and* $f^-(x)$ *could not be integrable by item 3 of Remark 2.27, and thus* $f(x)$ *could not be Lebesgue integrable.*

But $m(E) = m(E') = 0$ *does not assure Lebesgue integrability as the example* $f(x) = 1/x$ *for* $x \neq 0$ *and* $f(0) = \infty$, *demonstrates.*

Remark 2.47 (On two definitions of $(\mathcal{L}) \int_E |f(x)|\, dx$**)** *There is no conflict between the definition of* $(\mathcal{L}) \int_E |f(x)|\, dx$ *in 2.33 and that which would be produced directly by the prior section as applied to the nonnegative function* $|f(x)|$. *Indeed, for measurable* $f(x)$, *if* $|f(x)|$ *is Lebesgue integrable by the prior section then, since* $f^+(x) \le |f(x)|$ *and* $f^-(x) \le |f(x)|$, *both* $f^+(x)$ *and* $f^-(x)$ *are Lebesgue integrable by Corollary 2.32. Then since* $|f(x)| = f^+(x) + f^-(x)$, *2.33 follows by additivity of the integral in Proposition 2.31.*

Example 2.48 *The function*

$$f(x) = \begin{cases} 1/x^2, & x > 0, \\ \infty, & x = 0, \\ -1/x^2, & x < 0, \end{cases}$$

is Lebesgue integrable. This follows from $f^+(x) = 1/x^2$ *for* $x > 0$ *and is* 0 *otherwise, while* $f^-(x) = 1/x^2$ *for* $x < 0$ *and is* 0 *otherwise. Both are Lebesgue integrable as noted in Example 2.28, and:*

$$(\mathcal{L}) \int_0^\infty f^+(x)dx = (\mathcal{L}) \int_{-\infty}^0 f^-(x)dx = 1.$$

Consequently:

$$(\mathcal{L}) \int_{-\infty}^\infty f(x)dx = 0,$$

$$(\mathcal{L}) \int_{-\infty}^\infty |f(x)|\, dx = 2.$$

2.5.2 Properties of the Integral

We summarize the essential properties of the Lebesgue integral in the following proposition. We assume that both $f(x)$ and $g(x)$ are Lebesgue integrable functions, and not just Lebesgue measurable, to avoid definitional problems. For example, if stated assuming only Lebesgue measurability, then for item 2, $f(x) + g(x)$ may be integrable and indeed identically 0, with both $f(x)$ and $g(x)$ not integrable. Similarly, for item 6, measurable $f(x)$ could be integrable on a subcollection of the sets $E_i \subset E$ without being integrable on E.

Proposition 2.49 (Properties of the integral) *If* $f(x)$ *and* $g(x)$ *are Lebesgue integrable functions defined on a Lebesgue measurable set* $E \in \mathcal{M}_L(\mathbb{R}^n)$, *then suppressing the* $(\mathcal{L})$ *notation:*

1. *For any real* a:

$$\int_E af(x)dx = a \int_E f(x)dx.$$

2. *Arbitrarily defining* $f(x) + g(x)$ *on the set of Lebesgue measure* 0 *for which this sum may formally be equal to* $\infty - \infty$ *or* $-\infty + \infty$:

$$\int_E [f(x) + g(x)]dx = \int_E f(x)dx + \int_E g(x)dx.$$

3. *If* $f(x) = g(x)$ *a.e., then:*

$$\int_E f(x)dx = \int_E g(x)dx.$$

4. *If $f(x) \leq g(x)$ a.e., then:*

$$\int_E f(x)dx \leq \int_E g(x)dx.$$

5. *If $C_1 \leq f(x) \leq C_2$ a.e. and $m(E) < \infty$, then:*

$$C_1 m(E) \leq \int_E f(x)dx \leq C_2 m(E).$$

6. *If $E = \bigcup_i E_i$, a union of finitely or countably many disjoint Lebesgue measurable sets, then:*

$$\int_E f(x)dx = \sum_i \int_{E_i} f(x)dx. \tag{2.34}$$

7. *The **triangle inequality:***

$$\left| \int_E f(x)dx \right| \leq \int_E |f(x)| \, dx. \tag{2.35}$$

8. *If for all measurable E:*

$$\int_E f(x)dx = 0,$$

then $f(x) = 0$ a.e.

Proof. *Item 1 follows from Proposition 2.31 and (2.32).*

For item 2, the assumed integrability of $f(x)$ and $g(x)$ and the triangle inequality imply the integrability of $|f(x) + g(x)|$, and then of $f(x)+g(x)$ and this sum's positive and negative parts. The subtlety in this proof is that $[f(x)+g(x)]^+$ need not, and in general will not equal $f^+(x)+g^+(x)$. So we must investigate the implication of splitting $f(x)+g(x)$ two ways into positive and negative parts.

First by definition:

$$f(x) + g(x) = [f(x) + g(x)]^+ - [f(x) + g(x)]^-,$$

and then by splitting $f(x)$ and $g(x)$ separately:

$$f(x) + g(x) = [f^+(x) + g^+(x)] - [f^-(x) + g^-(x)].$$

These representations obtain:

$$[f(x) + g(x)]^+ + [f^-(x) + g^-(x)] = [f^+(x) + g^+(x)] + [f(x) + g(x)]^-,$$

All terms are integrable by the integrability of $f(x)$, $g(x)$, and the sum. Integrating and applying Proposition 2.31 produces:

$$\int_E [f(x) + g(x)]^+ dx + \int_E [f^-(x) + g^-(x)]dx$$
$$= \int_E [f^+(x) + g^+(x)]dx + \int_E [f(x) + g(x)]^- dx,$$

or

$$\int_E [f(x) + g(x)]dx = \int_E [f^+(x) + g^+(x)]dx - \int_E [f^-(x) + g^-(x)]dx.$$

The final result follows from another application of Proposition 2.31.

For parts 3 and 4, let:

$$g(x) = f(x) + [g(x) - f(x)],$$

and apply Proposition 2.31. Since $g(x) - f(x) = 0$ a.e. or $g(x) - f(x) \geq 0$ a.e., respectively, the results follow from items 1 and 2.

Item 5 is proved by noting that both $f(x) - C_1 \geq 0$ and $C_2 - f(x) \geq 0$ a.e., and thus have nonnegative integrals by item 4, and the result is completed by items 1 and 2 and (2.2).

For item 6, since $\chi_E = \sum_i \chi_{E_i}$ for disjoint sets:

$$
\begin{aligned}
\int_E f(x)dx &= \int_E f(x)\chi_E(x)dx \\
&= \int_E \sum_i f(x)\chi_{E_i}(x)dx \\
&= \int_E \sum_i f^+(x)\chi_{E_i}(x)dx - \int_E \sum_i f^-(x)\chi_{E_i}(x)dx,
\end{aligned}
$$

with the last step an application of parts 2 and 1. The final result now follows from Corollary 2.40.

Item 7 then follows from item 4 and the observation that $f(x) \leq |f(x)|$ and $-f(x) \leq |f(x)|$.

For item 8, assume that $f(x) \neq 0$ a.e. on E. Then $m(E') > 0$ for at least one of $E' = \{f > 0\}$ or $E' = \{f < 0\}$. We prove the result for the former set and leave as an exercise the latter set. There then exists n so that $m(E_n') > 0$ for $E_n' = \{f > 1/n\}$. Otherwise, since $\bigcup E_n' = E'$ and $E_n' \subset E_{n+1}'$, continuity from below (Proposition I.2.44) of m would obtain that $m(E') = 0$, a contradiction. Given such n, measurability of E_n', and item 4 obtain that:

$$\int_{E_n'} f(x)dx > m(E_n')/n > 0,$$

a contradiction. ■

Item 3 above is valid even if we do not assume that $g(x)$ is integrable.

Corollary 2.50 (On $g(x) = f(x)$ a.e.) *If $f(x)$ is Lebesgue integrable on a Lebesgue measurable set $E \in \mathcal{M}_L(\mathbb{R}^n)$, and $g(x) = f(x)$ a.e., then $g(x)$ is Lebesgue integrable on E and:*

$$\int_E g(x)dx = \int_E f(x)dx.$$

Proof. *The function $g(x)$ is Lebesgue measurable by Proposition I.3.16, and by Exercise 2.44, $g^+(x) = f^+(x)$ a.e. and $g^-(x) = f^-(x)$ a.e. Item 3 of Proposition 2.31 obtains that:*

$$\int_E g^\pm(x)dx = \int_E f^\pm(x)dx,$$

and thus $g(x)$ is Lebesgue integrable and item 3 of Proposition 2.49 applies. ■

Corollary 2.32 stated that every nonnegative Lebesgue integrable function can be used as a **test function** to identify other Lebesgue integrable functions, and it was noted in Remark 2.33 that this could be extended to general integrable functions. But we need to be careful here. If $g(x)$ is Lebesgue integrable and $f(x) \leq g(x)$ a.e., then in general we cannot predict that $f(x)$ is so integrable, since we have no lower bound on such $f(x)$.

Corollary 2.51 (Integrability comparison test) *Let $f(x)$, $g(x)$ be Lebesgue measurable functions and $E \in \mathcal{M}_L(\mathbb{R}^n)$.*

1. **If $g(x)$ is Lebesgue integrable on** E, and $|f(x)| \leq |g(x)|$ a.e. on E, then $f(x)$ is Lebesgue integrable on E with:

$$\int_E |f(x)|\, dx \leq \int_E |g(x)|\, dx.$$

2. **If $g(x)$ is not Lebesgue integrable on** E, and $|f(x)| \geq |g(x)|$ a.e. on E, then $f(x)$ is not Lebesgue integrable on E.

Proof. *Since $|f(x)|$ and $|g(x)|$ are Lebesgue measurable, the integral bound in item 1 follows from part 4 of Proposition 2.31. Then integrability of $|f(x)|$ implies integrability of $f^{\pm}(x)$ by (2.33), which in turn implies integrability of $f(x)$ by (2.32).*

Item 2 is a logical restatement of item 1. If $f(x)$ and thus $|f(x)|$ were Lebesgue integrable, then integrability of $g(x)$ would follow, which is a contradiction. ∎

2.5.3   Lebesgue's Dominated Convergence Theorem

This section derives **Lebesgue's dominated convergence theorem,** the final integration to the limit theorem for measurable function sequences, and again named after **Henri Lebesgue** (1875–1941). As was the case for the monotone convergence theorem, the result in (2.36) can be written as in (2.23) to emphasize that we are interchanging two limiting processes.

We suppress the $(\mathcal{L})$ notation.

Proposition 2.52 (Lebesgue's dominated convergence theorem) *Let $\{f_m(x)\}_{m=1}^{\infty}$ be a sequence of Lebesgue measurable functions on a Lebesgue measurable set $E \in \mathcal{M}_L(\mathbb{R}^n)$, and $f(x) = \lim_{m \to \infty} f_m(x)$ almost everywhere.*

If there is a Lebesgue measurable function $g(x)$, integrable on E, so that on E:

$$|f_m(x)| \leq g(x), \ \text{all } m,$$

then $f(x)$ is integrable on E and:

$$\int_E f(x)dx = \lim_{m \to \infty} \int_E f_m(x)dx. \tag{2.36}$$

Further,

$$\int_E |f_m(x) - f(x)|\, dx \to 0, \ \text{as } m \to \infty. \tag{2.37}$$

Proof. *The limit function $f(x)$ is measurable by Corollary I.3.48, and since $|f_m(x)| \leq g(x)$ implies that $|f(x)| \leq g(x)$ a.e., it follows from Corollary 2.51 that $f(x)$ is Lebesgue integrable.*

Also, $|f_m(x) - f(x)| \leq 2g(x)$ obtains that $2g(x) - |f_m(x) - f(x)|$ is nonnegative for all m and has pointwise limit $2g(x)$. Applying Fatou's lemma and Remark I.3.44 obtains:

$$
\begin{aligned}
\int_E 2g(x)dx \ &\leq \ \liminf \int_E [2g(x) - |f_m(x) - f(x)|]dx \\
&= \ \int_E 2g(x)dx + \liminf \int_E [-|f_m(x) - f(x)|]dx \\
&= \ \int_E 2g(x)dx - \limsup \int_E |f_m(x) - f(x)|\, dx.
\end{aligned}
$$

Subtracting $\int_E 2g(x)dx$, *which is finite by assumption, obtains:*

$$\limsup \int_E |f_m(x) - f(x)|\, dx \leq 0.$$

Since each integral is nonnegative, the limit superior of this integral sequence equals 0. It then follows that the limit inferior of this sequence is also 0 since $\liminf \leq \limsup$, *and thus so too is the limit, proving (2.37).*

Then by the triangle inequality in (2.35):

$$\left| \int_E [f_m(x) - f(x)]dx \right| \leq \int_E |f_m(x) - f(x)|\, dx,$$

and so (2.37) implies (2.36). ∎

Corollary 2.53 (Lebesgue's dominated convergence theorem) *Let* $\{h_j(x)\}_{m=1}^\infty$ *be a sequence of Lebesgue measurable functions on Lebesgue measurable set* $E \in \mathcal{M}_L(\mathbb{R}^n)$, *and assume that* $f(x) - \sum_{j=1}^\infty h_j(x)$ *a.e. If there is a Lebesgue measurable function* $g(x)$, *integrable on* E, *so that:*

$$\left| \sum_{j=1}^m h_j(x) \right| \leq g(x) \text{ for all } m,$$

then $f(x)$ *is integrable on* E *and:*

$$\int_E f(x)dx = \sum_{j=1}^\infty \int_E h_j(x)dx.$$

Further,

$$\int_E \left| \sum_{j=m}^\infty h_j(x) \right| dx \to 0 \text{ as } m \to \infty.$$

Proof. *Left as an exercise. Hint: Let* $f_m(x) = \sum_{j=1}^m h_j(x)$ *in Proposition 2.52.* ∎

2.5.4 Evaluating Integrals

We recall the introductory comments in Remark 2.23, and continue with results on evaluation. The definition of the integral $\int_E f(x)dx$ in (2.32) references the definition in (2.19), and thus reflects uncountably many bounded measurable functions in each of the supremum calculations implied by the component integrals.

Lebesgue's dominated convergence theorem gives a simplified way to specify the Lebesgue integral of a general measurable function, generalizing Corollary 2.41.

Corollary 2.54 (Lebesgue's dominated convergence theorem) *Let* $\{f_m(x)\}_{m=1}^\infty$ *be a sequence of Lebesgue measurable functions defined on a Lebesgue measurable set* $E \in \mathcal{M}_L(\mathbb{R}^n)$, *with* $|f_m(x)| \leq |g(x)|$ *on* E *for all* m *for some Lebesgue integrable* $g(x)$. *If* $f_m(x) \to f(x)$ *almost everywhere on* E, *then* $f(x)$ *is integrable and:*

$$\int_E f(x)dx = \lim_{m \to \infty} \int_E f_m(x)dx.$$

Proof. *Immediate from Proposition 2.52.* ∎

Example 2.55 *If* $f(x)$ *is known to be integrable, and* $\{f_m(x)\}_{m=1}^\infty$ *any sequence of simple functions with* $|f_m(x)| \leq |f(x)|$ *for all* m *and* $f_m(x) \to f(x)$ *almost everywhere on* E, *the above result follows. The simple function approximations of* $f^+(x)$ *and* $f^-(x)$ *implied by Example 2.42 provide such sequences.*

2.5.5   Absolute Riemann Implies Lebesgue

In this section, we generalize Proposition 2.18 which showed that if $f(x)$ is a bounded, Riemann integrable function on a bounded rectangle $R \equiv \prod_{j=1}^{n} [a_j, b_j]$, then $f(x)$ is measurable and Lebesgue integrable, and the values of the Lebesgue and Riemann integrals agree. Here we show that a locally bounded function $f(x)$, which is Riemann integrable and absolutely Riemann integrable on $\mathbb{R}^n$, is Lebesgue integrable on $\mathbb{R}^n$ and again the Riemann and Lebesgue integrals agree. By **locally bounded** is meant that $f(x)$ is bounded on bounded sets.

In the special case where $f(x)$ is nonnegative (or nonpositive) and Riemann integrable on $\mathbb{R}^n$, it is then also absolutely Riemann integrable. Thus this proposition applies and assures that $f(x)$ is also Lebesgue integrable on $\mathbb{R}^n$ and that the integrals agree.

For this statement, recall Exercises 1.25 and 1.74. Unlike Lebesgue integration, it is not redundant to state that $f(x)$ is both Riemann integrable and absolutely Riemann integrable, as the latter does not imply the former even on bounded domains. Analogously, while Riemann integrable implies absolute Riemann integrable on bounded domains by item 1 of Propositions 1.23 and 1.72, this does not extend to improper integrals as seen in Example 1.27.

Proposition 2.56 (Absolute Riemann $\Rightarrow$ Lebesgue integrable) *Let $f(x)$ be a real-valued, locally bounded function defined on $\mathbb{R}^n$ that is both Riemann integrable and absolutely Riemann integrable as an improper integral defined in (1.53).*

Then $f(x)$ is Lebesgue integrable and hence measurable, and:

$$(\mathcal{L}) \int_{\mathbb{R}^n} f(x) dx = (\mathcal{R}) \int_{\mathbb{R}^n} f(x) dx. \tag{2.38}$$

Proof. *Given m, let:*

$$R^{(m)} \equiv \prod_{j=1}^{n} [a_j^{(m)}, b_j^{(m)}],$$

where we assume that for all j, that $a_j^{(m)} \to -\infty$ and $b_j^{(m)} \to \infty$ monotonically as $m \to \infty$.
   Define:

$$f_m(x) = \begin{cases} f(x), & x \in R^{(m)} \text{ and } |f(x)| \leq m, \\ m, & x \in R^{(m)} \text{ and } f(x) > m, \\ -m, & x \in R^{(m)} \text{ and } f(x) < -m, \\ 0, & x \notin R^{(m)}. \end{cases}$$

Since $f(x)$ is bounded and Riemann integrable on each $R^{(m)}$ by assumption, it is continuous outside a set of measure 0 by Lebesgue's existence theorem. Thus $f_m(x)$ is also continuous outside such a set and bounded on $R^{(m)}$, and is again Riemann integrable by Lebesgue's result.
   By Proposition 2.18, $f_m(x)$ is Lebesgue measurable, integrable on $R^{(m)}$, and:

$$(\mathcal{L}) \int_{R^{(m)}} f_m(x) dx = (\mathcal{R}) \int_{R^{(m)}} f_m(x) dx. \tag{1}$$

The identity in (1) also holds for $|f_m(x)|$ by the same argument, and then also to $f_m^+(x) = (f_m(x) + |f_m(x)|)/2$, the positive part of $f_m(x)$. As $f_m(x)$ as 0 outside $R^{(m)}$, these identities hold as integrals over $\mathbb{R}^n$:

$$(\mathcal{L}) \int_{\mathbb{R}^n} f_m^+(x) dx = (\mathcal{R}) \int_{\mathbb{R}^n} f_m^+(x) dx. \tag{2}$$

Now $\{f_m^+(x)\}_{m=1}^\infty$ *is an increasing sequence of measurable functions that converges to* $f^+(x)$. *Thus by Lebesgue's monotone convergence theorem:*

$$(\mathcal{L}) \int_{\mathbb{R}^n} f^+(x)dx = \lim_{m\to\infty} (\mathcal{L}) \int_{\mathbb{R}^n} f_m^+(x)dx. \tag{3}$$

Since $f(x)$ *is Riemann and absolutely Riemann integrable,* $f^+(x)$ *is Riemann integrable, which obtains by (3) and (2):*

$$(\mathcal{L}) \int_{\mathbb{R}^n} f^+(x)dx = \lim_{m\to\infty} (\mathcal{R}) \int_{\mathbb{R}^n} f_m^+(x)dx = (\mathcal{R}) \int_{\mathbb{R}^n} f^+(x)dx.$$

The same argument applies to $f_m^-(x) = (|f_m(x)| - f_m(x))/2$, *and 2.38 is proved.* ∎

2.6 Summary of Convergence Results

Though requiring a bit of care from step to step, the development of the existence and properties of the Lebesgue integral largely proceeded without many surprises. The final reward is that the Lebesgue integral significantly expanded the class of functions that can be integrated compared with the Riemann approach, and similarly expanded the class of domains over which these integrals can be defined. Further, in cases where a given function could be integrated in both the Riemann and Lebesgue sense, the integrals almost always agreed.

The critical exception to this final statement had to do with the definitional constraint on a Lebesgue integral. In order for $f(x)$ to be defined as Lebesgue integrable, it is required that $|f(x)|$ be Lebesgue integrable, while no such condition is required on Riemann integrability. Thus for functions where $f(x)$ is Riemann integrable and $|f(x)|$ is not, the Lebesgue integral does not exist by definition.

The value of this definitional restriction is that if $f(x)$ is Lebesgue integrable over a given set E, of finite or infinite Lebesgue measure, then it is integrable over any Lebesgue measurable subset $E_j \subset E$. Further, if $E = \bigcup_{j=1}^\infty E_j$ as a disjoint union, then:

$$\int_E f(x)dx = \sum_{j=1}^\infty \int_{E_j} f(x)dx.$$

This result cannot be assumed within the Riemann theory, as was seen in Example 1.27.

But where the Lebesgue integration theory really shines is in the realm of limit theorems, which investigates $\left\{ \int_E f_m(x)dx \right\}_{m=1}^\infty$ when $\{f_m(x)\}_{m=1}^\infty$ is a function sequence that in some manner converges to $f(x)$. In the first chapter of this book, limitations in the Riemann theory were illustrated, with two being fundamental:

1. $f_m(x)$ may be Riemann integrable for all m, converge pointwise to $f(x)$, and yet $f(x)$ need not be Riemann integrable;

2. $f_m(x)$ may be Riemann integrable for all m, converge to $f(x)$, which is Riemann integrable, and yet $\int_E f_m(x)dx \nrightarrow \int_E f(x)dx$.

As we have seen above, the Lebesgue theory is more powerful in both of these respects, and we summarize the results developed in this chapter in the next section.

The final section of this chapter then revisits the Riemann examples of Chapter 1.

2.6.1   Lebesgue Integration to the Limit

In this section, we catalog the various situations in which it can be concluded that the pointwise limit function $f(x)$ is Lebesgue integrable, and, if so, how the value of this integral relates to the integral sequence $\int_E f_m(x)dx$. Such results are often categorized under the heading of **integration to the limit**.

In Book V, as part of a more general development of integration theory, another integration to the limit result will be developed that reflects the notion of **uniform integrability** of a function sequence. This notion will play an important role in the probability theory of Books IV and VI.

As this section is on Lebesgue integration, we suppress the $(\mathcal{L})$ notation.

1. **Bounded convergence theorem:** Let $\{f_m(x)\}_{m=1}^{\infty}$ be a sequence of uniformly bounded, Lebesgue measurable functions defined on a Lebesgue measurable set $E \in \mathcal{M}_L(\mathbb{R}^n)$ with $m(E) < \infty$. By uniformly bounded is meant that there exists M so that $|f_m(x)| \le M$ for all m and all $x \in E$.

   Then if $f_m(x) \to f(x)$ almost everywhere for $x \in E$:

   $$\int_E f(x)dx = \lim_{m \to \infty} \int_E f_m(x)dx. \tag{2.39}$$

   Comments: *The bounded convergence theorem requires uniformly bounded Lebesgue measurable functions $\{f_m(x)\}_{m=1}^{\infty}$, and a measurable set E of finite measure, but then guarantees both the integrability of the limit function $f(x)$, and the convergence of the integral sequence to the integral of $f(x)$.*

2. **Fatou's Lemma:** If $\{f_m(x)\}_{m=1}^{\infty}$ is a sequence of nonnegative Lebesgue measurable functions, and $f(x) \equiv \liminf f_m(x)$ almost everywhere on a Lebesgue measurable set $E \in \mathcal{M}_L(\mathbb{R}^n)$, then:

   $$\int_E f(x)dx \le \liminf \int_E f_m(x)dx. \tag{2.40}$$

   Comments: *Fatou's Lemma requires nonnegative Lebesgue measurable functions $\{f_n(x)\}_{m=1}^{\infty}$, and then for arbitrary measurable E provides an upper bound for the integral of the limit inferior of these functions in terms of the limit inferior of the integral sequence.*

   But note that:

   a. *Fatou's lemma does not in general guarantee that $f(x)$ is integrable.*

   b. *If the limit inferior of the integral sequence is finite, then $f(x)$ is integrable and has integral no bigger than this limit.*

   c. *If the limit inferior of the integral sequence is 0, then by nonnegativity of $f(x)$ we can conclude that $\int_E f(x)dx = 0$.*

   d. *Limit inferiors can be replaced by limits if they exist.*

   e. *This result applies to the sequence $\{|f_m(x)|\}_{m=1}^{\infty}$ for arbitrary Lebesgue measurable $\{f_m(x)\}_{m=1}^{\infty}$.*

3. **Lebesgue's monotone convergence theorem:** If $\{f_m(x)\}_{m=1}^{\infty}$ is an increasing sequence of nonnegative Lebesgue measurable functions on Lebesgue measurable $E \in \mathcal{M}_L(\mathbb{R}^n)$ and $f(x) = \lim_{m \to \infty} f_m(x)$ almost everywhere on E, then (2.39) holds.

Comments: *Lebesgue's monotone convergence theorem requires nonnegative Lebesgue measurable functions $\{f_m(x)\}_{m=1}^{\infty}$ that are monotonically increasing, $f_m(x) \leq f_{m+1}(x)$ for all m, or at least for all $m \geq N$ for some N. But then for arbitrary measurable E, (2.39) holds.*

Like Fatou's lemma, this result does not guarantee that $f(x)$ is Lebesgue integrable, but if the increasing sequence of integrals is bounded and thus has a limit, then the integral of $f(x)$ exists and equals this limit.

Though restricted to nonnegative functions, it again applies to the sequence $\{|f_m(x)|\}_{m=1}^{\infty}$ for arbitrary Lebesgue measurable functions $\{f_m(x)\}_{m=1}^{\infty}$, as long as this absolute value sequence is increasing.

4. **Lebesgue's dominated convergence theorem:** If $\{f_m(x)\}_{m=1}^{\infty}$ is a sequence of Lebesgue measurable functions on Lebesgue measurable set $E \in \mathcal{M}_L(\mathbb{R}^n)$ and $f(x) = \lim_{m \to \infty} f_m(x)$ almost everywhere on E, and if there is a Lebesgue integrable function $g(x)$ on E, so that on E:

$$|f_m(x)| \leq g(x) \text{ for all } m,$$

or at least for all $m \geq N$ for some N, then $f(x)$ is integrable on E and (2.39) holds. Further,

$$\int_E |f_m(x) - f(x)|\, dx \to 0 \text{ as } m \to \infty.$$

Comments: *Lebesgue's dominated convergence theorem applies to general Lebesgue measurable functions $\{f_m(x)\}_{m=1}^{\infty}$ defined on an arbitrary measurable set E, if they are dominated by an integrable function $g(x)$. The assumption that these functions are dominated by $g(x)$ assures both that each $f_m(x)$ is Lebesgue integrable, but more importantly assures that if this function sequence converges to $f(x)$ almost everywhere, then $f(x) \leq g(x)$, and thus the integrability of $f(x)$ is assured. In addition, the value of the integral of $f(x)$ is then given by the limit of the integral sequence.*

2.6.2　The Riemann Integral: A Discussion

Returning to Chapter 1, we now analyze the earlier examples from the perspective of the Lebesgue theory. This is justified because as should be verified, the examples there contained functions that were Lebesgue measurable, and for which the Riemann and Lebesgue integrals agree when they both exist.

But first, recall that there were positive results for limits of a function sequence in the Riemann theory which ensured both that the limit function $f(x)$ is Riemann integrable, and that (2.39) is satisfied. A more general statement was found in Proposition 1.48:

- If $\{f_m(x)\}_{m=1}^{\infty}$ *is a sequence of bounded Riemann integrable functions on a closed and bounded interval $[a, b]$, and there is a function $f(x)$ so that $f_m(x) \to f(x)$ uniformly, then $f(x)$ is Riemann integrable and (2.39) is valid on $E = [a, b]$.*

The more specialized result of Proposition 1.47 assumed that $\{f_m(x)\}_{m=1}^{\infty}$ was a sequence of continuous functions on $[a, b]$, which of necessity satisfied the general criteria of boundedness and Riemann integrability.

The general positive result of Proposition 1.48 is also confirmed true within the Lebesgue theory. This follows because bounded and Riemann integrable implies Lebesgue integrable by Proposition 2.18, and uniform convergence assures uniform boundedness, so this result also follows from the bounded convergence theorem.

We now turn to the other examples of Chapter 1.

Example 2.57 *For each of the examples in Section 1.2.2, we investigate the applicability of the Lebesgue results.*

1. **Pointwise convergence of continuous $f_n(x)$ on a bounded set $[a, b]$ with $f(x)$ integrable.**

   a. **Integrals converge:** *The example:*

   $$f_n(x) = \begin{cases} 1, & x \leq 0, \\ 1 - nx, & 0 < x \leq \frac{1}{n}, \\ 0, & x > \frac{1}{n}, \end{cases} \qquad f(x) = \begin{cases} 1, & x \leq 0, \\ 0, & x > 0, \end{cases}$$

   shows continuous $f_n(x)$ converging pointwise to a discontinuous $f(x)$, but where $f(x)$ is integrable over any compact $[a, b]$, and the series of Riemann integrals converges in the sense of (2.39).

   *The **bounded convergence theorem** predicts the results of this example when interpreted as Lebesgue integrals, as does the **dominated convergence theorem** with $g(x) = f(x)$.*

   b. **Integrals do not converge:** *The sequence:*

   $$f_n(x) = \begin{cases} 2^n, & 1/2^n \leq x \leq 1/2^{n-1}, \\ 0, & elsewhere, \end{cases}$$

   converges pointwise but not uniformly on $[0, 1]$ to a continuous function $f(x) \equiv 0$, and for all n:

   $$\int_0^1 f_n(x)dx = 1 \neq \int_0^1 f(x)dx.$$

   Consequently, pointwise convergence of almost everywhere continuous function on a compact interval does not assure convergence of Riemann integrals.

   Of course there cannot be any theorem in the Lebesgue (or any other) theory that predicts a result in conflict with the above conclusion. But it is still instructive to investigate the potential applicability of this theory.

   *We cannot apply either the **bounded convergence** or **Lebesgue's monotone convergence theorem** to this example since we lack boundedness and monotonicity.*

   *To investigate why **Lebesgue's dominated convergence theorem** does not apply, note that there is no integrable function $g(x)$ that dominates all $f_n(x)$. The logical guess would be $g(x) = \sum f_n(x)$, as this is the best possible upper bound for all x. But $g(x)$ is not Lebesgue integrable.*

   *However we can apply **Fatou's lemma,** noting that $f(x) = \liminf_{n \to \infty} f_n(x) = 0$ for all x. The Fatou conclusion is then,*

   $$0 = \int_0^1 \liminf f_n(x)dx \leq \liminf \int_0^1 f_n(x)dx = 1,$$

   consistent with this example's result of strict inequality.

2. **Uniform convergence of continuous $f_n(x)$ on an unbounded set $[a, \infty)$ with $f(x)$ integrable.**

a. Integrals converge: *Defining $f_n(x)$ on $[1, \infty)$ for $n \geq 1$:*

$$f_n(x) = x^{-(n+1)}/n,$$

provides a continuous function sequence that converges uniformly to the continuous function $f(x) \equiv 0$ and:

$$\int_1^\infty f_n(x)dx = 1/n^2 \to 0 = \int_1^\infty f(x)dx.$$

Because the set $E = [1, \infty)$ has infinite measure, we look to Fatou or one of Lebesgue's theorems. The **monotone convergence theorem** *does not apply because this sequence is decreasing, but the* **dominated convergence theorem** *predicts this positive result using $g(x) = f_1(x)$.*

b. Integrals do not converge: *Here:*

$$f_n(x) = x^{-(n+1)/n}/n.$$

Again $f_n(x)$ is continuous, $f_n(x) \to f(x) \equiv 0$ uniformly on $[1, \infty)$, and the integrals are well defined with $\int_1^\infty f_n(x)dx = 1$. However,

$$\int_1^\infty f_n(x)dx \to 1 \neq \int_1^\infty f(x)dx = 0.$$

Again there cannot be any theorem in the Lebesgue theory that predicts a result in conflict with the above conclusion, but we investigate nonetheless.

The **bounded convergence theorem** *does not apply despite the fact that the function sequence is uniformly bounded since $E = [1, \infty)$ does not have finite measure. But as a nonnegative measurable function sequence, we can again apply* **Fatou's lemma,** *which as in item 1.b confirms the possibility of this example's result of strict inequality.*

Lebesgue's **monotone convergence theorem** *is inapplicable because this function sequence is not monotonically increasing. For $x \in [1, \infty)$, the coefficient $1/n$ diminishes this function but the exponent increases it. Using calculus applied to the function $h(y) = x^{-(y+1)/y}/y$ obtains that for each $x > 1$, this function has $h'(y) = 0$ when $y_0 = \ln x$, and $h''(y_0) < 0$, indicating that y_0 is a maximum. Consequently, for any x the function sequence $\{f_n(x)\}_{n=1}^\infty$ increases until $n \approx \ln x$, then decreases to 0.*

Finally, we look for an integrable function $g(x)$ that dominates this function sequence and could justify the application of **Lebesgue's dominated convergence theorem.** *To have $g(x) \geq f_n(x)$ for all x and all n, it is clear that for $x \in [1, \infty)$ that we need $g(x) \geq x^{-(n+1)/n}/n$ for all n. Using the above analysis of the function $h(y) = x^{-(y+1)/y}/y$, $x > 1$, we find that this bounding function $g(x)$ can be defined by $g(x) = h(\ln x)$. In other words, for $x > 1$:*

$$g(x) = \frac{1}{x \ln x} x^{-1/\ln x} = \frac{1}{ex \ln x}.$$

Since the Lebesgue and Riemann integrals of $g(x)$ on the intervals $[1 + \epsilon, N]$ agree and are unbounded, this proves that $g(x)$ is not Lebesgue integrable on $(1, \infty)$, and hence Lebesgue's dominated convergence theorem is not applicable as expected.

3. Pointwise convergence of Riemann integrable $f_n(x)$ with $f(x)$ not integrable.

    a. Convergence on a bounded set $[a, b]$: *For any ordering $\{r_j\}_{j=1}^{\infty}$ of the rational numbers in $[0, 1]$, define:*

$$f_n(x) = \begin{cases} 1, & x = r_j, \ 1 \leq j \leq n, \\ 1/n, & \text{elsewhere.} \end{cases}$$

Then $f_n(x)$ is continuous except at n points, and $(\mathcal{R}) \int_0^1 f_n(x)dx = 1/n$. However, $f_n(x) \to f(x)$ pointwise, where:

$$f(x) = \begin{cases} 1, & x \text{ rational}, \\ 0, & x \text{ irrational}, \end{cases}$$

which is nowhere continuous and hence not Riemann integrable.

    *Taken as Lebesgue integrals, a positive result is achieved since $f(x)$ is Lebesgue integrable with integral 0. This result is then predicted by either the **bounded convergence theorem** or **Lebesgue's dominated convergence theorem**. **Monotone convergence** does not apply because this function sequence is not monotonically increasing.*

    b. Convergence on an unbounded set $[a, \infty)$: *For any ordering on the rational numbers $\{r_j\}_{j=1}^{\infty}$ in $[1, \infty)$, define:*

$$f_n(x) = \begin{cases} 1, & x = r_j, \ 1 \leq j \leq n, \\ x^{-(n+1)}, & \text{elsewhere.} \end{cases}$$

Then $f_n(x)$ is continuous except at n points and $(\mathcal{R}) \int_1^{\infty} f_n(x)dx = 1/n$. However, $f_n(x) \to f(x)$ pointwise, where for $x \geq 1$:

$$f(x) = \begin{cases} 1, & x \text{ rational}, \\ 0, & x \text{ irrational}, \end{cases}$$

which is nowhere continuous and hence not Riemann integrable.

    *As Lebesgue integrals, a positive result is achieved since $f(x)$ is Lebesgue integrable with integral 0. Because $E = [1, \infty)$ has infinite measure and this function sequence is not monotonically increasing, we can only attempt to apply **Lebesgue's dominated convergence theorem** or **Fatou's lemma**, and both apply.*

    First, $f_n(x)$ is dominated by the Lebesgue integrable function $f_1(x)$. Second, Fatou's lemma applies and states that:

$$\begin{aligned} \int_1^{\infty} f(x)dx & \leq \liminf \int_1^{\infty} f_n(x)dx \\ & = \liminf\{1/n\} \\ & = 0. \end{aligned}$$

    Since $f(x) \geq 0$ implies that $\int_1^{\infty} f(x)dx \geq 0$, Fatou's lemma also provides the positive result in (2.39) in this example.

3

Lebesgue Integration and Differentiation

In this chapter, we limit the discussion to the 1-dimensional Lebesgue measure space $(\mathbb{R}, \mathcal{M}_L, m)$, and study the relationships between differentiation and Lebesgue integration. As we have seen in Propositions 2.18 and 2.56, the value of a Lebesgue integral of a function equals the value of that function's Riemann integral in virtually any situation in which both are defined. Since there are intimate connections between integrals and derivatives in the Riemann theory when $n = 1$, it is natural to investigate to what extent these connections continue to exist within the Lebesgue theory, perhaps with "almost everywhere" qualifications.

To set the stage for this chapter's investigation, recall the results within the Riemann theory of Section 1.1.6. There it was seen that while there were various representations of the Fundamental Theorem of Calculus, these representations could be summarized into two categories:

- Derivative of a Riemann integral with variable upper limit of integration as in Proposition 1.33;

- Riemann integral of the derivative of a function as in Proposition 1.30.

Thinking ahead to potential Lebesgue counterparts to these results, the first observation is that Lebesgue integration requires only measurability of a function. While Lebesgue measurable functions are "nearly" continuous by Lusin's theorem of Proposition I.4.10, one predicts that the conclusion in (1.28), that $F'(x) = f(x)$, may be more than can reasonably be expected.

Indeed, we know by Proposition 2.49 that the integrand $f(y)$ can be changed on any set of measure 0 without changing the value of the Lebesgue integral in (1.27):

$$F(x) = F(a) + \int_a^x f(y)dy.$$

So if $F(x)$ is differentiable, it cannot be expected that $F'(x)$ will reproduce such an arbitrarily defined $f(x)$ everywhere. But we will see that for Lebesgue integrable functions $f(x)$, the conclusion of (1.28) will be true almost everywhere. This parallels the more general Riemann result noted earlier when $f(x)$ is only assumed to be continuous almost everywhere.

It turns out that generalizing (1.24) to a Lebesgue context is the more difficult task. Indeed, it is even difficult to generalize the Riemann result beyond the stated case of continuously differentiable $f(x)$. In seeking a generalization of this result to the Lebesgue context, we will look for a class of functions $f(x)$ that can be represented as Lebesgue integrals of their derivatives as in (1.26):

$$f(x) = f(a) + \int_a^x f'(y)dy.$$

This class of functions will include all continuously differentiable functions, but not include all differentiable functions. The final answer will fit neatly between continuously

DOI: 10.1201/9781003264590-3

differentiable functions and differentiable functions, and will be called **absolutely continuous functions.**

The final section of this chapter addresses Lebesgue integration by parts.

3.1   Derivative of a Lebesgue Integral

In this section, we address the generalization of Proposition 1.33 to a Lebesgue context. To do so we will first investigate when a real-valued function is differentiable almost everywhere, and prove that monotonic functions have this property. These results will then be applied to functions of the form:

$$F(x) = (\mathcal{L}) \int_a^x f(y) dy,$$

in pursuit of situations for which such $F(x)$ is differentiable a.e., and $F'(x) = f(x)$ a.e. This will then provide the Lebesgue version of this fundamental theorem of calculus.

To investigate differentiability of monotonic functions requires the introduction of the four **Dini derivates** of a function. In contrast to a derivative, the four derivates always exist, even if not finite, and the pursuit of differentiability is reduced to identifying when these derivates agree. For this, we will require a technical result called **Vitali's covering lemma**, which provides a useful approximation of a set, even if this set is not known to be Lebesgue measurable. In this general case, such approximations are defined in terms of Lebesgue outer measure.

To extend beyond monotonic functions, we will investigate a new class of functions, so-called **functions of bounded variation.** Somewhat surprisingly given their definition, such functions always equal a difference of monotonic functions and hence are then differentiable almost everywhere.

The bounded variation characterization proves to be quite applicable to the question at hand. The final section investigates the question: Is $F(x)$ as defined above of bounded variation, and thus differentiable almost everywhere? The answer to this question turns out to be "yes," and then the final investigation addresses the relationship between $F'(x)$ and $f(x)$.

3.1.1   Monotonic Functions

The goal of this section is to prove that a monotonic function $f(x)$, defined on an interval $[a, b]$, is differentiable almost everywhere. In addition, the resulting derivative $f'(x)$ is shown to be Lebesgue measurable and integrable on this interval.

While not the objective of this section's investigation, we will prove a result related to (1.24) on the integral of a derivative that will be useful in a later section. For monotonically increasing functions, $f(b) - f(a)$ will be seen to be an upper bound for the integral $(\mathcal{L}) \int_a^b f'(y) dy$, and a lower bound for monotonically decreasing functions. We will return to this investigation below and obtain equality under additional restrictions on $f(x)$.

To prove that monotonic $f(x)$ is differentiable a.e., we will need a more refined set of quantities related to derivatives and called **Dini derivates** defined below, named after **Ulisse Dini** (1845–1918). For this, we first recall the notation and definition of various one-sided limits.

Limits inferior and superior of a function sequence were originally encountered in Definition I.3.42. These limits were defined pointwise for a function sequence $\{f_n\}_{n=1}^{\infty}$, in terms of the collection of real-values $\{f_n(x)\}_{n=1}^{\infty}$. Below we have but one function f, and define these limits in terms of $\{f(x \pm h)\}$, the values of f near x. The reader should compare the definitions in these two contexts and confirm that the logic of these definitions is the same.

Definition 3.1 (One-sided limits) *Given $f(x)$, the various one-sided limits are defined as follows:*

1. Right limit at x:

$$\lim_{y \to x+} f(y) \equiv \lim_{y \to x, \, y > x} f(y). \tag{3.1}$$

2. Right limit superior at x:

$$\limsup_{y \to x+} f(y) \equiv \lim_{h \to 0+} \left[\sup\{f(y) | y \in (x, x+h)\} \right]. \tag{3.2}$$

3. Right limit inferior at x:

$$\liminf_{y \to x+} f(y) \equiv \lim_{h \to 0+} \left[\inf\{f(y) | y \in (x, x+h)\} \right]. \tag{3.3}$$

4. Left limit at x:

$$\lim_{y \to x-} f(y) \equiv \lim_{y \to x, \, y < x} f(y). \tag{3.4}$$

5. Left limit superior at x:

$$\limsup_{y \to x-} f(y) \equiv \lim_{h \to 0+} \left[\sup\{f(y) | y \in (x-h, x)\} \right]. \tag{3.5}$$

6. Left limit inferior at x:

$$\liminf_{y \to x-} f(y) \equiv \lim_{h \to 0+} \left[\inf\{f(y) | y \in (x-h, x)\} \right]. \tag{3.6}$$

Remark 3.2 *Note that the use of "limit" in the various $\limsup$ and $\liminf$ definitions is justified because the expressions in square brackets are monotonic in h.*

For example, $\sup\{f(y) | y \in (x, x+h)\}$ is monotonically decreasing as $h \to 0^+$, while $\inf\{f(y) | y \in (x-h, x)\}$ is monotonically increasing as $h \to 0^+$. Thus given f and x, the quantities in items 2, 3, 5, and 6 always exist, though need not be finite. In contrast, the quantities in items 1 and 4 need not exist, and as usual they exist if given $\epsilon > 0$ there exists δ, etc., where Definition 1.12 is adapted to reflect the restriction on y.

The left limits inferior and superior could have been defined in terms of left limits, for example:

$$\limsup_{y \to x-} f(y) \equiv \lim_{h \to 0-} \left[\sup\{f(y) | y \in (x+h, x)\} \right],$$

but the above convention of keeping $h > 0$ seems more natural notationally. This convention will be seen again in Definition 3.4.

Exercise 3.3 *Prove that $\lim_{y \to x+} f(y)$ exists if and only if $\limsup_{y \to x+} f(y) = \liminf_{y \to x+} f(y)$, and analogously for the existence of $\lim_{y \to x-} f(y)$.*

Definition 3.4 (Dini derivates of $f(x)$) *Given a real-valued function $f(x)$, the **four Dini derivates of f at x** are defined as follows, noting that these values need not be finite.*

1. Right upper derivate:

$$D^+ f(x) = \limsup_{h \to 0+} \frac{f(x+h) - f(x)}{h} \tag{3.7}$$

2. Right lower derivate:

$$D_+ f(x) = \liminf_{h \to 0+} \frac{f(x+h) - f(x)}{h} \tag{3.8}$$

3. Left upper derivate:

$$D^- f(x) = \limsup_{h \to 0+} \frac{f(x) - f(x-h)}{h} \tag{3.9}$$

4. Left lower derivate:

$$D_- f(x) = \liminf_{h \to 0+} \frac{f(x) - f(x-h)}{h} \tag{3.10}$$

Remark 3.5 (The derivative $f'(x)$) *Recall that the derivative of a function is defined by the following limit, when it exists:*

$$f'(x) = \lim_{h \to 0} \frac{f(x+h) - f(x)}{h}. \tag{3.11}$$

The four Dini derivates first split this limit into two groups:
- *$h < 0$ provides the left derivates,*
- *$h > 0$ provides the right derivates.*

Since such one-sided limits need not exist, each is then split into the upper and lower Dini derivates, defined in terms of the limits superior and inferior of these ratios as $h \to 0$. As noted above, all definitions have been structured to notationally maintain $h > 0$.

By definition of these limits:

$$D_+ f(x) \le D^+ f(x), \quad \text{and}, \quad D_- f(x) \le D^- f(x). \tag{3.12}$$

As will be proved in Exercise 3.7, $f'(x)$ exists if and only if all four derivates are equal. In general, however, there are no other predictable relationships between these four values.

Exercise 3.6 *Consider the function:*

$$f(x) = \begin{cases} x, & x \text{ rational} \\ -x, & x \text{ irrational}. \end{cases}$$

Show that $D^+ f(0) = D^- f(0) = 1$, $D_+ f(0) = D_- f(0) = -1$. Evaluate the Dini derivates of this function for $x \neq 0$, considering both rational and irrational x.

Generalize this example to a function for which all four Dini derivates differ at $x = 0$.

Exercise 3.7 *Prove that $f'(x)$ exists if and only if all four derivates are equal at x.*

We prove below that if $f(x)$ is monotonic on an open interval I, then $f'(x)$ exists almost everywhere, is Lebesgue measurable, and is Lebesgue integrable over any interval $[a, b] \subset I$. Further, the value of this integral is bounded above by $f(b) - f(a)$ when $f(x)$ is monotonically increasing and bounded below by this value when $f(x)$ is monotonically decreasing.

For the proof of this theorem, we require a technical result known as **Vitali's covering lemma,** named after **Giuseppe Vitali** (1875–1932). The version cited here is one of many that go by this name. First a definition:

Definition 3.8 (A cover in the sense of Vitali) *A collection of intervals $\mathcal{I} \equiv \{I_\alpha\}$ is said to **cover a set E in the sense of Vitali** if for any $x \in E$ and $\epsilon > 0$, there is an interval $I \in \mathcal{I}$ so that $x \in I$ and $|I| < \epsilon$, where $|I|$ denotes interval length.*

Remark 3.9 *Note that $\mathcal{I}$ is indeed a covering of E since by definition $E \subset \bigcup_\alpha I_\alpha$. The subscript α connotes that this collection need not be countable.*

Also $|I| = m(I)$, and thus equals Lebesgue measure whether I is open, closed, or semi-closed.

Vitali's covering lemma states that given such a covering of E with $m^*(E) < \infty$, one can choose **finite, disjoint** subcollections of $\mathcal{I}$ for which $m^* \left[E - \bigcup_{j=1}^N I_j \right]$ can be made arbitrarily small. Here m^* denotes Lebesgue outer measure of Definition I.2.25. An equivalent formulation of this result states that subsets of E can be identified, and specifically defined by $E' \equiv E \cap \left[\bigcup_{j=1}^N I_j \right]$, with outer measure arbitrarily close to the outer measure of E.

In the proof below that $f'(x)$ exists almost everywhere for monotonic $f(x)$, we will define such E-sets as well as natural coverings in the sense of Vitali. Instrumental to this proof is the selection of these finite disjoint subcollections and the construction of the associated E'-sets.

This proof must use m^* rather than m because at the outset we will not know if the E-sets so defined are Lebesgue measurable. No assumption on E is needed other than that this set has finite outer measure.

Proposition 3.10 (Vitali's Covering lemma) *Let $E \subset \mathbb{R}$ be a set with $m^*(E) < \infty$, and $\mathcal{I} \equiv \{I_\alpha\}$ a collection of intervals that cover E in the sense of Vitali.*

Then for any $\epsilon > 0$, there is a finite disjoint collection $\{I_j\}_{j=1}^N \subset \mathcal{I}$ so that:

$$m^* \left[E - \bigcup_{j=1}^N I_j \right] < \epsilon. \tag{3.13}$$

In addition, with $E' \equiv E \cap \bigcup_{j=1}^N I_j$:

$$m^* [E] - \epsilon < m^* [E'] \leq m^* [E]. \tag{3.14}$$

Proof. *Assume that (3.13) has been proved. Expressing:*

$$E = E \cap \left[\bigcup_{j=1}^N I_j \right] \cup E \cap \left[\bigcup_{j=1}^N I_j \right]^c,$$

where $A^c \equiv \widetilde{A}$ is the complement of A, then since m^* is subadditive by Proposition I.2.29 and $E \cap A^c \equiv E - A$, it follows from (3.13) that:

$$m^* [E] < m^* [E'] + \epsilon.$$

Since $m^*[E'] \leq m^*[E]$ by monotonicity of m^* by Proposition I.2.28, this obtains (3.14).

We will prove (3.13) with the closures $\{\bar{I}_\alpha\}$ of the intervals in $\mathcal{I}$. To see that this obtains the desired result, if $\{\bar{I}_j\}_{j=1}^N$ is chosen that satisfies (3.13), then:

$$\bigcup_{j=1}^N \bar{I}_j = \bigcup_{j=1}^N I_j \cup Z,$$

where $Z = \bigcup_{j=1}^N \bar{I}_j - \bigcup_{j=1}^N I_j$ is a finite collection of endpoints, and thus $m^*(Z) = 0$ by Proposition I.2.28.

Thus:

$$E - \bigcup_{j=1}^N I_j = \left[\left(E - \bigcup_{j=1}^N I_j\right) \cap Z\right] \cup \left[\left(E - \bigcup_{j=1}^N I_j\right) \cap Z^c\right],$$

and monotonicity of m^* obtains that:

$$m^*\left[\left(E - \bigcup_{j=1}^N I_j\right) \cap Z\right] = 0.$$

Also,

$$\left(E - \bigcup_{j=1}^N I_j\right) \cap Z^c = E - \bigcup_{j=1}^N \bar{I}_j,$$

and thus by subadditivity of m^*:

$$m^*\left(E - \bigcup_{j=1}^N I_j\right) \leq m^*\left(E - \bigcup_{j=1}^N \bar{I}_j\right) < \epsilon,$$

completing the proof of (3.13) with $\{I_j\}_{j=1}^N$.

We now suppress the closure notation and simply assume that $\mathcal{I}$ is a collection of closed intervals that cover E in the sense of Vitali. Because $m^*(E) < \infty$, there exists an open set G with $E \subset G$ and $m^*(G) < \infty$ by Corollary I.2.30. We claim that we can include in the collection $\mathcal{I}$ only such intervals with $I_\alpha \subset G$, that this subcollection retains the Vitali covering property, and that then by monotonicity:

$$m^*(\bigcup_{I_\alpha \subset G} I_\alpha) < \infty. \tag{1}$$

To prove this, recall that the open set G is a disjoint countable union of open intervals by Proposition I.2.12. Thus if $x \in E$ then $x \in (a,b) \subset G$ for some open interval. Choosing $\epsilon < \min\{b - x, x - a\}$, there exists $I_\alpha \in \mathcal{I}$ with $x \in I_\alpha$ and $|I_\alpha| < \epsilon$, and so $I_\alpha \subset G$. This proves the Vitali covering property for this subcollection, and thus we will assume:

$$E \subset \bigcup_\alpha I_\alpha \subset G. \tag{2}$$

Let $\epsilon > 0$ be given. We use induction to choose a disjoint collection of intervals for (3.13). Selecting any interval as I_1, assume that $I_1, I_2, ..., I_n$ have been chosen. If $E \subset \bigcup_{j=1}^n I_j$ then the proof of 3.13 is complete. Otherwise, let:

$$d_n \equiv \sup\{m^*(I_\alpha) \mid I_\alpha \cap \bigcup_{j=1}^n I_j = \emptyset\}.$$

Recalling that $m^*(I_\alpha) = |I_\alpha|$ for all intervals by Proposition I.2.28, it follows that $d_n \leq m^*(G) < \infty$ by (2), so choose I_{n+1} so that $I_{n+1} \cap \bigcup_{j=1}^n I_j = \emptyset$ and $m^*(I_{n+1}) \geq d_n/2$. This is possible since there is at least one such interval by definition of supremum.

Proceeding in this way, either the process eventually stops because $E \subset \bigcup_{j=1}^N I_j$ for some N and thereby completing the proof, or a countable collection of disjoint intervals $\{I_j\}_{j=1}^\infty$ is chosen. Since $\bigcup_{j=1}^\infty I_j \subset G$ by (2) and $m^*(G) < \infty$, monotonicity of m^* assures that $m^*\left(\bigcup_{j=1}^\infty I_j\right) < \infty$. Further, since these disjoint intervals are closed and thus Lebesgue measurable, it follows by Proposition I.2.39 that $m^* = m$ and:

$$m^*\left(\bigcup_{j=1}^\infty I_j\right) = \sum_{j=1}^\infty m^*(I_j) < \infty. \tag{3}$$

As a convergent series, choose N so that $\sum_{j=N+1}^{\infty} m^*(I_j) < \epsilon/5$ and define:

$$R = E - \bigcup_{j=1}^{N} I_j.$$

The proof will be complete by showing that $m^*(R) < \epsilon$.

Let $x \in R$. Since $\mathcal{I}$ covers E in the sense of Vitali, there exists $I \in \mathcal{I}$ with $x \in I$ and $m^*(I) < \delta$ for any δ. We choose δ small enough so that $I \cap \bigcup_{j=1}^{N} I_j = \emptyset$. This is possible because the I_j-union is a closed set, and thus since $x \in \left(\bigcup_{j=1}^{N} I \right)^c$, there exists an open interval with $x \in (a,b) \subset \left(\bigcup_{j=1}^{N} I \right)^c$. Now choose I with $x \in I \subset (a,b)$ as above.

Since $d_n \leq 2m^*(I_{n+1}) \to 0$ by construction and (3), choosing any n with $d_n < m^*(I)$ obtains that $I \cap \bigcup_{j=1}^{n} I_j \neq \emptyset$. Otherwise, since then also $I \cap \bigcup_{j=1}^{n-1} I_j = \emptyset$, the interval I would have been selected as I_n. Thus by selection of I, for any such $n > N$, $I \cap \bigcup_{j=N+1}^{n} I_j \neq \emptyset$. Choose the smallest m with $N+1 \leq m \leq n$ and $I \cap I_m \neq \emptyset$. Because now $I \cap \bigcup_{j=1}^{m-1} I_j = \emptyset$, the interval I was included in the supremum calculation for d_{m-1}, and thus $m^*(I) \leq d_{m-1} \leq 2m^*(I_m)$.

If y_m denotes the midpoint of I_m, then:

$$|x - y_m| \leq m^*(I) + 0.5m^*(I_m) \leq 2.5m^*(I_m).$$

This implies $I \subset J_m$, where the interval J_m has the same centerpoint as I_m, but five times the interval length: $m^*(J_m) = 5m^*(I_m)$.

Summarizing, for any $x \in R$, there are such intervals I and J_m with $x \in I \subset J_m$, and thus:

$$R \subset \bigcup_{m \geq N+1} J_m,$$

where this union is over all such $m \geq N+1$ produced as above.

As m^* is subadditive:

$$m^*(R) \leq \sum_{m \geq N+1} m^*(J_m) \leq 5 \sum_{j=N+1}^{\infty} m^*(I_j) < \epsilon.$$

∎

Remark 3.11 *In the fourth paragraph of the above proof was derived:*

$$m^* \left(E - \bigcup_{j=1}^{N} I_j \right) \leq m^* \left(E - \bigcup_{j=1}^{N} \bar{I}_j \right),$$

but in fact these outer measures are equal. This follows because $E - \bigcup_{j=1}^{N} \bar{I}_j \subset E - \bigcup_{j=1}^{N} I_j$, and so by monotonicity of m^*:

$$m^* \left(E - \bigcup_{j=1}^{N} \bar{I}_j \right) \leq m^* \left(E - \bigcup_{j=1}^{N} I_j \right).$$

With the machinery of the Vitali covering lemma in hand, the proof that monotonic functions are differentiable almost everywhere is now possible, even if a bit lengthy.

Proposition 3.12 (Monotonic $\Rightarrow$ differentiable a.e.) *Let $f(x)$ be a real-valued monotonic function defined on an open interval I. Then $f'(x)$ exists almost everywhere.*
Proof. *It suffices to prove the result for $f(x)$ monotonically increasing. Then if $f(x)$ is monotonically decreasing, $-f(x)$ is increasing and thus $-f'(x)$ exists almost everywhere.*

The existence of $f'(x)$ a.e. for monotonically increasing $f(x)$ will be proved by showing:

$$m^*\{x \in I | D^+ f(x) > D_- f(x)\} = m^*\{x \in I | D^- f(x) > D_+ f(x)\} = 0. \tag{1}$$

This then implies that these sets are Lebesgue measurable with measure zero by Propositions I.2.35 and I.2.39. As noted above, the use of outer measure is required here because it cannot be assumed that the sets in (1) *are Lebesgue measurable.*

Once (1) *is proved, this implies that:*

$$D_- f(x) \geq D^+ f(x) \text{ a.e., and, } D_+ f(x) \geq D^- f(x) \text{ a.e.}$$

This with (3.12):

$$D_+ f(x) \leq D^+ f(x), \text{ and, } D_- f(x) \leq D^- f(x),$$

are seen to combine to prove that all four derivates are equal a.e., and thus $f'(x)$ exists almost everywhere by Exercise 3.7.

We prove that $m^*\{x | D^+ f(x) > D_- f(x)\} = 0$ and leave the other demonstration as an exercise. For this we show that if $p > q$ are any rationals, that $m^*(E_{p,q}) = 0$ where:

$$E_{p,q} \equiv \{x \in I | D^+ f(x) > p > q > D_- f(x)\}.$$

For this latter demonstration, it is enough to prove $m^*(E_{p,q}^{(n)}) = 0$ for all $n \in \mathbb{Z}$, where:

$$E_{p,q}^{(n)} \equiv E_{p,q} \bigcap [n, n+1].$$

Because $\{x \in I | D^+ f(x) > D_- f(x)\}$ is a countable union of such $E_{p,q}^{(n)}$-sets over all p, q, n, the result follows by the countable subadditivity of m^* by Proposition I.2.29.

To simplify notation, we suppress the (n) designation for the remainder of the proof and assume that $E_{p,q} \subset [n, n+1]$ for some n.

Let $\epsilon > 0$ be given. Because $m^*(E_{p,q}) \leq 1$ by monotonicity of m^*, it follows from Corollary I.2.30 that there is an open set G with $E_{p,q} \subset G$ and:

$$m^* [G] \leq m^* [E_{p,q}] + \epsilon. \tag{2}$$

If $x \in E_{p,q}$, then by definition of $D_- f(x) < q$, there is a sequence $h_i(x) \to 0^+$ so that:

$$f(x) - f(x - h_i(x)) < q h_i(x).$$

Let $I_i(x) \equiv [x - h_i(x), x]$ and:

$$\mathcal{I} \equiv \{I_i(x) \mid x \in E_{p,q} \text{ and } I_i(x) \subset G\}.$$

Then since $h_i(x) \to 0^+$ for all x, $\mathcal{I}$ is a covering of $E_{p,q}$ in the sense of Vitali and thus Proposition 3.10 applies.

Choose disjoint $\{I_j\}_{j=1}^N \equiv \{[x_j - h_{i_j}(x_j), x_j]\}_{j=1}^N \subset \mathcal{I}$ so that by (3.14):

$$m^* [E'_{p,q}] > m^* [E_{p,q}] - \epsilon, \tag{3}$$

with $E'_{p,q} \equiv E_{p,q} \bigcap \bigcup_{j=1}^N I_j^o$ and $I_j^o \equiv (x_j - h_{i_j}(x_j), x_j)$. This is possible by Remark 3.11 since (3) *is satisfied with $E'_{p,q}$ defined as $E_{p,q} \bigcap \bigcup_{j=1}^N I_j$ by Proposition 3.10. Since $\{I_j\}_{j=1}^N$ are disjoint and $\bigcup_{j=1}^N I_j \subset G$, it follows from $m^* (I_j) = h_{i_j}(x_j)$ that $\sum_{j=1}^N h_{i_j}(x) < m^* [G]$, and then by* (2):

$$\sum_{j=1}^N [f(x_j) - f(x_j - h_{i_k}(x))] < q \sum_{j=1}^N h_{i_j}(x)$$
$$< q [m^* [E_{p,q}] + \epsilon]. \tag{4}$$

If $y \in E'_{p,q}$ defined as above as $E'_{p,q} \equiv E_{p,q} \bigcap \bigcup_{j=1}^{N} I_j^o$, then by definition of $D^+ f(y) > p$, there is a sequence $k_i(y) \to 0^+$ so that:

$$f(y + k_i(y)) - f(y) > pk_i(y).$$

Since $y \in \bigcup_{j=1}^{N} I_j^o$ by definition, it follows that for $k_i(y)$ small enough that $J_i(y) \equiv [y, y + k_i(y)] \subset I_j$ for some j. Define:

$$\mathcal{I}' \equiv \{J_i(y) \mid y \in E'_{p,q} \text{ and } J_i(y) \subset I_j \text{ for some } j\}.$$

Then again since $k_i(y) \to 0^+$ for all y, $\mathcal{I}'$ is a covering of $E'_{p,q}$ in the sense of Vitali and Proposition 3.10 applies.

Choose disjoint $\{J_j\}_{j=1}^{M} \subset \mathcal{I}'$ so that with $E''_{p,q} \equiv E'_{p,q} \bigcap \bigcup_{j=1}^{M} J_j$ we have (3.14). Then with the bound in (3) for $m^* \left[E'_{p,q} \right]$:

$$m^* \left[E''_{p,q} \right] > m^* \left[E'_{p,q} \right] - \epsilon > m^* \left[E_{p,q} \right] - 2\epsilon. \tag{5}$$

As $J_j \equiv \left[y_j - k_{i_j}(y_j), y_j \right]$ are disjoint and $E''_{p,q} \subset \bigcup_{j=1}^{M} J_j$, it follows from $m^* (J_j) = k_{i_j}(y_j)$ that $\sum_{j=1}^{M} k_{i_j}(x) > m^* \left[E''_{p,q} \right]$. Thus by (5):

$$\sum_{j=1}^{M} \left[f(y_j + k_{i_j}(y_j)) - f(y_j) \right] > p \sum_{j=1}^{M} k_{i_j}(x)$$
$$> p \left[m^* \left[E_{p,q} \right] - 2\epsilon \right]. \tag{6}$$

Now $f(x)$ is increasing by assumption, and since by construction each $J_j \subset I_i$ for some i:

$$\sum_{j=1}^{M} \left[f(y_j + k_{i_j}(y_j)) - f(y_j) \right] \leq \sum_{j=1}^{N} \left[f(x_j) - f(x_j - h_{i_k}(x)) \right].$$

Combining with the estimates in (4) and (6) yields:

$$p \left[m^* \left[E_{p,q} \right] - 2\epsilon \right] < q \left[m^* \left[E_{p,q} \right] + \epsilon \right].$$

But $q < p$, so this inequality can only hold for all $\epsilon > 0$ if $m^* \left[E_{p,q} \right] = 0$. ∎

Corollary 3.13 (Sign of $f'(x)$) *Given a real-valued monotonic function $f(x)$, then for any x for which $f'(x)$ exists:*

- *$f'(x) \geq 0$ for monotonically increasing functions;*

- *$f'(x) \leq 0$ for monotonically decreasing functions.*

Proof. *This follows from the assumed existence of the limit in (3.11). For a monotonically increasing function, $f(x + h) \geq f(x)$ for $h > 0$ and $f(x + h) \leq f(x)$ for $h < 0$, and so:*

$$[f(x + h) - f(x)] / h \geq 0 \text{ for all } h.$$

The same argument applies in the monotonically decreasing case. ∎

We next prove that $f'(x)$ of Proposition 3.12 is in fact Lebesgue measurable on I.

Corollary 3.14 ($f'(x)$ Lebesgue measurable) *Let $f(x)$ be a real-valued monotonic function defined on an open interval I. Then $f'(x)$ is Lebesgue measurable.*
Proof. *Proposition 3.12 states that $f'(x)$ exists a.e. on (a, b), so adding the endpoints a, b to this exceptional set obtains that $f'(x)$ exists and is given by (3.11) for almost all $x \in [a, b]$. Let $E \subset [a, b]$ denote the set where $f'(x)$ exists.*

For each positive integer n, define the function $g_n(x)$ on $[a, b]$ by:

$$g_n(x) = n[f(x + 1/n) - f(x)],$$

where $f(x+1/n) \equiv f(b)$ if $x+1/n > b$. Since $f(x)$ is monotonic and thus measurable, $g_n(x)$ is measurable for all n.

Further, $g_n(x) \to f'(x)$ on E, which means this convergence is almost everywhere, and thus $f'(x)$ is Lebesgue measurable by Corollary I.3.48. ■

Remark 3.15 (On Lebesgue integrability of $f'(x)$) *For Lebesgue integrability, we cannot apply Proposition 2.15 to conclude that Lebesgue measurable $f'(x)$ is integrable on $[a, b] \subset I$ since we do not know that $f'(x)$ is bounded. However, Lebesgue integrability will be established below in Proposition 3.19.*

Proposition 3.16 provides an important corollary result, that integrals of Lebesgue integrable functions with variable upper limits of integration are differentiable almost everywhere. This conclusion supports the hope that Version II of the fundamental theorem in (1.28) will be satisfied once we develop the tools to value such a derivative.

Proposition 3.16 (Differentiability of a Lebesgue integral) *If $f(x)$ is a Lebesgue integrable function on $[a, b]$, then:*

$$F(x) \equiv (\mathcal{L}) \int_a^x f(y) dy,$$

is differentiable a.e. for $x \in [a, b]$.

Proof. *Since $f(x)$ is Lebesgue integrable, (2.32) obtains:*

$$\int_a^x f(y) dy = \int_a^x f^+(y) dy - \int_a^x f^-(y) dy \equiv F^+(x) - F^-(x),$$

with $f^+(x)$ and $f^-(x)$ nonnegative. For any interval $[x, x+h] \subset [a, b]$, Proposition 2.31 obtains:

$$\begin{aligned} F(x+h) - F(x) &= \int_x^{x+h} f^+(y) dy - \int_x^{x+h} f^-(y) dy \\ &= \left[F^+(x+h) - F^+(x) \right] - \left[F^-(x+h) - F^-(x) \right]. \end{aligned}$$

Since $f^+(x) \geq 0$ and $f^-(x) \geq 0$, each integral is nonnegative for all x, h by Proposition 2.31. Consequently, $F^+(x+h) \geq F^+(x)$ and $F^-(x+h) \geq F^-(x)$ for all x, h. Thus both $F^+(x)$ and $F^-(x)$ are monotonically increasing and differentiable almost everywhere by Proposition 3.12, and then so too is $F(x) = F^+(x) - F^-(x)$. ■

Remark 3.17 (Continuous a.e.) *Proposition 3.12 proved that the property of monotonicity is powerful enough to ensure that such a function is differentiable except on a set of measure zero, which could be the empty set. This implies that such functions are also continuous almost everywhere:*

$$\lim_{h \to 0} [f(x+h) - f(x)] = \lim_{h \to 0} \left[\frac{f(x+h) - f(x)}{h} \right] h = 0 \ a.e.,$$

because the bracketed term converges to the finite limit $f'(x)$ a.e.

This is not a new result. Without all the above machinery it was proved in Proposition I.5.8 that every increasing function is continuous except on a set of measure zero. This result naturally applies to decreasing functions.

But it should be noted that sets of measure zero can be anything but simple to characterize outside of the requirement of Definition 1.18 that they can be covered by a countable collection of intervals of arbitrarily small total measure. Indeed, in addition to singleton sets $\{x\}$ and finite collections of singletons, $\{x_i\}_{i=1}^n$, sets of measure zero can also be countably infinite like the integers $\mathbb{Z}$, but also countably infinite and dense in $\mathbb{R}$ like the rationals $\mathbb{Q}$, and even uncountable (see Example 3.50).

Example 3.18 (On demonstrating differentiability) *Let $\{r_n\}_{n=1}^\infty$ be an arbitrary enumeration of the rationals in $[0, 1]$ and define the monotonically increasing function:*

$$f(x) = \begin{cases} 0, & x < 0, \\ \sum_{r_n \leq x} 1/2^n, & 0 \leq x \leq 1, \\ 1, & x > 1. \end{cases} \tag{3.15}$$

In other words, for any $x \in [0, 1]$, we include in the summation every term $1/2^n$ for which there is a rational $r_n \leq x$. This function is discontinuous at every rational $x = r_n$, at which point it has a "jump" discontinuity equal to $1/2^n$. It is an exercise to show that while $f(x)$ is discontinuous from the left at every rational, it is continuous from the right.

This function is also continuous at every irrational x. This follows from the observation that while any interval $[x - h, x + h]$ contains infinitely many rationals, any such interval has a "largest contributing" rational, by which is meant, the rational r_n with the smallest value of n and hence the largest value of $1/2^n$. This largest contributor can also be decreased by decreasing h.

More formally, given irrational x and $\epsilon > 0$, there are only finitely many rationals with associated contribution $1/2^n \geq \epsilon$. Thus we can choose h so that the largest contributing rational in $[x - h, x + h]$ is $1/2^N < \epsilon$. Then for $y \in [x - h, x + h]$:

$$\begin{aligned} |f(y) - f(x)| &\leq \sum_{x-h \leq r_n < x+h} 1/2^n \\ &\leq \sum_{n=N}^\infty 1/2^n < 2\epsilon. \end{aligned}$$

It is worth a moment of thought to understand why this demonstration fails when x is rational.

The above proposition now asserts that this function is in fact differentiable except on a set of measure zero. To be sure, this exceptional set must contain the rationals since differentiability implies continuity, and thus discontinuity implies nondifferentiability. But it seems quite difficult to directly show that this function is differentiable anywhere. Unlike the continuity analysis, differentiability requires not just that $|f(x + h) - f(x)|$ is small when h is small, but that it is small enough relative to h so that $[f(x + h) - f(x)]/h$ converges as $h \to 0$.

Even more generally, an exceptional set of measure 0 need not be countable. An example is the **Cantor set** below, named after **Georg Cantor** (1845–1918) who first recognized that there were infinite sets of different orders of magnitude, or more formally, of different cardinalities. In his work, two sets are deemed to have the same **cardinality** if their elements can be put into one-to-one correspondence, or "paired," without missing elements or double counting any elements. By this definition an uncountable set is one that cannot be "paired" with the integers.

Cantor discovered that while the rationals can be so paired with the integers, the reals could not and hence they are "uncountable."

More surprising perhaps than the existence of an uncountable set is Cantor's discovery that an uncountable set can have Lebesgue measure 0.

For the next application of Proposition 3.12, we now demonstrate that for monotonic functions, $f'(x)$ is Lebesgue integrable. Recall Remark 3.15 that this conclusion did not follow from Proposition 2.15.

Further, there is a predictable relationship between the Lebesgue integral of $f'(x)$ over $[a, b]$ and $f(b) - f(a)$. While this is a long way from the version of the fundamental theorem in (1.24), it is an indication that these quantities are related when integration is defined in a Lebesgue context. Equally important presently, the next proposition will play an instrumental role later in this section.

Proposition 3.19 ($f'(x)$ **is integrable for monotonic functions**) *Let $f(x)$ be a real-valued monotonic function defined on an interval I.*

Then $f'(x)$ is Lebesgue integrable on $[a, b] \subset I$, and:

- *If $f(x)$ is monotonically increasing:*

$$(\mathcal{L}) \int_a^b f'(x)dx \leq f(b) - f(a). \tag{3.16}$$

- *If $f(x)$ is monotonically decreasing:*

$$(\mathcal{L}) \int_a^b f'(x)dx \geq f(b) - f(a). \tag{3.17}$$

Proof. *By Corollary 3.14, $f'(x)$ is Lebesgue measurable. We provide the proof of the integral bound of (3.16) assuming $f(x)$ is monotonically increasing. For decreasing functions, the inequality in (3.17) then follows from (3.16) when applied to increasing $-f(x)$.*

For each positive integer n, define the function $g_n(x)$ on $[a, b]$ by:

$$g_n(x) = n[f(x + 1/n) - f(x)],$$

where $f(x + 1/n) \equiv f(b)$ if $x + 1/n > b$. Since $f(x)$ is increasing, $g_n(x) \geq 0$ for all n. Further, $g_n(x) \to f'(x)$ almost everywhere.

By Fatou's lemma of Proposition 2.34, and suppressing the $(\mathcal{L})$ notation:

$$\int_a^b f'(x)dx \leq \liminf \int_a^b g_n(x)dx. \tag{1}$$

Although the integral of $g_n(x)$ is defined as a Lebesgue integral in Fatou's lemma, $f(x)$ is increasing and differentiable almost everywhere, and thus $f(x)$ is bounded and continuous almost everywhere.

Hence, the following integrals of $f(x)$ can be manipulated as Riemann integrals by Propositions 1.22 and 2.18. Recalling that $f(x) \equiv f(b)$ if $x > b$:

$$\int_a^b g_n(x)dx = \int_a^b n[f(x + 1/n) - f(x)]dx$$

$$= n \int_{a+1/n}^{b+1/n} f(x)dx - n \int_a^b f(x)dx$$

$$= n \int_b^{b+1/n} f(x)dx - n \int_a^{a+1/n} f(x)dx$$

$$= f(b) - n \int_a^{a+1/n} f(x)dx. \tag{2}$$

Since $f(x)$ is an increasing function, it follows from item 5 of Proposition 2.49 that:

$$-f(a+1/n) \leq -n \int_a^{a+1/n} f(x)dx \leq -f(a),$$

and so

$$\liminf_n \left[-n \int_a^{a+1/n} f(x)dx\right] \leq -f(a). \tag{3}$$

Combining (1), (2), and (3) obtains:

$$\int_a^b f'(x)dx \leq f(b) - f(a).$$

∎

The results in (3.16) and (3.17) are perhaps initially surprising because of the inequalities. The fundamental theorem of calculus in (1.24) reinforces the expectation that integrating a continuous derivative obtains $f(b) - f(a)$.

These expectations from calculus are fundamentally changed when we move from the Riemann world of "continuously differentiable" to the Lebesgue world of "differentiable almost everywhere."

Example 3.20 (Inequality in (3.16) and (3.17)) *Even for Riemann integrals, the expected equality in (1.24) can be corrupted by small violations in the assumptions.*

1. *On $[0,2]$ let $f(x) = x$ on $[0,1)$ and $x+1$ on $[1,2]$. Then $f'(x) = 1$ a.e., and defined as a Riemann or Lebesgue integral:*

$$\int_0^2 f'(x)dx = 2 < f(2) - f(0) = 3.$$

Of course, changing the sign of $f(x)$ produces a decreasing function and example for (3.17).

2. *On $[0,n]$ define $F(x) = a_j$ on $[j, j+1)$ for $j = 0, 1, 2..., n-1$, and $F(n) = a_n$. Then independent of the choice of $\{a_j\}_{j=0}^n$ it follows that $F'(x) = 0$ a.e. and so $\int_0^n F'(x)dx = 0$ whether defined as a Riemann or Lebesgue integral. Choosing $\{a_j\}$ to be an increasing sequence provides an example of strict inequality in (3.16), while a decreasing sequence exemplifies (3.17).*

The increasing example with $a_j = (j+1)/n$ for $j < n$ and $a_n = 1$ obtains a discrete probability distribution function of Book II for which discrete probabilities are assigned to the integers $\{j\}_{j=0}^{n-1}$ and defined by $\Pr[j] = 1/n$. Then $F(x)$ is the distribution function defined on $\mathbb{R}$ by:

$$F(x) = \sum_{j \leq x} \Pr[j].$$

These examples are somewhat transparent because in all cases the failure of (3.16) and (3.17) to hold with equality is due to jump discontinuities in the given functions. So, it is natural to speculate that except for discontinuities, perhaps Version I of the fundamental theorem of calculus in (1.24) would be true after all.

But this is not the case. There are even more surprising examples ahead in the next section. It will be shown that there exist increasing functions for which $f'(x) = 0$ a.e. and thus $\int_E f'(x)dx = 0$, but these functions are in fact continuous **everywhere**.

The examples above also illustrate that for general functions, which need not be monotonic, the value of $\int_a^b f'(x)dx$ can be entirely unpredictable *vis-a-vis* the value of $f(b) - f(a)$. So, if we are to have any hope of recovering a Version I of the fundamental theorem for Lebesgue integrals that looks like (1.24), additional restrictions will be needed beyond the existence of $f'(x)$ a.e. We will return to this question in Section 3.2.

For the rest of this section, we continue the development of Version II of the fundamental theorem.

3.1.2 Functions Differentiable a.e.

The prior section demonstrated that every monotonic, real-valued function is differentiable a.e., and this naturally implies that if $\{f_j(x)\}_{j=1}^n$ is any finite collection of such functions on $[a, b]$, then:

$$f(x) \equiv \sum_{j=1}^n f_j(x),$$

is differentiable a.e. This follows because $f'(x)$ then exists at least outside the finite union of the exceptional sets of $\{f_j(x)\}_{j=1}^n$, and this union has measure 0.

A moment's thought reveals that this summation of monotonic functions can be characterized into three cases:

1. $f(x)$ is increasing, which happens at least in cases when all $f_j(x)$ are increasing;

2. $f(x)$ is decreasing, which happens at least in cases when all $f_j(x)$ are decreasing;

3. $f(x) = I(x) + D(x)$, with $I(x)$ increasing and $D(x)$ decreasing, or equivalently:

$$f(x) = I_1(x) - I_2(x),$$

with increasing functions $I_1(x) = I(x)$ and $I_2(x) = -D(x)$.

Case 3 can also be expressed as a difference of decreasing functions, $f(x) = (-I_2(x)) - (-I_1(x))$, but the convention is to view this case as a difference of monotonically increasing functions. Moreover, case 3 is the only conclusion actually needed for such sums of monotonic functions since case 1 corresponds to $I_2(x) = 0$ and case 2 corresponds to $I_1(x) = 0$.

Conclusion 3.21 *Any finite summation of monotonic functions produces a function $f(x)$ that is differentiable a.e., and that can be expressed as $f(x) = I_1(x) - I_2(x)$, a difference of increasing functions.*

We have seen such a decomposition before. In Proposition 3.16, it was proved that if $f(x)$ is Lebesgue integrable, then $F(x)$ defined by:

$$F(x) = (\mathcal{L}) \int_a^x f(y)dy,$$

is differentiability almost everywhere. And indeed, we proved this precisely because $F(x)$ could be expressed as a difference of monotonically increasing functions:

$$F(x) = \int_a^x f^+(y)dy - \int_a^x f^-(y)dy.$$

With this introduction, we will next study functions of the form $I_2(x) - I_2(x)$, and attempt to characterize when a given function can be expressed in this way. Such a characterization would be valuable since any such function would be differentiable almost everywhere. Quite remarkably, functions $f(x)$ that can be represented as $f(x) = I_1(x) - I_2(x)$ can be characterized by a property known as **bounded variation**, which will initially seem to have nothing at all to do with this characterization.

Notation 3.22 ($[X]^+$, $[X]^-$) *To define bounded variation, we use the notational convention that given any numerical expression X:*

$$[X]^+ \equiv \max[X, 0], \qquad [X]^- \equiv \max[-X, 0]. \tag{3.18}$$

This notational convention has been seen before in Definition 2.43.

Definition 3.23 (Variations of a function) *Given a real-valued function $f(x)$ on an interval $[a, b]$, and a partition $\Pi = \{x_0, x_1, ..., x_n\}$ where:*

$$a = x_0 < x_1... < x_n = b,$$

let:

$$v^+ \equiv \sum_{i=1}^{n} [f(x_i) - f(x_{i-1})]^+,$$
$$v^- \equiv \sum_{i=1}^{n} [f(x_i) - f(x_{i-1})]^-,$$
$$t \equiv \sum_{i=1}^{n} |f(x_i) - f(x_{i-1})|$$
$$= v^+ + v^-.$$

Then define:

Positive Variation of f :	$P = \sup_\Pi v^+$,
Negative Variation of f :	$N = \sup_\Pi v^-$,
Total Variation of f :	$T = \sup_\Pi t$,

where $\sup_\Pi$ denotes the supremum over all such partitions.
*We say that $f(x)$ **is of bounded variation,** and abbreviated $f \in B.V.$, if $T < \infty$.*

Notation 3.24 *It is common to add various qualifiers to the above notation when it is necessary to either identify the interval used, for example T_a^b, or the function, $T(f)$, or both, $T_a^b(f)$. Similarly, the notation $f(x) \in B.V.[a, b]$ is common.*

Exercise 3.25 (On $\mu \to 0$) *Recall the definition of mesh size μ of a partition Π given in (1.2). Let Π_μ denote a partition of mesh size μ. Prove that the supremum in the definition of T is necessarily obtained as $\mu \to 0$ by proving that:*

$$T = \lim_{\mu \to 0} \sup_{\Pi_\mu} t,$$

where $\sup_{\Pi_\mu}$ denotes the supremum over all partitions of mesh size μ. Hint: Given any two partitions, $P_1 = \{x_i\}_{i=0}^n$ and $P_2 = \{x_j'\}_{j=0}^m$, show that with $P = \{x_i\}_{i=0}^n \bigcup \{x_j'\}_{j=0}^m$ and obvious notation:

$$\max\{t_{P_1}, t_{P_2}\} \leq t_P.$$

*The partition P is call the **common refinement** of P_1 and P_2.*

Example 3.26 (Monotonic functions) *$B.V.[a, b]$ contains all real-valued monotonic functions. If monotonically increasing, then for every partition $v^- = 0$ by definition, and $v^+ = f(b) - f(a)$ since $[f(x_i) - f(x_{i-1})]^+ = f(x_i) - f(x_{i-1})$. Hence, $T = P = f(b) - f(a)$, while $N = 0$. For monotonically decreasing functions, $T = N = -[f(b) - f(a)]$, while $P = 0$.*

It is shown below as Corollary 3.31 that $B.V.[a, b]$ is a vector space over $\mathbb{R}$, meaning that if $f, g \in B.V.[a, b]$, then so too are $f + g$ and cf for $c \in \mathbb{R}$. See also Definition 4.33.

The next result provides some tools for working with these variations when $T < \infty$.

Proposition 3.27 (Variation relations) *If $f(x)$ is of bounded variation on $[a, b]$, then:*

$$P - N = f(b) - f(a), \tag{3.19}$$

$$P + N = T. \tag{3.20}$$

Proof. *For any given partition Π:*

$$[f(x_i) - f(x_{i-1})]^+ - [f(x_i) - f(x_{i-1})]^- = f(x_i) - f(x_{i-1}),$$

which obtains by summation:

$$v^+ = v^- + f(b) - f(a). \tag{1}$$

Taking suprema on v^- then v^+ provides,

$$\begin{aligned} v^+ &\leq N + f(b) - f(a), \\ P &\leq N + f(b) - f(a). \end{aligned}$$

The same steps applied to $v^- = v^+ + f(a) - f(b)$ obtains:

$$N \leq P + f(a) - f(b).$$

Since both P and N are bounded by T and $T < \infty$, subtraction is justified and obtains:

$$f(b) - f(a) \leq P - N \leq f(b) - f(a),$$

which is (3.19).

For (3.20), first note that $T \leq P + N$ since any partition allowed in the T calculation can be used for both P and N, but different partitions can also be used for the latter calculations. From the definition, (1) and (3.19):

$$T \geq v^+ + v^- = [v^- + f(b) - f(a)] + v^- = 2v^- + P - N,$$

and taking suprema in v^- provides $T \geq P + N$ and the result follows. ∎

Exercise 3.28 ($f(x) \in B.V. \Rightarrow |f(x)| \in B.V.$) *Prove that if $f(x)$ is of bounded variation on $[a, b]$ with total variation T, then $|f(x)|$ is of bounded variation on $[a, b]$ with total variation $T' \leq T$. Hint: Prove that for all real numbers:*

$$||c| - |d|| \leq |c - d|.$$

Quite surprisingly perhaps, the following proposition states that the class $B.V.[a, b]$ is identical to the class of functions that can be expressed as $f(x) = I_1(x) - I_2(x)$. The decomposition of a $B.V.$ function into a difference of increasing functions is not unique because we can add any given increasing function to the component functions to achieve a new decomposition.

Two explicit decompositions are provided below that reflect the function's variations.

Proposition 3.29 (Characterization of $B.V.[a,b]$) *A function $f(x)$ defined on $[a,b]$ is of bounded variation if and only if $f(x) = I_1(x) - I_2(x)$ for increasing functions $I_1(x)$ and $I_2(x)$.*

In particular, two such decompositions are:

$$f(x) = f(a) + P_a^x - N_a^x, \tag{3.21}$$

$$f(x) = f(a) + T_a^x - 2N_a^x, \tag{3.22}$$

where T_a^x is the total variation over $[a,x] \subset [a,b]$, and similarly for P_a^x and N_a^x.
Proof. *If $f(x) \in B.V.[a,b]$, then $f(x) \in B.V.[a,x]$ for any $x \in [a,b]$. Thus (3.19) applied to the interval $[a,x]$ obtains (3.21):*

$$f(x) = f(a) + P_a^x - N_a^x.$$

Both $I_1(x) \equiv f(a) + P_a^x$ and $I_2(x) \equiv N_a^x$ are increasing by definition. In addition, these functions are bounded above by $f(a) + T_a^b$ and T_a^b, respectively, and bounded below by $f(a)$ and 0, respectively. The decomposition in (3.22) then follows from $T_a^x = P_a^x + N_a^x$ by (3.20).
Conversely, if $f(x) = I_1(x) - I_2(x)$, then for any partition:

$$\begin{aligned} t &= \sum_{i=1}^{n} [I_1(x_i) - I_1(x_{i-1})] + \sum_{i=1}^{n} [I_2(x_i) - I_2(x_{i-1})] \\ &= I_1(b) - I_1(a) + I_2(b) - I_2(a). \end{aligned}$$

Thus:

$$T = I_1(b) - I_1(a) + I_2(b) - I_2(a),$$

and $f(x) \in B.V.[a,b]$. ∎

Remark 3.30 (On $f(x) = I_1(x) - I_2(x)$) *Note that for any decomposition that $I_1(a) - I_2(a) - f(a)$, and thus if we define $\Delta_x f = f(x) - f(a)$, and similarly for $\Delta_x I_1$ and $\Delta_x I_2$, then for all $x \in [a,b]$:*

$$\Delta_x f = \Delta_x I_1 - \Delta_x I_2.$$

This will play a role in the chapter on Riemann-Stieltjes integration, and in particular, Proposition 4.23.

Corollary 3.31 ($B.V.[a,b]$ is a vector space) *If $\{f_j(x)\}_{j=1}^{n} \subset B.V.[a,b]$ and $\{a_j\}_{j=1}^{n} \subset \mathbb{R}$, then $f(x) \equiv \sum_{i=1}^{n} a_j f_j(x) \in B.V.[a,b]$. Thus $B.V.[a,b]$ is a vector space over $\mathbb{R}$ generated by monotonically increasing functions.*
Proof. *Using the triangle inequality and apparent notation, for any partition Π:*

$$t \le \sum_{i=1}^{n} |a_j| t_j \le \sum_{i=1}^{n} |a_j| T_j.$$

Thus

$$T \le \sum_{i=1}^{n} |a_j| T_j,$$

and $B.V.[a,b]$ is closed under finite linear combinations and is hence a vector space.
Using the characterization of Proposition 3.29 provides another proof of this vector space structure, plus proves that this vector space is generated by monotonically increasing functions. ∎

Corollary 3.32 ($B.V.[a,b]$ functions are differentiable a.e.) *If $f(x) \in B.V.[a,b]$, then $f'(x)$ exists almost everywhere.*
Proof. *Immediate from Proposition 3.12, since $f(x) = I_1(x) - I_2(x)$ by Proposition 3.29.* ∎

Remark 3.33 (Differentiable a.e. $\not\Rightarrow$ B.V.) *While $f(x) \in B.V.[a,b]$ is sufficient to ensure that $f'(x)$ exists almost everywhere, it is by no means necessary.*

An example on $[0,\pi]$ is:

$$f(x) = \sin(1/x),$$

which is differentiable almost everywhere, and indeed everywhere except at $x = 0$. But this function is not of bounded variation on this interval.

Exercise 3.34 *Prove that $f(x)$ in Remark 3.33 is not of bounded variation on any interval $[0,b]$, but is of bounded variation on every interval $[a,b]$ with $a > 0$. Hint: Focus on relative maximums and minimums.*

3.1.3   The Lebesgue FTC – Version II

In this section, we achieve the desired result of generalizing Version II of the fundamental theorem in Proposition 1.33 to a Lebesgue context. The first step is to show that for Lebesgue integrable functions $f(x)$, the associated integrals with variable upper limits are continuous and of bounded variation. We already know that such integrals are differentiable almost everywhere by Proposition 3.16, but this new characterization will allow us to prove below that $F'(x) = f(x)$ a.e.

Proposition 3.35 ($\displaystyle\int_a^x f(y)dy$ continuous, B.V.) *Let $f(x)$ be Lebesgue integrable on $[a,b]$. Then the Lebesgue integral:*

$$F(x) = (\mathcal{L}) \int_a^x f(y)dy, \tag{3.23}$$

is a continuous function of bounded variation on $[a,b]$, with:

$$T_a^b \le (\mathcal{L}) \int_a^b |f(y)|\, dy. \tag{3.24}$$

Proof. *Since $f(x)$ is Lebesgue integrable, then suppressing the $(\mathcal{L})$ notation:*

$$\int_a^x f(y)dy = \int_a^x f^+(y)dy - \int_a^x f^-(y)dy, \tag{1}$$

with $f^+(x)$ and $f^-(x)$ nonnegative. Thus as in the proof of Proposition 3.16, $F(x)$ is a difference of increasing functions and is of bounded variation by Proposition 3.29.

For any partition of $[a,b]$, Proposition 2.49 obtains:

$$\begin{aligned}
t &= \sum_{i=1}^n |F(x_i) - F(x_{i-1})| \\
&= \sum_{i=1}^n \left| \int_{x_{i-1}}^{x_i} f(y)dy \right| \\
&\le \sum_{i=1}^n \int_{x_{i-1}}^{x_i} |f(y)|\, dy \\
&= \int_a^b |f(y)|\, dy.
\end{aligned}$$

Taking a supremum obtains (3.24).

By (1), *it is enough to prove continuity for increasing $f(x)$. If $[x, x + h] \subset [a, b]$, then since $f(x)$ is increasing:*

$$|F(x + h) - F(x)| = \int_x^{x+h} f(y) dy.$$

If $f(x) \leq M$, then by Proposition 2.49,

$$|F(x + h) - F(x)| \leq Mh, \tag{2}$$

and continuity follows.

Otherwise, define $f_n(x) \equiv \min[n, f(x)]$, and note that $\{f_n(x)\}_{n=1}^\infty$ is monotonically increasing for each x, and $f_n(x) \to f(x)$. By Lebesgue's monotone convergence theorem:

$$\int_a^b f_n(y) dy \to \int_a^b f(y) dy, \text{ and, } \int_a^b [f(y) - f_n(y)] dy \to 0.$$

So for any $\epsilon > 0$, there is an N such that for $n \geq N$:

$$\int_a^b [f(y) - f_n(y)] dy < \epsilon.$$

Since $f(y) - f_n(y) \geq 0$, this implies that for all $[x, x + h] \subset [a, b]$:

$$\int_x^{x+h} [f(y) - f_n(y)] dy < \epsilon. \tag{3}$$

By (2) and (3):

$$\int_x^{x+h} f(y) dy - \int_x^{x+h} f_N(y) dy + \int_x^{x+h} [f(y) - f_N(y)] dy < Nh + \epsilon.$$

Choosing $h < \epsilon/N$ yields $|F(x + h) - F(x)| < 2\epsilon$, and continuity of $F(x)$ follows. ∎

Before turning to the main result of this section, we need one more technical result. Recall from Proposition 2.49 that if measurable $f(x)$ satisfies $f(x) = 0$ a.e., then $\int_E f(x) dx = 0$ for any measurable set E. Conversely, by this proposition, if $\int_E f(x) dx = 0$ for **all** measurable E, then $f(x) = 0$ a.e.

Our goal is to investigate if $\int_E f(x) dx = 0$ for a given more limited collection of measurable sets, also implies that $f(x) = 0$ a.e.

A single set E is clearly inadequate as $f(x) = x$ on $E \equiv [-1, 1]$ demonstrates. Similarly, this example shows that one can also have $\int_E f(x) dx = 0$ for a finite or countable or uncountable collection of sets E and still the result fails.

The following provides a sufficient condition for the desired conclusion when the collection of sets are of the form $E \equiv [a, x]$, recalling the notational convention in Definition 2.9.

Proposition 3.36 ($F(x) = 0$ all $x \Rightarrow f(x) = 0$ a.e.) *If $f(x)$ is Lebesgue integrable on $[a, b]$ and $F(x) = 0$ **for all** $x \in [a, b]$, where $F(x)$ is defined in (3.23), then $f(x) = 0$ a.e.*

Proof. *Assume that there is a set E of positive Lebesgue measure on which $f(x) > 0$. By Proposition I.2.42, there is a closed set $F \subset E \subset [a, b]$ with positive measure. Denoting $I \equiv [a, b]$ obtains from Proposition 2.49:*

$$0 = \int_I f(y)dy = \int_F f(y)dy + \int_{I-F} f(y)dy. \tag{1}$$

Since $\widetilde{F} \equiv \mathbb{R} - F$ is open, it equals an at most countable union of disjoint open intervals by Proposition I.2.12:

$$\mathbb{R} - F = \bigcup_{k=1}^{\infty} J_k.$$

Hence, $I - F = \bigcup_{k=1}^{\infty} I_k$ with $I_k \equiv I \bigcap J_k$, and this is a union of disjoint intervals.

Now $\int_F f(y)dy > 0$ by Proposition 2.49 since $f(x) > 0$ on F and $m(F) > 0$, and since $\{I_k\}_{k=1}^{\infty}$ are disjoint, it follows from (1) and this same proposition:

$$\int_{I-F} f(y)dy = \sum_{k=1}^{\infty} \int_{I_k} f(y)dy < 0.$$

Thus there is at least one interval I_k on which

$$\int_{I_k} f(y)dy < 0. \tag{2}$$

Let this interval be $I_k = \langle a_k, b_k \rangle$, denoting that this interval can be open, closed or semi-closed. Then by assumption:

$$\int_a^{a_k} f(y)dy = \int_a^{b_k} f(y)dy = 0, \tag{3}$$

while by Proposition 2.49:

$$\int_a^{b_k} f(y)dy = \int_a^{a_k} f(y)dy + \int_{I_k} f(y)dy.$$

Thus $\int_{I_k} f(y)dy = 0$ by (3), in contradiction to (2). Hence, $m(E) = 0$.

The same argument applies if $f(x) < 0$ on E, and so $f(x) = 0$, a.e. ∎

Remark 3.37 (On $F(x) = 0$, all x) *In stark contrast to many results in the Lebesgue theory, the above conclusion requires that $F(x) = 0$ **for all** $x \in [a, b]$, and not, $F(x) = 0$ **for almost all** $x \in [a, b]$. This was used in the proof to conclude that (3) was valid no matter which $I_k = \langle a_k, b_k \rangle$ had the given property in (3).*

*However, this result can be generalized to the assumption that $F(x) = 0$ **for a dense set of** $x \in [a, b]$, since by continuity of $F(x)$ as given by Proposition 3.35, it would then follow that $F(x) = 0$ for all x.*

We finally arrive at the goal of this section to show that Version II of the fundamental theorem of calculus of Proposition 1.33 is true almost everywhere for a function $f(x)$ that is Lebesgue integrable on $[a, b]$. To simplify this proof, we first develop this result for a bounded measurable function and then generalize.

Proposition 3.38 (FTC Version II, bounded $f(x)$) *Let $f(x)$ be a bounded, Lebesgue measurable function defined on $[a, b]$, and define for arbitrary $F(a)$:*

$$F(x) = F(a) + (\mathcal{L}) \int_a^x f(y) dy. \tag{3.25}$$

Then, $F'(x)$ exists almost everywhere on $[a, b]$, and $F'(x) = f(x)$ a.e.

Proof. *The existence of $F'(x)$ a.e. follows from Proposition 3.16, and $F'(x)$ is Lebesgue measurable by Corollary 3.14.*

Set $f(x) = 0$ for $x > b$, and define for $x \in [a, b]$:

$$\begin{aligned} F_n(x) &= n\left[F(x + 1/n) - F(x)\right] \\ &= n \int_x^{x+1/n} f(y) dy. \end{aligned}$$

As $f(x)$ is bounded, say $|f(x)| \leq M$, it follows that $|F_n(x)| \leq M$ by Proposition 2.49.

Thus $\{F_n(x)\}_{n=1}^\infty$ are uniformly bounded and measurable on $[a, b]$, and $F_n(x) \to F'(x)$ everywhere $F'(x)$ exists, which is almost everywhere.

To show that $F'(x) = f(x)$ a.e., we first prove that **for all** $x \in [a, b]$ that defined as Lebesgue integrals:

$$\int_a^x f(y) dy = \int_a^x F'(y) dy. \tag{1}$$

We prove this for $x < b$, then use continuity of these integrals by Proposition 3.35 to extend to $x = b$.

Because $F(x)$ is continuous by Proposition 3.35, the Lebesgue integral $\int_a^x F(y) dy$ exists **for all** $x \in [a, b]$ and equals the associated Riemann integral by Proposition 2.18. For any $x < b$, the bounded convergence theorem, and then a change of variables for the Riemann integrals obtains:

$$\begin{aligned} \int_a^x F'(y) dy &= \lim_{n \to \infty} \int_a^x F_n(y) dy \\ &= \lim_{n \to \infty} \left[n \int_a^x F(y + 1/n) dy - n \int_a^x F(y) dy \right] \\ &= \lim_{n \to \infty} \left[n \int_x^{x+1/n} F(y) dy - n \int_a^{a+1/n} F(y) dy \right] \\ &= F(x) - F(a) \\ &\equiv \int_a^x f(y) dy. \end{aligned}$$

Note that the second to last step of this derivation is justified by continuity of $F(x)$. By continuity and Exercise 1.13, $F(x)$ achieves a maximum M_n and a minimum m_n on each interval of integration $[d, d + 1/n]$, and Proposition 2.49 obtains:

$$m_n \leq n \int_d^{d+1/n} F(y) dy \leq M_n.$$

Since $m_n \to F(d)$ and $M_n \to F(d)$ for $d \in [a, b)$ by continuity, the above results follow for $d = x < b$ or $d = a$.

Hence, (1) obtains for all $x \in [a, b]$:

$$\int_a^x [F'(y) - f(y)] dy = 0,$$

and by Proposition 3.36, $F'(y) = f(y)$ a.e. ∎

The final step is the generalization to integrable $f(x)$, removing the boundedness assumption.

Proposition 3.39 (FTC Version II: Derivative of an integral) *Let $f(x)$ be a Lebesgue integrable function on $[a, b]$ and define for arbitrary $F(a)$:*

$$F(x) = F(a) + (\mathcal{L}) \int_a^x f(y) dy. \tag{3.26}$$

Then $F'(x)$ exists almost everywhere on $[a, b]$, and:

$$F'(x) = f(x), \text{ a.e.} \tag{3.27}$$

Proof. *Because $f(x) = f^+(x) - f^-(x)$ for nonnegative Lebesgue integrable functions $f^+(x)$ and $f^-(x)$, it follows from Proposition 2.49 that:*

$$\begin{aligned} F(x) &= F(a) + \int_a^x f^+(y) dy - \int_a^x f^-(y) dy \\ &\equiv F_1(x) - F_2(x). \end{aligned}$$

Thus if (3.27) is proved assuming that $f(x)$ is nonnegative, the general result follows.

Define $f_n(x) = \min[f(x), n]$ for nonnegative $f(x)$. Then for each x the sequence $\{f_n(x)\}_{n=1}^\infty$ is monotonically increasing and $f_n(x) \to f(x)$. As $f(x) - f_n(x) \geq 0$ for all n:

$$F_n(x) \equiv \int_a^x [f(y) - f_n(y)] dy,$$

is monotonically increasing, differentiable a.e. by Proposition 3.12, and:

$$F_n(x) = F(x) - F(a) - \int_a^x f_n(y) dy. \tag{1}$$

By monotonicity and Corollary 3.13, $F_n'(x) \geq 0$ anywhere it exists, which is almost everywhere. As $f_n(y)$ is bounded and measurable, it is integrable by Proposition 2.15, and Proposition 3.38 obtains:

$$\frac{d}{dx} \int_a^x f_n(y) dy = f_n(x) \text{ a.e.}$$

Thus by (1):

$$\begin{aligned} F'(x) &= [F_n(x) + F(a)]' + \frac{d}{dx} \int_a^x f_n(y) dy \\ &\geq f_n(x) \text{ a.e.} \end{aligned}$$

This inequality is true for all n, and so:

$$F'(x) \geq f(x), \text{ a.e.} \tag{2}$$

Integrating this inequality:

$$\int_a^b F'(y) dy \geq \int_a^b f(y) dy \equiv F(b) - F(a).$$

However, since $f(x) \geq 0$ implies that $F(x)$ is monotonically increasing, Proposition 3.19 obtains:

$$\int_a^b F'(y) dy \leq F(b) - F(a),$$

and hence:

$$\int_a^b F'(y)dy = F(b) - F(a) = \int_a^b f(y)dy.$$

This implies that:

$$\int_a^b [F'(y) - f(y)]dy = 0.$$

But, as $F'(x) \geq f(x)$ a.e. by (2), it follows that $F'(x) = f(x)$ a.e. by item 8 of Proposition 2.31. ∎

Remark 3.40 (Lebesgue vs. Riemann FTC) *Recall that the general statement in Proposition 1.33 in the Riemann context is that if $f(x)$ is bounded and Riemann integrable on $[a, b]$, then the function $F(x)$ in (1.27) is differentiable almost everywhere on (a, b), and $F'(x) = f(x)$ at each continuity point of $f(x)$. Because bounded and Riemann integrable on $[a, b]$ implies continuous except on a set of measure 0 by Proposition 1.22, the conclusion of the last statement can be expressed as $F'(x) = f(x)$ almost everywhere.*

Hence, the Lebesgue result is a generalization of the general Riemann result. Under either Riemann or Lebesgue integrability of bounded $f(x)$, the associated function $F(x)$ is differentiable a.e., and $F'(x) = f(x)$ a.e. In the Lebesgue case, this conclusion holds even without the boundedness condition.

In the Riemann case, we get the additional insight that $F'(x) = f(x)$ at all continuity points of $f(x)$. Since a Lebesgue integrable function can be discontinuous everywhere, for example, the characteristic function of the irrationals in $[0, 1]$, such a characterization does not seem feasible.

Exercise 3.41 *Show that for $f(x)$ defined as the characteristic function of the irrationals in $[0, 1]$, that $F'(x) \equiv 1$ on $[0, 1]$. Thus $F'(x) = f(x)$ on the irrationals.*

3.2   Lebesgue Integral of a Derivative

In this section, we investigate Version I of the fundamental theorem of calculus, and that is the representation of a function as the indefinite Lebesgue integral of its derivative as in Proposition 1.32, but with $(\mathcal{L})$ instead of $(\mathcal{R})$. Specifically, we seek conditions on a function $f(x)$ defined on $[a, b]$ say, so that $f'(x)$ exists almost everywhere, is Lebesgue integrable, and for at least almost all $x \in [a, b]$:

$$f(x) = f(a) + (\mathcal{L})\int_a^x f'(y)dy. \tag{3.28}$$

If $f'(x)$ is continuous, then (3.28) follows from (1.26), since the Lebesgue and Riemann integrals agree for all x by Proposition 2.18. The goal of this section is to seek weaker conditions on $f(x)$ that assure (3.28). To set the stage, we first investigate behaviors of functions that cause this identity to fail.

3.2.1   Examples of FTC Failures

Before we turn to the theoretical development of this section, we investigate a variety of examples that provide insights as to why this generalization of the Riemann result is such a challenge.

Claim 3.42 *Continuity of $f(x)$ is necessary for (3.28) to hold, but is not sufficient.*

Proposition 3.35 asserts that if $f'(x)$ is Lebesgue integrable, which must be assumed if the above expression is to be meaningful, then $\int_a^x f'(y)dy$ is continuous and hence so too is $f(x)$.

That continuity is not sufficient may be a surprise since discontinuity was at the heart of earlier illustrations in Example 3.20 as to how $f'(x)$ might not integrate to $f(x)$. It would be natural to expect that by eliminating the "jump discontinuities" in $f(x)$, that an affirmative conclusion could be reached. But the illustrated functions there all had well-defined derivatives except at isolated points, and from these examples it would be natural to think that continuity implies differentiability, at least almost everywhere.

*And indeed this was the belief of most mathematicians before 1872 when **Karl Weierstrass (1815–1897)** discovered and published a **continuous, nowhere differentiable function,** whereby "nowhere" is meant, not differentiable at any x. This function came to be called the **Weierstrass function.***

Example 3.43 (Weierstrass function) *With appropriate restrictions on a, b, the Weierstrass function is defined on a given interval $[c, d]$ by:*

$$f(x) = \sum_{n=0}^{\infty} a^n \cos(b^n \pi x). \tag{3.29}$$

For $0 < a < 1$, this summation of continuous functions converges absolutely since $|\cos(b^n \pi x)| \leq 1$, and also converges uniformly since:

$$\left| \sum_{n=0}^{\infty} a^n \cos(b^n \pi x) \right| \leq \frac{1}{1-a}, \text{ for all } x.$$

By Exercise 1.44, this summation converges to a continuous function $f(x)$.

Proving that $f(x)$ is differentiable nowhere is more of a challenge, but the intuition is easier to appreciate. Because the period of $\cos(x)$ is 2π, meaning $\cos(x + 2\pi) = \cos(x)$ for all x, the period of $\cos(b^n \pi x)$ is $2/b^n$. So, if $b > 1$, each additional summand provides more and more oscillation to $f(x)$, and intuitively oscillates too much for the difference quotient:

$$[f(x + h) - f(x)]/h,$$

to converge as $h \to 0$.

To prove nondifferentiability at a given x, one can prove a general statement about such limits, such as it does not exist for any limit $h \to 0$, or a more limited statement, such as limits exist but differ for different choices of $h \to 0$, or even a more limited statement, that the limit does not exist for one carefully chosen sequence $h_n \to 0$.

Since the introduction of Weierstrass' example, many other examples have been illustrated. In fact, it is known that differentiable functions are rare in the space of continuous functions in a similar way that rational numbers are rare in the space of real numbers. Specifically, on the space of real-valued continuous functions defined on $[0, 1]$, one can define a measure so that the collection of functions that are differentiable **at even one point** form a subset of measure 0, just as the rationals are a subset of $\mathbb{R}$ of Lebesgue measure 0. We will not formally pursue this result since we have no application for it beyond the existence of such examples. But the intuition is accessible.

Continuity at x requires that $f(x + h) - f(x) \to 0$ for any $h \to 0$, and it is completely irrelevant how fast this difference converges. To be differentiable requires that the speed of convergence is proportional to h, or notationally, $O(h)$. If the speed is slower than h, say equal to h^a for $0 < a < 1$, then the difference quotient explodes, while if faster than h, say

h^a for $a > 1$, the quotient converges to 0. It can be shown that only constant functions can have this very fast rate of convergence of $a > 1$ for all x.

So differentiability requires a very specific rate of convergence, that $f(x + h) - f(x) = O(h)$. But this alone is not sufficient as is seen with $f(x) = |x|$ at $x = 0$, where the speed of convergence is just right, but the limit of the quotient depends on the sign of h.

We now investigate a more recent and accessible example.

Example 3.44 (McCarthy function) *In Exercise 3.45, the **continuous, nowhere differentiable function** introduced by **John McCarthy** in 1953 is developed in detail. Structurally, the McCarthy function $f(x)$ is similar to the Weierstrass function in that there is a damping factor a^n in the terms of the series, which assures uniform convergence to continuous $f(x)$. But this series possesses a simpler periodic oscillatory function than $\cos(b^n \pi x)$, which makes the proof of nondifferentiability far easier.*

The McCarthy function is given by:

$$f(x) = \sum_{n=1}^{\infty} 2^{-n} g(2^{2^n} x), \qquad (3.30)$$

where $g(x)$ is defined on $[-2, 2]$ by:

$$g(x) = \begin{cases} 1 + x, & -2 \le x \le 0, \\ 1 - x, & 0 \le x \le 2, \end{cases}$$

and extended periodically to $\mathbb{R}$ with period 4. Thus $g(x)$ is a piecewise linear version of $\cos x$ on $[-\pi, \pi]$, scaled to $[-2, 2]$. Note that $2^{2^n} \equiv 2^{(2^n)}$, and not $\left((2)^2\right)^n = 2^{2n}$.

Unlike the Weierstrass oscillatory functions, the terms in this summation are not differentiable everywhere. Specifically, $g(2^{2^n} x)$ has period $4/2^{2^n}$ and is not differentiable on the set $E_n = \{\pm 2k/2^{2^n} | k \in \mathbb{N}\}$, where here the natural numbers $\mathbb{N}$ contains 0. While each summand adds more nondifferentiable points, the countable union of these countable sets is countable. Thus the everywhere nondifferentiability conclusion is stronger than what can be explained only by the E_n-points. In this respect, the McCarthy function is similar to the Weierstrass function, for which the analogously defined $E_n = \emptyset$ for all n.

Exercise 3.45 (McCarthy function) *Prove that for the McCarthy function in (3.30):*

1. $f(x)$ is continuous.

2. The nondifferentiable exceptional points of each summand satisfy $E_n \subset E_{n+1}$ for all n.

3. The union $\bigcup_{n=1}^{\infty} E_n$ is dense in $\mathbb{R}$. Hint: If $x = 2^y > 0$, say, we need $\{k_m\}_{m=1}^{\infty}$ so that $2k_m/2^{2^m} \to 2^y$ as $j \to \infty$. Define $k_m = \lfloor 2^{y-1+2^m} \rfloor$, where $\lfloor x \rfloor$ denotes the "greatest integer function" defined by $\lfloor x \rfloor = \min\{n \in \mathbb{Z} | n \le x\}$, and so $x - 1 < \lfloor x \rfloor \le x$.

4. Given x, prove that with $g_n(x) \equiv g(2^{2^n} x)$ and $h_k = 2^{-2^k}$, that:

a. $g_n(x \pm h_k) - g_n(x) = 0$ if $n > k$ for either choice of $\pm h_k$. Hint: Recall that $g_n(x)$ is periodic with period $4/2^{2^n}$.

b. $|g_n(x \pm h_n) - g_n(x)| = 1$ for at least one choice of $\pm h_n$.

c. With the above notation:

$$\left| \sum_{n=1}^{k-1} [g_n(x \pm h_k) - g_n(x)] \right| \le (k-1) \max_n |[g_n(x \pm h_k) - g_n(x)]|$$
$$\le (k-1) 2^{2^{k-1}} 2^{-2^k}$$
$$< 2^k 2^{-2^{k-1}},$$

for any choice of $\pm h_k$.

5. *From part 4, show that as $k \to \infty$ and thus $h_k \to 0$:*

$$\left| \frac{f(x + h_k) - f(x)}{h_k} \right| . \geq 2^{-k} 2^{2^k} - 2^k 2^{2^{k-1}} \to \infty,$$

and hence $f(x)$ is not differentiable at x.

Claim 3.46 *If $f(x) \notin B.V.$ (Definition 3.23), then (3.28) need not hold even if $f'(x)$ exists a.e.*
   This example of failure is caused by the lack of integrability of $f'(x)$.
   Let:

$$f(x) = \begin{cases} x^2 \sin\left(1/x^2\right), & 0 < x \leq 1, \\ 0, & x = 0. \end{cases}$$

That $f(x) \notin B.V$ follows by choosing the partition:

$$\Pi = \{x | 1/x^2 \in \{2n\pi, \pi/2 + 2n\pi, \pi + 2n\pi, 3\pi/2 + 2n\pi, 2(N+1)\pi | 1 \leq n \leq N\}\}.$$

Then the total variation t of Definition 3.23 is given:

$$t = \sum_{n=1}^{N} 2/(\pi/2 + 2n\pi) + \sum_{n=1}^{N} 2/(3\pi/2 + 2n\pi).$$

This summation is unbounded, being proportional to the harmonic series, and hence $f(x) \notin B.V.$
   This function is continuous and differentiable for $0 < x \leq 1$ and hence a.e. on $[0, 1]$, with:

$$f'(x) = 2x \sin\left(1/x^2\right) - 2\cos\left(1/x^2\right)/x.$$

However, $f'(x)$ is not Lebesgue integrable on $[0, 1]$ since $|f'(x)|$ is not integrable. So, (3.28) fails on this interval because the integral is not even defined.
   To see this, note that by continuity a.e. of $f'(x)$ that this integral can be interpreted as a Riemann integral. Substituting $y = 1/x^2$ obtains:

$$\int_0^1 f'(x)dx = \int_1^\infty \left[\frac{\sin y}{y^2} - \frac{\cos y}{y}\right] dy.$$

The first integral is finite since $|(\sin y)/y^2| \leq 1/y^2$ which is integrable. It is an exercise that $(\cos y)/y$ is not absolutely Riemann integrable, and hence not Lebesgue integrable on $[1, \infty)$. Hint: Consider the period of this function, and construct subordinate triangles within these cycles.

Claim 3.47 *If $f(x) \in B.V.$, (3.28) need not hold even if $f'(x)$ is Lebesgue integrable.*
   This is an easy conclusion since B.V. does not imply continuous, so a step function is the counterexample.

Claim 3.48 *If $f(x) \in B.V.$ and is continuous, (3.28) need not hold even if $f'(x)$ is Lebesgue integrable.*
   This may be a very surprising result because the assumptions appear to reflect all the right ingredients. Being of bounded variation assures that $f'(x)$ exists almost everywhere by Corollary 3.32, and Lebesgue integrability avoids the absolute convergence problems exemplified above. In addition, continuity assures that (3.28) will not fail because of the previously illustrated problem with jumps in $f(x)$. The source of the problem here, and perhaps a problem even more surprising and remarkable than the existence of Weierstrass-like functions, is as follows.

There exist continuous, increasing functions $f(x)$, which of necessity are of bounded variation, but with the property that $f'(x) = 0$ a.e.

These are called **singular functions** and once shown to exist, it will be clear that (3.28) must fail. Specifically, a singular function satisfies $f(b) > f(a)$ by definition, but the increase $f(b) - f(a)$ cannot possibly be replicated by the integral of $f'(x)$ that has the well defined value of 0 a.e.

Definition 3.49 (Singular function) *A function $f(x)$ is **singular** on the interval $[a, b]$ if $f(x)$ is continuous, monotonically increasing with $f(b) > f(a)$, and $f'(x) = 0$ almost everywhere.*

Recall that "monotonically increasing" means that $f(x) \geq f(y)$ if $x > y$, while "strictly monotonically increasing" means that $f(x) > f(y)$ if $x > y$. So, a singular function is allowed to have "flat" spots for which $f(x) = f(y)$ for $x > y$, but it must strictly increase over the full interval $[a, b]$.

Intuition struggles to grasp how a function can have $f'(x) = 0$ "almost everywhere," and yet truly increase between $x = a$ and $x = b$ in a continuous way. The struggle in envisioning this behavior stems from the combination of properties of continuity and that $f(b) > f(a)$. Eliminating either of these properties allows easy examples like a step function, which lacks continuity, and constant functions, which lack a genuine increase.

The classic and original example of a singular function is the **Cantor function**, named after **Georg Cantor** (1845–1918), the discoverer of the **Cantor ternary set** on which this function is defined.

Example 3.50 (Cantor ternary set) *The Cantor ternary set K is a subset of the interval $[0, 1]$ that is closed, uncountable, and has Lebesgue measure 0.*

It is constructed as the intersection of a countable number of closed sets $\{F_n\}_{n=1}^{\infty}$, for which $F_{n+1} \subset F_n$ for all n. Thus:

$$K \equiv \bigcap_{n=1}^{\infty} F_n,$$

so K is closed, and Lebesgue measurable as the intersection of measurable sets. For this, recall that closed sets are Borel measurable by Definition I.2.13, and hence Lebesgue measurable by Proposition I.2.38.

Each successive closed set is defined to equal the prior set but with the open "middle third" intervals removed. For example:

$$
\begin{aligned}
F_0 &= [0, 1], \\
F_1 &= [0, 1] - (1/3, 2/3), \\
F_2 &= F_1 - \left\{ (1/9, 2/9) \bigcup (7/9, 8/9) \right\},
\end{aligned}
$$

$$\vdots$$

The total length of the open intervals removed is 1, the length of the original interval $[0, 1]$. This can be derived by noting that in the first step one interval of length one-third is removed, then two intervals of length one-ninth, then four intervals of length one-twenty-seventh, and so forth. The total length of the intervals removed is thus:

$$\sum_{n=0}^{\infty} 2^n / 3^{n+1} = \frac{1}{3} \sum_{n=0}^{\infty} (2/3)^n = 1.$$

Note that this conclusion is justified by the measurability of K, and hence that of its complement, and that the measure of this complement is the sum of the measures above by countable additivity. Because the complement of the Cantor ternary set in $[0,1]$ has length 1, the Cantor ternary set is thus a set of measure 0 by finite additivity.

The Cantor ternary set is also uncountable, and this is demonstrated using the base-3 expansion of numbers in the interval $[0,1]$:

$$x_{(3)} \equiv 0.a_1a_2a_3a_4.....$$
$$= \sum_{j=1}^{\infty} a_j/3^j, \text{ where } a_j = 0,1,2.$$

Removal of the "middle thirds" at each step of the construction is equivalent to eliminating all points with $a_j = 1$ for some j. So, the Cantor ternary set is made up of all numbers in $[0,1]$ with base-3 expansions using only 0s and 2s.

This may at first appear counterintuitive because $1/3 \in K$ by construction and yet the base-3 expansion of $1/3$ is simply $0.1_{(3)}$. The same is true for the left endpoint of each of the left-most intervals removed at each step, which are all numbers of the form $1/3^j$. But these can all be rewritten as:

$$1/3^j = \sum_{n=j+1}^{\infty} 2/3^n,$$

in the same way that in base-10, $0.1 = 0.09999...$ Consequently, each $x \in K$ has a base-3 expansion using only $a_j \in \{0,2\}$.

That K is uncountable now follows by identifying each $x \in K$ with a unique real number $b(x) \in [0,1]$. If $x = \sum_{j=1}^{\infty} a_j/3^j \in K$, define:

$$b(x) = \sum_{j=1}^{\infty} (a_j/2)/2^j.$$

It is left as an exercise to show that the mapping $b : K \to [0,1]$ is 1:1 and onto, and hence K is uncountable.

The Cantor function is defined as follows.

Definition 3.51 (Cantor function) *If $x_{(3)} = 0.a_1a_2a_3a_4.....a_na_{n+1}...$ in base-3, and N is the first index with $a_N = 1$, so $a_j = 0$ or 2 for $j \leq N-1$, define:*

$$f(x_{(3)}) \equiv \sum_{j=1}^{N-1} (a_j/2)/2^j + 1/2^N. \tag{3.31}$$

If there is no such first digit, and thus $x_{(3)} \in K$, define $N = \infty$ and so:

$$f(x_{(3)}) \equiv \sum_{j=1}^{\infty} (a_j/2)/2^j. \tag{3.32}$$

Exercise 3.52 *Check that in the definition of the Cantor function, if x has two base-3 expansions:*

$$x_{(3)} = 0.a_1a_2a_3a_4.....a_{N-1}1000... = 0.a_1a_2a_3a_4.....a_{N-1}0222...,$$

that $f(x_{(3)})$ is well defined. Also check that $f(0) = 0$, $f(1) = 1$ and $f(x_{(3)}) \leq 1$ for all $x_{(3)}$.

Proposition 3.53 (Cantor function is singular) *The Cantor function $f(x)$ is monotonically increasing on $[0,1]$ with $f(0) = 0$ and $f(1) = 1$, is continuous, and $f'(x) = 0$ a.e. Thus $f(x)$ is singular on $[0,1]$.*

Proof. *We first claim that given digits $a_1, a_2, ..., a_n$ with all $a_j \in \{0, 2\}$, $f(x_{(3)})$ is constant within the interval:*

$$0.a_1 a_2 ..... a_n 1 \leq x_{(3)} \leq 0.a_1 a_2 ..... a_n 2. \tag{1}$$

If such $x_{(3)} < 0.a_1 a_2 ..... a_n 2$, then $x_{(3)} = 0.a_1 a_2 ..... a_n 1 a_{n+2}...$ for arbitrary digits $\{a_j\}_{j=n+2}^{\infty}$ including all 0s. Then by (3.31), $f(x_{(3)}) \equiv \sum_{j=1}^{n} (a_j/2)/2^j + 1/2^{n+1}$, which equals $f(0.a_1 a_2 ..... a_n 2)$.

When $n = 0$, there is one interval as in (1) : $0.1 \leq x_{(3)} \leq 0.2$, which in decimal units is $1/3 \leq x \leq 2/3$. For $n = 1$, the intervals are $0.01 \leq x_{(3)} \leq 0.02$ and $0.21 \leq x_{(3)} \leq 0.22$, which in decimal units are $1/9 \leq x \leq 2/9$ and $7/9 \leq x \leq 8/9$, and so forth. For each n there are 2^n such intervals. The intervals in (1) are the closures of the "middle third" intervals removed from $[0,1]$ in the construction of K. With this identification, it follows that $f'(x) = 0$ for all x in the union of these intervals, which has measure 1 as noted above. Thus $f'(x) = 0$ almost everywhere.

Now $f(0) = 0$ and $f(1) = 1$ by Exercise 3.52. To verify that $f(x)$ is monotonically increasing over $[0, 1]$, let:

$$x_{(3)} \equiv 0.a_1 a_2 a_3 a_4 ..... a_n a_{n+1}... < 0.b_1 b_2 b_3 b_4 ..... b_n b_{n+1}... \equiv y_{(3)},$$

and note that we seek to prove that $f(x_{(3)}) \leq f(y_{(3)})$.

If $x_{(3)} = 0$ and/or $y_{(3)} = 1$, then Exercise 3.52 assures this conclusion. Given $0 < x_{(3)} < y_{(3)} < 1$, there is a first digit k so that $a_k < b_k$ and $a_i = b_i$ for $i < k$. We assume that $a_i, b_i \in \{0, 2\}$ for $i < k$, since otherwise there is a first $i < k$ with $a_i = b_i = 1$ and then $f(x_{(3)}) = f(y_{(3)})$ by (3.31). It is an exercise to explicitly evaluate the three cases with $(a_k, b_k) \in \{(0, 1), (0, 2), (1, 2)\}$ to confirm that $f(x_{(3)}) \leq f(y_{(3)})$, and thus $f(x)$ is monotonically increasing.

For continuity, let $x_{(3)}, y_{(3)}$ be given. If $x_{(3)}, y_{(3)} \in K$, then using the above notation let k be the first digit so that $a_k \neq b_k$ and $a_i = b_i$ for $i < k$. If $a_k > b_k$, then $a_k = 2$ and $b_k = 0$, so by (3.32):

$$x_{(3)} - y_{(3)} = 2/3^k + \sum_{i=k+1}^{\infty} (a_i - b_i)/3^i \geq 1/3^k.$$

This follows since $a_i - b_i \geq -2$ for $i \geq k + 1$. The same lower bound is true for $y_{(3)} - x_{(3)}$ if $a_k < b_k$ and so $|x_{(3)} - y_{(3)}| \geq 1/3^k$.

Hence, if $x_{(3)}, y_{(3)} \in K$ and $|x_{(3)} - y_{(3)}| < 1/3^k$, then $a_i = b_i$ at least for $i \leq k$. Thus since $|a_i - b_i| \leq 2$ for all i:

$$\left| f(x_{(3)}) - f(y_{(3)}) \right| \leq \sum_{i=k+1}^{\infty} |a_i - b_i|/2^{i+1} \leq 1/2^k.$$

This statement is equivalent to the standard $\epsilon - \delta$ definition, and it follows that $f(x)$ is continuous on K. Further, $f(x)$ is constant and hence continuous on the intervals in (1) above, the right end points of which are elements of K. Thus left to prove is that $f(x)$ is left continuous at the left endpoints of such intervals.

To this end, let $x_{(3)} = 0.a_1 a_2 a_3 a_4 ..... a_m 1$ be such a left endpoint, an element of K as noted above, whereas in (1) each a_i is 0 or 2. If $y^{(n)} < x_{(3)}$ and $y^{(n)} \to x_{(3)}$, we can assume infinitely many $y^{(n)} \notin K$ since otherwise $f(y^{(n)}) \to f(x_{(3)})$ by continuity of f on K. Thus each $y^{(n)} = 0.b_1^{(n)} b_2^{(n)} b_3^{(n)} .... \notin K$ has a first digit $b_k^{(n)} = 1$ with $b_i^{(n)}$ equal to 0 or 2 for $i < k$. Now $y^{(n)} < x_{(3)}$ does not assure that $k \leq m + 1$. For example, if $y^{(n)} = 0.a_1 a_2 a_3 ..... a_m 0 b_{m+2}^{(n)}...$, then $\{b_j^{(n)}\}_{j=m+2}^{\infty}$ are arbitrary, and thus k can be any value.

To simplify the analysis, choose N so that $\left|y^{(n)} - x_{(3)}\right| < 3^{-m}$ *for* $n \geq N$. *For any such* n *it then follows from* $y^{(n)} < x_{(3)}$ *that* $b_j^{(n)} = a_j$ *for all* $j \leq m$, *and either* $b_{m+1}^{(n)} = 1$ *or* $b_{m+1}^{(n)} = 0$. *In the first case* $f(x_{(3)}) = f(y^{(n)})$ *and there is nothing to prove. In the second case,* $b_j^{(n)}$ *can be arbitrary for* $j > m + 1$. *Thus since* $b_j^{(n)} = a_j$ *for all* $j \leq m$ *and* $b_{m+1}^{(n)} = 0$, *we have with* k *the first digit with* $b_k^{(n)} = 1$:

$$f(x_{(3)}) - f(y^{(n)}) = 1/2^{m+1} - \sum\nolimits_{j=m+2}^{k-1} b_j^{(n)}/2^{j+1} - 1/2^k. \tag{2}$$

Now since $y^{(n)} \to x_{(3)}$, *from:*

$$x_{(3)} = 0.a_1 a_2 a_3 a_4 \ldots .. a_m 1,$$
$$y^{(n)} = 0.a_1 a_2 a_3 \ldots .. a_m 0 b_{m+2}^{(n)} \ldots,$$

we obtain that as $n \to \infty$.

$$\sum\nolimits_{j=m+2}^{\infty} b_j^{(n)}/3^j \to 1/3^{m+1} = \sum\nolimits_{j=m+2}^{\infty} 2/3^j.$$

As $b_j^{(n)} \in \{0, 1, 2\}$, *it follows that* $b_j^{(n)} \to 2$ *for all* $j \geq m + 2$. *This assures in* (2) *that* $k \to \infty$ *and* $b_j^{(n)} \to 2$ *and thus:*

$$1/2^{m+1} - \sum\nolimits_{j=m+2}^{k-1} b_j^{(n)}/2^{j+1} - 1/2^k \to 0.$$

Thus $f(y^{(n)}) \to f(x_{(3)})$ *and the proof of continuity is complete.* ∎

3.2.2   The Lebesgue FTC – Version I

The examples of the prior section compel the conclusion that if there is a Lebesgue Version I of the fundamental theorem of calculus of Proposition 1.33, the functions $f(x)$ to which it applies will need to be "special" in some way. Apparently, none of the properties of continuity, uniform continuity, monotonicity, or of bounded variation are adequate given that the Cantor function has all four properties. On the other hand, this theorem applies if $f(x)$ is differentiable with a continuous derivative.

So we return to an investigation of properties of functions that can be represented as integrals of Lebesgue integrable functions:

$$F(x) = (\mathcal{L}) \int_a^x f(y) dy, \tag{3.33}$$

seeking to identify this special property.

So far we have identified the following properties of such functions $F(x)$ when $f(x)$ is Lebesgue integrable on $[a, b]$:

- $F(x)$ is continuous on $[a, b]$ by Proposition 3.35, and hence uniformly continuous by Exercise 1.14.

- $F(x) \in B.V.[a, b]$ by Proposition 3.35.

- $F'(x)$ exists almost everywhere, and $F'(x) = f(x)$ a.e. by Proposition 3.39.

Thus, any function $F(x)$ that can be represented as a Lebesgue integral as in (3.33) must be at least:

- uniformly continuous,

- of bounded variation, and,

- differentiable almost everywhere.

It turns out that such indefinite integrals satisfy another property, stronger than those above, and we will see below that for all functions with this property that (3.28) holds. The property is called **absolute continuity** and is defined as follows.

Definition 3.54 (Absolute continuity) *A real-valued function $f(x)$ defined on $[a, b]$ is* ***absolutely continuous*** *if for any $\epsilon > 0$ there is a δ so that:*

$$\sum_{i=1}^{n} |f(x_i) - f(x_i')| < \epsilon,$$

for any finite collection of disjoint subintervals $\{(x_i', x_i)\}_{i=1}^{n} \subset [a, b]$ with:

$$\sum_{i=1}^{n} |x_i - x_i'| < \delta.$$

Exercise 3.55 *Prove that $\sum_{j=1}^{n} a_j f_j(x)$, a finite linear combination of absolutely continuous functions, is absolutely continuous. Hint: The triangle inequality.*

Remark 3.56 (On absolute continuity) *This definition may initially appear to be only a bit more restrictive than that for continuity of Definition 1.12. Here's a summary of results.*

1. **Absolute continuity implies uniform continuity,** *because the former with $n = 1$ is exactly the definition of the latter by Definition 1.12.*

2. **Absolute continuity implies Lebesgue measurability** *by item 1 and Proposition 1.3.11.*

3. **Absolute continuity implies bounded variation, and a Lebesgue integrable derivative:** *This is proved below in Proposition 3.59.*

4. **Absolute continuity is stronger than bounded variation:** *As an example, the Cantor function is B.V. since increasing, but it is not absolutely continuous as proved in Example 3.57.*

5. **Continuously differentiable implies absolutely continuous:** *If $f(x)$ is continuously differentiable then $|f'(x)| \leq M$ on $[a, b]$ for some M, and by the mean value theorem:*

$$\sum_{i=1}^{n} |f(x_i) - f(x_i')| \leq M \sum_{i=1}^{n} |x_i - x_i'|.$$

Hence, for any ϵ we choose $\delta = \epsilon/M$.

6. **Absolutely continuous does not imply continuously differentiable:** *It will be proved in Proposition 3.59 that absolutely continuous functions are differentiable almost everywhere, and by Propositions 3.39 and 3.58 it follows that such functions need not have more than this. So absolute continuity is in fact weaker than continuously differentiable.*

7. **The integral in (3.33) of a Lebesgue integrable function is absolutely continuous:** *Proposition 3.58 proves this result.*

Example 3.57 *(The Cantor function is not absolutely continuous): Absolute continuity is a strong condition because there is no limit on how many disjoint intervals can be used. The Cantor function is continuous and hence uniformly continuous on $[0,1]$, but it is not absolutely continuous.*

The intuitive explanation is that the Cantor function is constant on a set of measure 1, while on the Cantor set K of measure 0 this function increases from $f(0) = 0$ to $f(1) = 1$. This implies that with sufficiently many small intervals that contain a subset of K, we can make $\sum_{i=1}^{n} |f(x_i) - f(x_i')| > \epsilon$ and at the same time make $\sum_{i=1}^{n} |x_i - x_i'|$ as small as we desire.

To make this precise, define for any N:

$$x_N = 0.a_1 a_2 a_3 a_4 \ldots a_N 022222\ldots,$$

where we assume that all $a_i \in \{0, 2\}$. There are 2^N such numbers, all members of K by definition. For any such x_N and any integer $M > 1$, define $x_{N,M} \in K$ by:

$$x_{N,M} = 0.a_1 a_2 a_3 a_4 \ldots a_N 022..2022\ldots,$$

where the $(N + M)$th digit of $x_{N,M}$ is 0, and otherwise all digits agree with x_N.

Thus $|x_N - x_{N,M}| = 1/3^{N+M}$, and by definition $|f(x_N) - f(x_{N,M})| = 1/2^{N+M}$. Fixing M and considering all 2^N such x_N obtains:

$$\sum_{i=1}^{2^N} |f(x_N) - f(x_{N,M})| = 1/2^M, \quad \sum_{i=1}^{2^N} |x_N - x_{N,M}| = \left(\frac{2}{3}\right)^N /3^M. \qquad (1)$$

Given $\epsilon > 0$, choose M so that $\epsilon < 1/2^M$. For any δ choose N so that $\left(\frac{2}{3}\right)^N /3^M < \delta$. Then (1) obtains a disjoint collection $\{(x_{N,M}, x_N)\}_{i=1}^{2^N}$ with $\sum_{i=1}^{2^N} |x_N - x_{N,M}| < \delta$ and $\sum_{i=1}^{2^N} |f(x_N) - f(x_{N,M})| > \epsilon$.

We next prove that the Lebesgue integral function in (3.33), with Lebesgue measurable $f(x)$, is absolutely continuous. By Proposition 2.15, this result also applies with bounded, Lebesgue measurable integrands.

Proposition 3.58 ($F(x)$ in (3.33) is absolutely continuous) *Let $f(x)$ be Lebesgue integrable on $[a, b]$ and $F(a)$ arbitrary. Then:*

$$F(x) = F(a) + (\mathcal{L}) \int_a^x f(y) dy \qquad (3.34)$$

is absolutely continuous on this interval.
Proof. *By Proposition 2.49, $f(y)$ is Lebesgue integrable on all $[a, x] \subset [a, b]$, and so $F(x)$ is well defined.*

If $f(x)$ is bounded, $|f(x)| \leq M$, and $E \equiv \bigcup_{j=1}^{n}(x_j, y_j) \cdot \subset [a, b]$, a disjoint union with $m(E) < \delta$, then by Proposition 2.49:

$$\left| \sum_{j=1}^{n} [F(y_j) - F(x_j)] \right| \leq \sum_{j=1}^{n} \int_{x_j}^{y_j} |f(y)| \, dy$$
$$\leq M\delta.$$

Given $\epsilon > 0$, choose $\delta = \epsilon/M$ and $F(x)$ is absolutely continuous by Definition 3.54.

For general $f(x)$, define:

$$f_n(x) = \begin{cases} \min[f(x), n], & f(x) \geq 0, \\ \max[f(x), -n], & f(x) \leq 0. \end{cases}$$

Each $f_n(x)$ is bounded, $|f_n(x)| \leq n$, and $f_n(x) \to f(x)$ pointwise. Also, $|f_n(x)| \leq |f(x)|$, and since $f(x)$ is integrable, Lebesgue's dominated convergence theorem of Proposition 2.54 obtains:

$$\int_a^b f_n(x)dx \to \int_a^b f(x)dx, \qquad \int_a^b |f(x) - f_n(x)|\, dx \to 0.$$

Thus for any $\epsilon > 0$ there in an N with both:

$$\left| \int_a^b f_N(x)dx - \int_a^b f(x)dx \right| < \epsilon/2, \qquad \int_a^b |f(x) - f_N(x)|\, dx < \epsilon/2.$$

Since $|f(x) - f_N(x)| \geq 0$, it follows from item 6 of Proposition 2.31 that for any measurable $E \subset [a, b]$:

$$\int_E |f(x) - f_N(x)|\, dx < \epsilon/2.$$

Now let $E \equiv \bigcup_{j=1}^n (x_j, y_j) \subset [a, b]$, a disjoint union with $m(E) < \delta \equiv \epsilon/2N$. Then by Proposition 2.49 and the above result for a bounded function:

$$\begin{aligned}
\left| \sum_{j=1}^n [F(y_j) - F(x_j)] \right| &\leq \sum_{j=1}^n \int_{x_j}^{y_j} |f(y)|\, dy \\
&\leq \sum_{j=1}^n \int_{x_j}^{y_j} |f(y) - f_N(x)|\, dy + \sum_{j=1}^n \int_{x_j}^{y_j} |f_N(x)|\, dx \\
&= \int_E |f(x) - f_N(x)|\, dx + \int_E |f_N(x)|\, dx \\
&< \epsilon/2 + Nm(E) \\
&< \epsilon.
\end{aligned}$$

∎

We next investigate properties of absolutely continuous functions. The first result states that such functions are of bounded variation and hence differentiable almost everywhere by Corollary 3.32.

Proposition 3.59 (Absolute continuity $\Rightarrow$ B.V.) *If $f(x)$ is absolutely continuous on $[a, b]$, then $f(x) \in B.V.[a, b]$. Hence, $f'(x)$ exists a.e. and is Lebesgue integrable.*
Proof. *Given ϵ and associated δ in the definition of absolute continuity, define integer $N \geq (b - a)/\delta$ and intervals $I_k = [a + (k - 1)\delta, a + k\delta]$ so that $[a, b] \subset \bigcup_{k=1}^N I_k$.*
Given any partition of the interval $[a, b]$ with:

$$a = x_0 < x_1 ... < x_n = b,$$

the definition of absolute continuity yields that for any k:

$$\sum_{\{x_j\} \in I_k} |f(x_{j+1}) - f(x_j)| < \epsilon, \text{ since } \sum_{\{x_j\} \in I_k} |x_{j+1} - x_j| < \delta.$$

If $x_j \in I_k$ and $x_{j+1} \in I_{k+1}$, then since $|x_{j+1} - x_j| < 2\delta$ there exists y_j so that both $|x_{j+1} - y_j| < \delta$ and $|y_j - x_j| < \delta$, and thus by the triangle inequality:

$$|f(x_{j+1}) - f(x_j)| \leq |f(x_{j+1}) - f(y_j)| + |f(y_j) - f(x_j)| < 2\epsilon.$$

Consequently:

$$t \equiv \sum_j |f(x_{j+1}) - f(x_j)| < 3N\epsilon,$$

and so $T \leq 3N\epsilon$ and $f(x)$ is B.V.

Differentiability a.e. then follows from Corollary 3.32, and Lebesgue integrability of $f'(x)$ follows from Propositions 3.19 and 3.29. ∎

To motivate the next result, recall from calculus that if $f(x)$ is continuous on $[a, b]$, and differentiable on (a, b) with $f'(x) = 0$ for all x, then $f(x)$ is a constant function by an application of the mean value theorem. If only $f'(x) = 0$ a.e., the result can be quite different. Step functions are nonconstant with $f'(x) = 0$ a.e., and the Cantor function exemplifies this same conclusion even though continuous.

In contrast, while absolutely continuous functions are in general only differentiable almost everywhere, these functions cannot exhibit the singular behavior of the Cantor function. Specifically, if an absolutely continuous function satisfies $f'(x) = 0$ a.e., then $f(x)$ must again be the constant function.

In this respect, absolutely continuous functions are "closer" to differentiable functions than to functions that are continuous, of bounded variation, and differentiable almost everywhere. And we will see below that they are close enough to differentiable functions to ensure that $f(b) - f(a)$ can be recovered from the Lebesgue integral of $f'(x)$ on $[a, b]$, which is (3.28).

Proposition 3.60 (Absolutely continuity and $f'(x) = 0$ a.e. $\Rightarrow f(x) = c$) *If $f(x)$ is absolutely continuous on $[a, b]$ and $f'(x) = 0$ a.e., then $f(x)$ is a constant function.*
Proof. *We seek to prove that $f(c) = f(a)$ for any $c \in [a, b]$. Because $f'(x) = 0$ a.e., let $E \subset [a, c]$ be the set of measure $m(E) = c - a$ on which $f'(x) = 0$. Then for arbitrary $\eta > 0$ and any $x \in E$, there is an interval $(x - h_x, x + h_x)$, so that:*

$$|f(x + k) - f(x)| < \eta k, \ for \ |k| \leq h_x. \tag{1}$$

For arbitrary $\epsilon > 0$, let δ be chosen as in the definition of absolute continuity, and choose a closed set $F \subset E$ with $m(E) - m(F) < \delta$. The set F exists by Proposition I.2.42 with $m(E - F) < \delta$, and this characterization follows from finite additivity. Since F is closed and bounded and hence compact, and $\{(x - h_x, x + h_x)\}$ is an open cover of E and hence F, there is a finite subcover $\{(x_j - h_j, x_j + h_j)\}_{j=1}^n$ of F by the Heine-Borel theorem of Proposition I.2.27. We make this collection disjoint by replacing with $\{(x_j - h_j', x_j + h_j'')\}_{j=1}^n$ where $h_j', h_j'' \leq h_j$, so that now $F \subset \bigcup_{j=1}^n [x_j - h_j', x_j + h_j''] \subset [a, c]$.
Relabel these disjoint open intervals $\{(y_j, z_j)\}_{j=1}^n$, so that:

$$z_0 \equiv a \leq y_1 < z_1 \leq y_2 < z_2 \leq ... \leq y_n < z_n \leq c \equiv y_{n+1}.$$

We claim that for the collection of intervals defined as

$$[a, c] - \bigcup_{j=1}^n (y_j, z_j) = \bigcup_{j=0}^n [z_j, y_{j+1}],$$

that:

$$\sum_{j=0}^n |y_{j+1} - z_j| < \delta. \tag{2}$$

To see this, note that:

$$F \subset \bigcup_{j=1}^n [y_j, z_j] = \bigcup_{j=1}^n (y_j, z_j) \bigcup Z,$$

where Z is the collection of endpoints of measure 0. Thus by De Morgan's laws:

$$\left[\bigcup_{j=1}^{n}(y_j, z_j)\right]^c \bigcap Z^c \subset F^c,$$

and so:

$$\left[\bigcup_{j=1}^{n}(y_j, z_j)\right]^c \bigcap E \bigcap Z^c \subset E \bigcap F^c.$$

Since $m(Z) = 0$, finite additivity obtains:

$$m\left[E \bigcap \left[\bigcup_{j=1}^{n}(y_j, z_j)\right]^c\right] \leq m\left[E \bigcap F^c\right].$$

Recalling that $F \subset E \subset [a, c]$ and $m(E) = c - a$, it follows from this last inequality that:

$$
\begin{aligned}
m\left[[a, c] - \bigcup_{j=1}^{n}(y_j, z_j)\right] &\equiv m\left[E \bigcap \left[\bigcup_{j=1}^{n}(y_j, z_j)\right]^c\right] \\
&\leq m\left[E \bigcap F^c\right] \\
&\equiv m\left[E - F\right] \\
&= m(E) - m(F) < \delta,
\end{aligned}
$$

which is (2).

By absolute continuity of $f(x)$ and (2):

$$\sum_{j=0}^{n} |f(y_{j+1}) - f(z_j)| < \epsilon. \tag{3}$$

while by construction of $\{(y_j, z_j)\}_{j=1}^{n}$ and (1):

$$\sum_{j=1}^{n} |f(z_j) - f(y_j)| < \eta \sum_{j=1}^{n}(z_j - y_j) < \eta(c - a). \tag{4}$$

Combining (3) and (4):

$$
\begin{aligned}
|f(c) - f(a)| &= \left|\sum_{j=0}^{n}[f(y_{j+1}) - f(z_j)] + \sum_{j=1}^{n}[f(z_j) - f(y_j)]\right| \\
&< \eta(c - a) + \epsilon.
\end{aligned}
$$

Since η and ϵ are arbitrary, it follows that $f(c) = f(a)$. ∎

Remark 3.61 (On the proof) *It should be noted that **almost all** of this proof is possible for an arbitrary function $f(x)$ for which $f'(x) = 0$ a.e. In general, we can always derive the estimate in (4) of $\sum_{j=1}^{n} |f(z_j) - f(y_j)|$ over the constructed collection of disjoint intervals $\bigcup_{j=1}^{n}(y_j, z_j)$, and this estimate can be made arbitrarily small exactly because $f'(x) = 0$ a.e. This construction can be implemented with any step function, or the Cantor function, for example.*

However, we must then be able to link these estimates with estimates of $|f(y_{j+1}) - f(z_j)|$, the changes in $f(x)$ "between" intervals. This is not possible for a general function with $f'(x) = 0$ a.e., despite the fact that it will be again true that $\sum |y_{j+1} - z_j| < \delta$.

This problem is apparent with a step function. The intervals (y_j, z_j) would exist within the steps, while the intervals (z_j, y_{j+1}) would span steps and thus will be impossible to obtain the estimates in (3). This is also seen with the Cantor function, whereby the intervals (y_j, z_j) would be contained within the constant steps of $f(x)$, and the intervals (z_j, y_{j+1}) would exist between steps. The sum of $|f(y_{j+1}) - f(z_j)|$ terms cannot be made small as was seen in Example 3.57.

Thus the key role of absolute continuity was to obtain the estimate in (3). Then with the estimate in (4) due to $f'(x) = 0$ a.e., we could complete the proof that $f(x)$ was the constant function.

The property of absolute continuity is key to the desired result of Version I of a Lebesgue fundamental theorem of calculus.

Proposition 3.62 (FTC Version I: Integral of a Derivative) *A function $f(x)$ is defined by a Lebesgue integral on $[a, b]$ with Lebesgue integrable integrand $g(x)$:*

$$f(x) = f(a) + (\mathcal{L}) \int_a^x g(y)dy,$$

if and only if $f(x)$ is absolutely continuous.

Any such function is then equal on $[a, b]$ to the integral of its derivative:

$$f(x) = f(a) + (\mathcal{L}) \int_a^x f'(y)dy, \tag{3.35}$$

and thus also:

$$(\mathcal{L}) \int_a^b f'(y)dy = f(b) - f(a). \tag{3.36}$$

Proof. *Since (3.35) implies (3.36), we address the former.*

If $f(x)$ can be defined as above with $g(x)$, then $f(x)$ is absolutely continuous by Proposition 3.58. Then, by Proposition 3.59, $f'(x)$ exists almost everywhere, and $f'(x) = g(x)$ almost everywhere by Proposition 3.39. By Proposition 2.49, $f'(x)$ is Lebesgue integrable with,

$$\int_a^x f'(y)dy = \int_a^x g(y)dy$$

for all x, and $f(x)$ has representation as in (3.35).

Next, assume $f(x)$ is absolutely continuous. Then $f(x) \in B.V.$ by Proposition 3.59 and so by Proposition 3.29 equals the difference of monotonically increasing functions, $f(x) = I_1(x) - I_2(x)$. All three functions are differentiable a.e. by Proposition 3.12, and since $I_j'(x) \geq 0$ by Corollary 3.13:

$$|f'(x)| \leq I_1'(x) + I_2'(x).$$

Then by Propositions 2.31 and 3.19:

$$\int_a^b |f'(y)|\, dy \leq I_1(b) + I_2(b) - I_1(a) - I_2(a),$$

and so $f'(x)$ is Lebesgue integrable on $[a, b]$.

Define

$$g(x) = \int_a^x f'(y)dy.$$

Then $g(x)$ is differentiable a.e. with $g'(x) = f'(x)$ a.e. by Proposition 3.39. In addition, $g(x)$ is absolutely continuous as noted above, and thus by Exercise 3.55, $f(x) - g(x)$ is absolutely continuous and differentiable a.e. Since $f'(x) - g'(x) = 0$ a.e., it follows that $f(x) - g(x)$ is a constant by Proposition 3.60, say $f(x) - g(x) = f(a)$, and 3.35 follows. ∎

3.3 Lebesgue Integration by Parts

Recall integration by parts for the Riemann integral in (1.30). Given continuously differentiable functions $f(x)$ and $g(x)$:

$$(\mathcal{R}) \int_a^b f'(x)g(x)dx = f(b)g(b) - f(a)g(a) - (\mathcal{R}) \int_a^b f(x)g'(x)dx. \tag{3.37}$$

It follows from (1.24) and continuity of $(f(x)g(x))'$ that:

$$(\mathcal{R})\int_a^b (f(x)g(x))'dx = f(b)g(b) - f(a)g(a),$$

and $(f(x)g(x))' = f'(x)g(x) + f(x)g'(x)$ completes the derivation. There is a similar development in the Lebesgue theory, but one based on (3.36).

To begin with, we must show that if $f(x)$ and $g(x)$ are absolutely continuous on $[a, b]$, then so too is $f(x)g(x)$.

Exercise 3.63 *Prove that if $f(x)$ and $g(x)$ are absolutely continuous on $[a, b]$, then so too is $f(x)g(x)$. Hint: Bound $\sum_{i=1}^n |f(x_i)g(x_i) - f(x_i')g(x_i')|$ in terms of the associated $f(x)$ and $g(x)$ differences, recalling that each function is continuous and hence bounded.*

Given this, we have the following.

Proposition 3.64 (Lebesgue Integration by Parts) *Let $f(x)$ and $g(x)$ be absolutely continuous functions on $[a, b]$. Then:*

$$(\mathcal{L})\int_a^b f'(x)g(x)dx = f(b)g(b) - f(a)g(a) - (\mathcal{L})\int_a^b f(x)g'(x)dx. \tag{3.38}$$

Proof. *Since $f(x)g(x)$ is an absolutely continuous function on $[a, b]$ by Exercise 3.63, (3.36) obtains:*

$$f(b)g(b) = f(a)g(a) + (\mathcal{L})\int_a^b (f(x)g(x))'dx.$$

Now $(f(x)g(x))'$ exists almost everywhere by Proposition 3.59, as do $f'(x)$ and $g'(x)$, and thus:

$$(f(x)g(x))' = f'(x)g(x) + f(x)g'(x), \quad a.e.$$

Then by Proposition 2.49:

$$(\mathcal{L})\int_a^b (f(x)g(x))'dx = (\mathcal{L})\int_a^b (f'(x)g(x) + f(x)g'(x))dx. \tag{1}$$

Since $f(x)$ and $g(x)$ are absolutely continuous they are continuous and thus bounded on $[a, b]$, while $f'(x)$ and $g'(x)$ are Lebesgue integrable by Proposition 3.59. This assures that $f'(x)g(x)$ and $f(x)g'(x)$ are Lebesgue integrable. In detail, if $|f(x)| \leq M$ on $[a, b]$, then $|f(x)g'(x)| \leq M|g'(x)|$ and $f(x)g'(x)$ is integrable by Corollary 2.51. The same argument applies to $f'(x)g(x)$.

Propositions 2.49 and 2.62 now justify rewriting the integral in (1) as in (3.38). ∎

Remark 3.65 (Further generalizations) *In the next chapter, we study an integration approach that again generalizes the Riemann integral, known as the Riemann-Stieltjes integral, and will generalize Riemann integration by parts to this context. Then in Book V, this Lebesgue result will be generalized to Lebesgue-Stieltjes integrals, which are integrals defined relative to Borel measures.*

4

Stieltjes Integration

4.1 Introduction

The notion of a **Stieltjes integral** is named after **Thomas Joannes Stieltjes** (1856–1894) who introduced it in a paper published in 1894. This paper was his 1886 thesis and written under two advisors, **Jean-Gaston Darboux** (1842–1917) who was introduced in Chapter 1, and **Charles Hermite** (1822–1901), whose mathematical contributions are largely outside the topics of these books. Stieltjes' original idea generalized **Riemann's integral** introduced in 1868, but preceded the measure-theoretic approach underlying **Lebesgue's integral** that was introduced in 1904.

In the development of ideas since that time, Stieltjes' original approach has come to be known by the name **Riemann-Stieltjes integration**. A related approach studied in Book V analogously generalizes the Lebesgue integral, and has come to be known by the name **Lebesgue-Stieltjes integration**. These approaches are summarized next.

4.1.1 The Riemann-Stieltjes Integral

The Riemann-Stieltjes integral of a bounded function $g(x)$ begins as does the Riemann integral, with an arbitrary **partition** $P \equiv \{x_i\}_{i=0}^n$ of the interval $[a, b]$ into subintervals $\{[x_{i-1}, x_i]\}_{i=1}^n$ as in (1.1):

$$a = x_0 < x_1 < \cdots < x_{n-1} < x_n = b.$$

The mesh size μ is again defined as in (1.2) by $\mu \equiv \max_{1 \le i \le n}\{x_i - x_{i-1}\}$.

The Riemann-Stieltjes integral of $g(x)$ also requires an **integrator function** $F(x)$, and is then initially estimated with a type of Riemann sum, now called a Riemann-Stieltjes sum:

$$\sum\nolimits_{i=1}^n g(\widetilde{x}_i)\Delta F_i, \tag{4.1}$$

with **subinterval tags** $\widetilde{x}_i \in [x_{i-1}, x_i]$ and $\Delta F_i \equiv F(x_i) - F(x_{i-1})$.

Notation 4.1 ($g(x)$ **vs.** $F(x)$) *As will be seen in Proposition 4.14, the roles of $g(x)$ as integrand and $F(x)$ as integrator are often reversible. The motivation for capitalizing the integrator function is that for the primary applications of this theory in these books, $F(x)$ will be the distribution function of a random variable or random vector X defined on some probability space. In textbooks without such a "biased" point of view, both integrand and integrator functions may be lowercase, and also subject to other notational conventions.*

Remark 4.2 (**On** ΔF_i) *When $F(x)$ is an increasing function, so $\Delta F_i \geq 0$, the sum in (4.1), looks like the Riemann sum in (1.3), but where we have replaced the standard interval length $\Delta x_i \equiv x_i - x_{i-1}$ with ΔF_i. This latter value resembles the F-length in equation I.(5.4), but there is a subtle difference.*

DOI: 10.1201/9781003264590-4

In the Riemann development, partition subintervals are taken to be closed, but if we deemed them to be open or semi-closed, the interval length would not be changed. That convention is carried over to the Riemann-Stieltjes construction, with all subintervals taken as closed.

In contrast, the general Borel measure theory in Chapter I.5 derived from F-length, and similar developments elsewhere in that book, required partitions to create disjoint sets. There, as will be seen for the Lebesgue-Stieltjes integral below, ΔF_i is defined as the F-measure of the right semi-closed interval $(x_{i-1}, x_i]$.

Since $g(x)$ is assumed bounded, when $F(x)$ is an increasing function, the above sum can be bounded as in (1.4):

$$\sum_{i=1}^{n} m_i \Delta F_i \leq \sum_{i=1}^{n} g(\tilde{x}_i) \Delta F_i \leq \sum_{i=1}^{n} M_i \Delta F_i. \tag{4.2}$$

Recalling (1.5), m_i denotes the **greatest lower bound** or **infimum** of $g(x)$, and M_i the **least upper bound** or **supremum** of $g(x)$, both defined on the subinterval $[x_{i-1}, x_i]$.

For increasing $F(x)$, if the limits of the upper and lower bounding sums agree as $\mu \to 0$, the **Riemann-Stieltjes integral** can be defined in terms of these limits, or by:

$$\int_a^b g(x) dF \equiv \lim_{\mu \to 0} \sum_{i=1}^{n} g(\tilde{x}_i)(F(x_i) - F(x_{i-1})). \tag{4.3}$$

This limit is then independent of the choice of interval tags $\{\tilde{x}_i\}_{i=1}^{n}$, since $m_i \leq g(\tilde{x}_i) \leq M_i$ for all choices.

This theory can also be developed for other than increasing integrators $F(x)$, but then the above approach needs to be modified since (4.2) fails to be satisfied when $F(x_i) - F(x_{i-1})$ is no longer assumed nonnegative. One generalization pursued here is to integrators of bounded variation. Results on increasing integrators will provide the necessary ingredients for this study since, from Proposition 3.29, functions of bounded variation can be expressed as a difference of increasing functions.

But thinking ahead to the distribution functions of probability theory, the most common application in these books for the development below is to the case where $F(x)$ is a distribution function, and thus is increasing. In this case, the Riemann-Stieltjes integral becomes in effect a type of Riemann integral, but where the measure of the "length" of an interval $[x_{i-1}, x_i]$ is generalized from $\Delta x_i \equiv x_i - x_{i-1}$ to $\Delta F_i \equiv F(x_i) - F(x_{i-1})$.

The central question of course is existence, or more specifically, conditions on $g(x)$ and $F(x)$ that assure existence. The existence question is pursued below along with the analogous development for the n-dimensional version of this integral. The central applications of this integral will be found in the Books IV and VI developments of probability theory, where $F(x)$ will be a distribution function.

This theory will also be recalled in the introductory discussions in Book VIII, and there will motivate the need for a new **stochastic integral**.

4.1.2   The Lebesgue-Stieltjes Integral

For, completeness, it is perhaps worthwhile to introduce the notion of the **Lebesgue-Stieltjes integral,** even though the details of this development are deferred to Book V. Focusing on the case of an increasing integrator function $F(x)$ defined on $\mathbb{R}$, but now also assumed to be right continuous, recall the construction in Chapter I.5, of a Borel measure μ_F induced by $F(x)$. The measure μ_F is "induced" in the sense that on the semi-algebra of right semi-closed intervals $\{(a, b]\}$, μ_F is defined:

$$\mu_F[(a, b]] \equiv F(b) - F(a).$$

This set function definition is then extended to a measure on the Borel sigma algebra $\mathcal{B}(\mathbb{R})$, and to a complete sigma algebra $\mathcal{M}_{\mu_F}(\mathbb{R})$ with $\mathcal{B}(\mathbb{R}) \subset \mathcal{M}_{\mu_F}(\mathbb{R})$.

As in Definition 2.2, a **simple function** $g(x)$ defined on the measure space $(\mathbb{R}, M_{\mu_F}(\mathbb{R}), \mu_F)$ is a bounded function given by:

$$g(x) = \sum_{i=1}^m a_i \chi_{A_i}(x),$$

where:

1. $\{A_i\}_{i=1}^m \subset M_{\mu_F}(\mathbb{R})$ are disjoint μ_F-measurable sets with $\mu_F(\bigcup_{i=1}^m A_i) < \infty$.

2. $\chi_{A_i}(x)$ is the characteristic function or indicator function for A_i, defined in (1.8).

3. $\{a_i\}_{i=1}^m \subset \mathbb{R}$.

With a change to $\{A_i\}_{i=1}^m \subset B(\mathbb{R})$, the same definition applies to simple functions $g(x)$ defined on the measure space $(\mathbb{R}, B(\mathbb{R}), \mu_F)$.

Given a simple function $g(x)$, the **Lebesgue-Stieltjes integral** is defined as in Definition 2.4:

$$\int g(x) d\mu_F \equiv \sum_{i=1}^m a_i \mu_F(A_i).$$

Comparing to the first step in the development of the Lebesgue integral in (2.2), observe that the above integral simply replaces $m(A_i)$, the Lebesgue measure of A_i, with the Borel measure of this set, $\mu_F(A_i)$. The development of this integral is then largely the same, whereby the integral of more general μ_F-measurable functions is derived step by step, from simpler to more general functions.

This integration theory can also be generalized to right continuous integrator functions $F(x)$ of bounded variation, but then the induced μ_F will in general be a "signed" measure. A signed measure has all of the properties of a measure except nonnegativity. Signed measures will be studied in Book V.

The Lebesgue-Stieltjes theory can be developed on $\mathbb{R}^n$ using the measure theory of Chapter I.8, or on any measure space X. See Book V for details.

4.2 Riemann-Stieltjes Integral on $\mathbb{R}$

4.2.1 Definition of the Integral

In this section, we formally define the Riemann-Stieltjes integral, both in terms of limits of Riemann-Stieltjes sums, and also in terms of infima and suprema of bounding Darboux sums. This latter approach applies well to increasing integrators, while the Riemann-Stieltjes sums approach applies more generally. In the next section, we prove definitional equivalence for increasing integrators with the aid of the Cauchy criterion, as was seen in Chapter 1.

We begin with the Riemann-Stieltjes sums approach.

Definition 4.3 (Riemann-Stieltjes integral: General F) *Assume that $g(x)$ and $F(x)$ are defined on $[a, b]$ and that $g(x)$ is bounded.*
Given the partition $P \equiv \{x_i\}_{i=0}^n$ of $[a, b]$:

$$a = x_0 < x_1 < \cdots < x_{n-1} < x_n = b, \tag{4.4}$$

and subinterval tags $\{\widetilde{x}_i\}_{i=1}^n$ with $\widetilde{x}_i \in [x_i, x_{i-1}]$, define the **Riemann-Stieltjes summation** *$R(g, F; P)$ by:*

$$R(g, F; P) \equiv \sum_{i=1}^n g(\widetilde{x}_i) \Delta F_i, \tag{4.5}$$

where $\Delta F_i \equiv F(x_i) - F(x_{i-1})$.
If:

$$\lim_{\mu \to 0} R(g, F; P) = I < \infty, \tag{4.6}$$

meaning that for any $\epsilon > 0$, there is a δ so that for all **tagged partitions** *P of mesh size $\mu \le \delta$:*

$$|R(g, F; P) - I| < \epsilon, \tag{4.7}$$

then define the **Riemann-Stieltjes integral of $g(x)$ with respect to $F(x)$ over $[a, b]$:**

$$\int_a^b g(x)dF = I. \tag{4.8}$$

If (4.6) is not satisfied, we say that the Riemann-Stieltjes integral of $g(x)$ with respect to $F(x)$ over $[a, b]$ does not exist.
If $g(x)$ is Riemann-Stieltjes integrable with respect to $F(x)$, one sometimes uses the notation $g \in \mathcal{R}(F)$.

Notation 4.4 (Tagged partition) *The collections $\{x_i\}_{i=0}^n$ and $\{\widetilde{x}_i\}_{i=1}^n$ are collectively referred to as a* **tagged partition** *and sometimes denoted $P \equiv \{\widetilde{x}_i, [x_{i-1}, x_i]\}_{i=1}^n$.*

For the Darboux sums approach, given bounded $g(x)$ and partition $P \equiv \{x_i\}_{i=0}^n$ of $[a, b]$ as in (4.4), define the **lower and upper bounding step functions** as in (1.13):

$$\varphi(x) \equiv \sum_{i=1}^n m_i \chi_{[x_{i-1}, x_i]}(x), \quad \psi(x) \equiv \sum_{i=1}^n M_i \chi_{[x_{i-1}, x_i]}(x),$$

where $m_i \equiv \inf\{g(x)|x \in [x_{i-1}, x_i]\}$ and $M_i \equiv \sup\{g(x)|x \in [x_{i-1}, x_i]\}$. When the mesh size μ of the partition need be identified, these step functions will be denoted $\varphi_\mu(x)$ and $\psi_\mu(x)$.

Exercise 4.5 *($\int_a^b \varphi(x)dF$, $\int_a^b \psi(x)dF$) If $F(x)$ is an* **increasing, continuous function** *on $[a, b]$, so $F(x) \le F(y)$ if $x < y$, use (4.2) to prove that:*

$$\int_a^b \varphi(x)dF \equiv \sum_{i=1}^n m_i \Delta F_i, \quad \int_a^b \psi(x)dF \equiv \sum_{i=1}^n M_i \Delta F_i, \tag{4.9}$$

where $\Delta F_i \equiv F(x_i) - F(x_{i-1})$ and these integrals are defined as in (4.6).
Prove that this same result is obtained for increasing $F(x)$ assuming only continuity at the partition points $\{x_i\}_{i=0}^n$ underlying the definition of $\varphi(x)$ and $\psi(x)$.

For Riemann-Stieltjes integrals on $\mathbb{R}$ (and later $\mathbb{R}^n$), it is conventional to suppress the framework of bounding step functions for the integrand $g(x)$. Instead, as was seen in the Riemann development on $\mathbb{R}^n$, we define as our elementary objects the upper and lower **Darboux sums.** When $F(x) \equiv x$, these are the Darboux sums for the Riemann integral on $\mathbb{R}$ implicit in (1.14), and defined explicitly in (1.44) with $n = 1$.

Definition 4.6 (Darboux sums) *Given a* **bounded function** *$g(x)$ and an* **increasing function** *$F(x)$ defined on the compact interval $[a, b]$ with partition $P \equiv \{x_i\}_{i=0}^n$, define the* **lower Darboux sum** *$L(g, F; P)$, and the* **upper Darboux sum** *$U(g, F; P)$, by:*

$$L(g, F; P) \equiv \sum_{i=1}^n m_i \Delta F_i, \quad U(g, F; P) \equiv \sum_{i=1}^n M_i \Delta F_i, \tag{4.10}$$

where $m_i \equiv \inf\{g(x)|x \in [x_{i-1}, x_i]\}$, $M_i \equiv \sup\{g(x)|x \in [x_{i-1}, x_i]\}$, and $\Delta F_i \equiv F(x_i) - F(x_{i-1})$.

Because the integrator function $F(x)$ is assumed to be increasing, it is apparent by definition that for any given partition P that:

$$L(g, F; P) \leq U(g, F; P).$$

This relationship is less apparent when different partitions are used for the upper and lower sums. The following proposition clarifies this with an affirmative result, but first a definition.

Definition 4.7 (Refinement of a partition) *Given $[a, b]$, a partition $P' \equiv \{x'_j\}_{j=0}^m$ of $[a, b]$ is a **refinement** of a partition $P \equiv \{x_i\}_{i=0}^n$ if $\{x_i\}_{i=0}^n \subset \{x'_j\}_{j=0}^m$.*

Proposition 4.8 $(L(g, F; P_1) \leq U(g, F; P_2))$ *If $g(x)$ is bounded and $F(x)$ is increasing on $[a, b]$, then for any partitions P_1 and P_2:*

$$L(g, F; P_1) \leq U(g, F; P_2). \tag{4.11}$$

Proof. Let P be a partition of $[a, b]$ defined by all the points of both P_1 and P_2, and thus P is a **common refinement** of both P_1 and P_2. We claim that:

$$L(g, F; P_1) \leq L(g, F; P) \leq U(g, F; P) \leq U(g, F; P_2), \tag{4.12}$$

which proves (4.11).

By definition $L(g, F; P) \leq U(g, F; P)$, so the proof will be complete by demonstrating that $L(g, F; P_1) \leq L(g, F; P)$ and $U(g, F; P) \leq U(g, F; P_2)$. By induction, it is enough to prove this assuming that P contains only one more point than P_1 or P_2.

To this end, assume $P_1 \equiv \{x_i\}_{i=0}^n$ and that P has an additional point y with say $x_0 < y < x_1$. Then, with $m'_0 \equiv \inf\{g(x)|x \in [x_0, y]\}$, $m'_1 \equiv \inf\{g(x)|x \in [y, x_1]\}$, and m_1 defined as above relative to $[x_0, x_1]$:

$$
\begin{aligned}
&L(g, F; P) - L(g, F; P_1) \\
=\ &m'_0(F(y) - F(x_0)) + m'_1(F(x_1) - F(y)) - m_1(F(x_1) - F(x_0)) \\
=\ &[m'_1 - m_1](F(x_1) - F(y)) + [m'_0 - m_1](F(y) - F(x_0)).
\end{aligned}
$$

Since $m_1 = \min\{m'_0, m'_1\}$, it follows that $L(g, F; P) - L(g, F; P_1) \geq 0$.

That $U(g; P_2, F) - U(g; P, F) \geq 0$ is proved similarly. ∎

We next define the lower and upper Darboux integrals, reflecting the sums in Definition 4.9:

Definition 4.9 (Upper/lower Darboux integrals) *If $g(x)$ is bounded and $F(x)$ is increasing on $[a, b]$, the **lower Darboux integral of g with respect to F** over $[a, b]$ is defined:*

$$\underline{\int_a^b} g(x)dF \equiv \sup_P L(g, F; P), \tag{4.13}$$

*and the **upper Darboux integral of g with respect to F** over $[a, b]$ is defined:*

$$\overline{\int_a^b} g(x)dF \equiv \inf_P U(g, F; P), \tag{4.14}$$

where the supremum and infimum are defined over all partitions of $[a, b]$.

The lower and upper Darboux integrals always exist and are finite. First, the associated Darboux sums are monotonic with respect to partition refinements by Proposition 4.8. Further, these sums have an upper bound of $M[F(b) - F(a)]$ and lower bound of $m[F(b) - F(a)]$ if $m \leq g(x) \leq M$.

The following definition states that a function $g(x)$ is Riemann-Stieltjes integrable with respect to increasing $F(x)$ over $[a, b]$ when the upper and lower Darboux integrals agree.

Definition 4.10 (Riemann-Stieltjes integral: Increasing $F(x)$) *Let $g(x)$ be bounded and $F(x)$ increasing on $[a, b]$, and assume that:*

$$\sup_P L(g, F; P) = \inf_P U(g, F; P) = I < \infty, \tag{4.15}$$

where the supremum and infimum are taken over all partitions P of $[a, b]$.

*Then the **Riemann-Stieltjes integral of $g(x)$ with respect to $F(x)$ over** $[a, b]$ is defined by:*

$$\int_a^b g(x)dF = I. \tag{4.16}$$

If (4.15) is not satisfied then we say that the Riemann-Stieltjes integral of $g(x)$ with respect to $F(x)$ over $[a, b]$ does not exist.

If $g(x)$ is Riemann-Stieltjes integrable with respect to $F(x)$, one sometimes uses the notation $g \in \mathcal{R}(F)$.

Remark 4.11 (On $\mu \to 0$) *Recalling Remark 1.8 and Exercise 1.61, it is again the case that by virtue of (4.12), the integrability condition of (4.15) can be restated as:*

$$\lim_{\mu \to 0} \sup_{P_\mu} L(g, F; P_\mu) = \lim_{\mu \to 0} \inf_{P_\mu} U(g, F; P_\mu) = I, \tag{4.17}$$

where P_μ denotes a partition of mesh size μ.

4.2.2   Riemann and Darboux Equivalence

In this section, we prove that the Riemann-Stieltjes sums Definition 4.3 and Darboux sums Definition 4.10 provide equivalent criteria for the existence of the Riemann-Stieltjes integral of a bounded function $g(x)$ for an **increasing integrator** $F(x)$. In other words, we prove that for an increasing integrator, a bounded function is Riemann-Stieltjes integrable in terms of Riemann-Stieltjes sums if and only if it is Riemann-Stieltjes integrable in terms of Darboux sums, and further that the integrals agree.

We do this by showing that each is equivalent to (4.18), which is known as the **Cauchy criterion for the existence of the Riemann-Stieltjes integral** and named after **Augustin-Louis Cauchy** (1789–1857). The Cauchy criterion here is identical to this criterion for the Riemann integral in (1.16), but restated with the notation of (4.10) as was seen in (1.50) for the Riemann integral in $\mathbb{R}^n$.

Proposition 4.12 (Cauchy criterion for the Riemann-Stieltjes integral) *A bounded function $g(x)$ is Riemann-Stieltjes integrable with respect to increasing $F(x)$ on $[a, b]$ by the Riemann-Stieltjes sums criterion of Definition 4.3, if and only if it is Riemann-Stieltjes integrable on this interval by the Darboux sums criterion of Definition 4.10, and then the integrals agree.*

In either case, $g(x)$ is so integrable if and only for any $\epsilon > 0$ there is a partition P so that:

$$0 \leq U(g, F; P) - L(g, F; P) < \epsilon, \tag{4.18}$$

with $U(g, F; P)$ and $L(g, F; P)$ given in (4.10).

Further, if (4.18) is satisfied for P, it is also satisfied for every refinement of P.

Proof. *As noted above, we show that the requirement of either definition of integrability is equivalent to the above Cauchy criterion, and thus each is equivalent to the other.*

Given $\epsilon > 0$, assume that (4.18) is satisfied. It follows by (4.12) and the definition of infimum and supremum that for such P:

$$L(g, F; P) \leq \sup_{P'} L(g, F; P') \leq \inf_{P'} U(g, F; P') \leq U(g, F; P).$$

Thus:

$$\inf_{P'} U(g, F; P') - \sup_{P'} L(g, F; P') < \epsilon. \tag{1}$$

Since this is true for all ϵ, it follows that g is Riemann-Stieltjes integrable with respect to increasing $F(x)$ over $[a, b]$ by the Darboux sums criterion of Definition 4.10.

Now by (4.17), there exists μ' so that for every partition P_1 with mesh size $\mu_1 \leq \mu'$:

$$\sup_{P'} L(g, F; P') \leq L(g, F; P_1) + \epsilon/2.$$

Similarly, there exists μ'' so that for every partition P_2 with mesh size $\mu_2 \leq \mu''$:

$$\inf_{P'} U(g, F; P') \geq U(g, F; P_2) - \epsilon/2.$$

Thus by (1), given any partition P_3 with mesh size $\mu_3 \leq \min\{\mu', \mu''\}$:

$$U(g, F; P_3) - L(g, F; P_3) < 2\epsilon,$$

and given this partition and any interval tags $\{\widetilde{x}_i\}_{i=1}^n$:

$$L(g, F; P_3) \leq \sum_{i=1}^{n} g(\widetilde{x}_i) \Delta F_i \leq L(g, F; P_3) + 2\epsilon. \tag{2}$$

As (2) is satisfied for every partition P_3 with mesh size $\mu_3 \leq \min\{\mu', \mu''\}$, it is thus true taking a limit as $\mu_3 \to 0$. By (4.17), with I denoting the integral using Darboux sums:

$$I \leq \lim_{\mu \to 0} \sum_{i=1}^{n} g(\widetilde{x}_i) \Delta F_i \leq I + 2\epsilon. \tag{3}$$

Since ϵ is arbitrary, g is Riemann-Stieltjes integrable with respect to increasing $F(x)$ on $[a, b]$ by the Riemann-Stieltjes sums criterion of Definition 4.3.

Further by (3), the Riemann-Stieltjes integrals of these definitions agree.

Conversely, if g is Riemann-Stieltjes integrable with respect to increasing $F(x)$ on $[a, b]$ by the Darboux sums Definition 4.10 and $\epsilon > 0$ is given, then by definition of extrema there exist partitions P_1 and P_2 so that:

$$0 \leq U(g, F; P_1) - \int_a^b g(x) dF < \epsilon/2,$$

$$0 \leq \int_a^b g(x) dF - L(g, F; P_2) < \epsilon/2.$$

Thus:

$$0 \leq U(g, F; P_1) - L(g, F; P_2) < \epsilon. \tag{4}$$

It then follows from (4.12) that (4) is satisfied with both $U(g, F; P_1)$ and $L(g, F; P_2)$ defined relative to the common refinement of these partitions, and this is (4.18).

Similarly, if g is Riemann-Stieltjes integrable with respect to increasing $F(x)$ on $[a, b]$ by the Riemann-Stieltjes sums Definition 4.3 and $\epsilon > 0$ is given, there is a δ so that:

$$\left| \sum_{i=1}^{n} g(\widetilde{x}_i) \Delta F_i - \int_a^b g(x) dF \right| < \epsilon/3,$$

for all tagged partitions $P \equiv \{\widetilde{x}_i, [x_{i-1}, x_i]\}_{i=1}^n$ of mesh size $\mu \leq \delta$. Thus for any such partition and any two sets of tags $\{\widetilde{x}_i\}_{i=1}^n$ and $\{\widetilde{x}_i'\}_{i=1}^n$:

$$\left| \sum\nolimits_{i=1}^n g(\widetilde{x}_i)\Delta F_i - \sum\nolimits_{i=1}^n g(\widetilde{x}_i')\Delta F_i \right| < 2\epsilon/3.$$

By definition of M_i and m_i, we can choose such tags so that these Riemann-Stieltjes sums are within $\epsilon/6$ of $U(g, F; P)$ and $L(g, F; P)$, respectively, and this obtains (4.18).

Finally, if (4.18) is satisfied for a given partition, it is also satisfied for every refinement by (4.12). ∎

Below we turn to existence results, and then on to the matter of actually evaluating certain Riemann-Stieltjes integrals. But first we consider two examples.

Example 4.13

1. Let $g(x)$ be bounded. Define $F(x) = \chi_{[1,\infty)}(x)$, the increasing and right continuous characteristic function of $[1, \infty)$, so $F(x) = 1$ for $x \geq 1$ and $F(x) = 0$ otherwise, and consider the existence of $\int_0^2 g(x)dF$.

   For any partition of $[0, 2]$, it follows that $F(x_i) - F(x_{i-1}) = 0$ for all i except for the interval for which $x_{i-1} < 1 \leq x_i$, and then $F(x_i) - F(x_{i-1}) = 1$. Considering all partitions, all such intervals can be refined by $[x_{i-1}, x_i] = [x_{i-1}, 1] \bigcup [1, x_i]$, and thus the only intervals of interest are $[x_{i-1}, 1]$.

   For Definition 4.3:

   $$R(g, F; P) \equiv g(\widetilde{x}_i),$$

   with $\widetilde{x}_i \in [x_{i-1}, 1]$.

   For Definition 4.10:

   $$L(g, F; P) = \inf_I g(x), \qquad U(g, F; P) = \sup_I g(x),$$

   where the infimum and supremum are defined over $x \in I \equiv [x_{i-1}, 1]$.

   These definitions agree for integrable $g(x)$ by Proposition 4.12, so we investigate the details.

   a. **Definition 4.3 Criterion:** Since $R(g, F; P)$ must converge as $\mu \to 0$ for every partition sequence, this translates to $g(\widetilde{x}_i)$ converging as $x_{i-1} \to 1$, and this implies $\widetilde{x}_i \to 1$. Thus $\int_0^2 g(x)dF$ exists by Definition 4.3 if and only if $g(x)$ is left continuous at $x = 1$, and then $\int_0^2 g(x)dF = g(1^-)$. If $g(x)$ is not left continuous at $x = 1$, then $\int_0^2 g(x)dF$ does not exist.

   It is an exercise to check that:

   i. If we redefine $F(1) \equiv 0$, making $F(x)$ left continuous, then $\int_0^2 g(x)dF$ exists by Definition 4.3 if and only if $g(x)$ is right continuous at $x = 1$, and this integral then equals $g(1^+)$.

   ii. If $F(1)$ is redefined with $0 < F(1) < 1$, making $F(x)$ discontinuous at $x = 1$, then $\int_0^2 g(x)dF$ exists by Definition 4.3 if and only if $g(x)$ is continuous at $x = 1$, and this integral then equals $g(1)$.

   b. **Definition 4.10 Criterion:** By Proposition 4.12, the criterion is now that $U(g, F; P) - L(g, F; P)$ can be made arbitrarily small for given partitions. It follows that with $I \equiv [x_{i-1}, 1]$:

   $$U(g, F; P) - L(g, F; P) = \sup_I g(x) - \inf_I g(x).$$

Thus $\int_0^2 g(x)dF$ exists by Definition 4.10 if and only if $g(x)$ is left continuous at $x = 1$, and then $\int_0^2 g(x)dF = g(1^-)$. If $g(x)$ is not left continuous then the integral does not exist.

Redefining $F(1) = 0$ or $0 < F(1) < 1$ again produces the result that $\int_0^2 g(x)dF$ exists by Definition 4.10 if and only if $g(x)$ is right continuous, respectively continuous, at $x = 1$, and then $\int_0^2 g(x)dF = g(1^+)$, respectively $g(1)$.

2. Let $g(x)$ be bounded, $F(x) = \ln x$, and consider the existence of $\int_1^2 g(x)dF$.

Define the partition P of $[1, 2]$ with $x_i = 2^{i/n}$ for $0 \le i \le n$, so $F(x_i) - F(x_{i-1}) = \frac{1}{n}\ln 2$. Now:

$$U(g, F; P) - L(g, F; P) = \frac{1}{n}\ln 2 \sum_{i=1}^n [M_i - m_i],$$

where m_i and M_i are defined as the infimum and supremum of $g(x)$ over $[x_{i-1}, x_i]$. In this case, discontinuities in $g(x)$ need not create an existence problem due to the $1/n$ factor, but this conclusion is subject to an investigation into the growth of $\sum_{i=1}^n [M_i - m_i]$ with n.

If $g(x)$ is increasing, then $M_i - m_i = g(x_i) - g(x_{i-1})$ and so this summation equals $g(2) - g(1)$. It then follows from (4.18) of Proposition 4.12 that $\int_1^2 g(x)dF$ is well defined simply by choosing n large. The same result applies if g is decreasing, or is of bounded variation since such functions equal a difference of increasing functions by Proposition 3.29. For this, we will also need that Riemann-Stieltjes integrals are linear in the integrand, proved below in Proposition 4.24.

On the other hand, if $g(x) = \sin\left(\frac{1}{x-1}\right)$, a function that is not of bounded variation on $[1, 2]$ as can be verified, then $\sum_{i=1}^n [M_i - m_i] \approx 2n$ as $n \to \infty$. Thus the difference between upper and lower Darboux sums cannot be made small with this partition sequence by letting $n \to \infty$. However, the existence of $\int_1^2 g(x)dF$ by (4.18) requires only that $U(g, F; P) - L(g, F; P)$ can be made arbitrarily small for some partitions, not all partitions.

For this example, the existence of this integral is not easily confirmed with (4.18), or any of the above criteria. But we will see in Proposition 4.17 that $\int_{1+\epsilon}^2 g(x)dF$ exists for all $\epsilon > 0$, and then by Proposition 4.28 that:

$$\int_{1+\epsilon}^2 g(x)dF = (\mathcal{R})\int_{1+\epsilon}^2 \frac{1}{x}\sin\left(\frac{1}{x-1}\right)dx,$$

where we have converted this to a Riemann integral with $dF \equiv F'(x)dx$. The substitution $y = \frac{1}{x-1}$ obtains:

$$\int_{1+\epsilon}^2 g(x)dF = (\mathcal{R})\int_1^{1/\epsilon} \frac{\sin y}{y(1+y)}dy.$$

This integrand is absolutely integrable and has a well-defined limit as $\epsilon \to 0$:

$$\left|\int_{1+\epsilon}^2 g(x)dF\right| \le (\mathcal{R})\int_1^{1/\epsilon} \frac{dy}{y^2} \to 1.$$

Thus $g(x) = \sin\left(\frac{1}{x-1}\right)$ is Riemann-Stieltjes integrable with respect to $F(x) = \ln x$ over $[1, 2]$ after all.

4.2.3   On Existence of the Integral

The first existence result is quite remarkable, since there is nothing in the definition of the Riemann-Stieltjes integral that suggests that the roles of $g(x)$ and $F(x)$ are reversible. To highlight the symmetry of this result, we temporarily abandon the notational convention discussed in Notation 4.1 and represent both integrand and integrator functions in lowercase. Importantly, note that neither of the functions is assumed to be increasing, and thus the notion of integrability is that of the Riemann-Stieltjes sums Definition 4.3.

See Remark 4.29 for an explanation of how the result in (4.19) is indeed an integration by parts formula, despite the absence of any visible derivatives.

Proposition 4.14 (Integration by Parts) *If a bounded function $g(x)$ is Riemann-Stieltjes integrable with respect to $f(x)$ over $[a, b]$, then $f(x)$ is Riemann-Stieltjes integrable with respect to $g(x)$, and:*

$$\int_a^b g(x)df + \int_a^b f(x)dg = f(b)g(b) - f(a)g(a). \tag{4.19}$$

Proof. *Given an arbitrary partition $P \equiv \{x_i\}_{i=0}^n$ of $[a, b]$:*

$$f(b)g(b) - f(a)g(a) = \sum_{i=1}^n (f(x_i)g(x_i) - f(x_{i-1})g(x_{i-1})),$$

and so for any collection of tags $\{\widetilde{x}_i\}_{i=1}^n$:

$$f(b)g(b) - f(a)g(a) - \sum_{i=1}^n f(\widetilde{x}_i)(g(x_i) - g(x_{i-1}))$$
$$= \sum_{i=1}^n g(x_i)(f(x_i) - f(\widetilde{x}_i)) + \sum_{i=1}^n g(x_{i-1})(f(\widetilde{x}_i) - f(x_{i-1})).$$

Subtracting $\int_a^b g(x)df$:

$$\left| f(b)g(b) - f(a)g(a) - \int_a^b g(x)df - \sum_{i=1}^n f(\widetilde{x}_i)(g(x_i) - g(x_{i-1})) \right| \tag{1}$$

$$= \left| \sum_{i=1}^n g(x_i)(f(x_i) - f(\widetilde{x}_i)) + \sum_{i=1}^n g(x_{i-1})(f(\widetilde{x}_i) - f(x_{i-1})) - \int_a^b g(x)df \right|.$$

The last pair of sums in (1) reflects a Riemann-Stieltjes sum for $\int_a^b g(x)df$ using the partition $P' \equiv \{x_i\}_{i=0}^n \bigcup \{\widetilde{x}_i\}_{i=1}^n$ and endpoint interval tags.

By assumption on the integrability of $g(x)$ with respect to $f(x)$, Definition 4.3 obtains that given $\epsilon > 0$ there is a δ so that:

$$\left| \sum_{i=1}^n g(\widetilde{y}_i)(f(y_i) - f(y_{i-1})) - \int_a^b g(x)df \right| < \epsilon, \tag{2}$$

for any partition $P'' \equiv \{y_i\}_{i=0}^n$ of $[a, b]$ of mesh size $\mu'' \leq \delta$, and arbitrary tags $\widetilde{y}_i \in [y_i, y_{i-1}]$.

If the mesh size of P is $\mu \leq \delta$, the mesh size of the above refinement P' also satisfies $\mu' \leq \delta$, and by (1) and (2):

$$\left| f(b)g(b) - f(a)g(a) - \int_a^b g(x)df - \sum_{i=1}^n f(\widetilde{x}_i)(g(x_i) - g(x_{i-1})) \right| < \epsilon. \tag{3}$$

Thus (3) is true for all partitions of mesh size $\mu \leq \delta$ and all tags. Since $\epsilon > 0$ is arbitrary, it follows by Definition 4.3 that $f(x)$ is Riemann-Stieltjes integrable with respect to $g(x)$ and:

$$\int_a^b f(x)dg = f(b)g(b) - f(a)g(a) - \int_a^b g(x)df.$$

∎

Item 1 of Example 4.13 suggests that the existence of $\int_a^b g(x)dF$ does not require continuity of either $g(x)$ or $F(x)$, but illustrates what can happen when discontinuities coincide. This generalizes to the following necessary condition for the existence of $\int_a^b g(x)dF$ for increasing $F(x)$, and bounded $g(x)$.

Proposition 4.15 ($\int_a^b g(x)dF$ exists $\Rightarrow$ No common discontinuities) *If $F(x)$ is increasing, $g(x)$ is bounded, and $\int_a^b g(x)dF$ exists, then $g(x)$ and $F(x)$ have no common discontinuity points.*
Proof. *By contradiction, assume that $g(x)$ and $F(x)$ have a common discontinuity point at $x' \in [a, b]$, considering first $x' \in (a, b)$.*

Given any partition $P \equiv \{x_i\}_{i=0}^n$ of $[a, b]$ with $x' \in (x_{i-1}, x_i)$ for some i:

$$U(g, F; P) - L(g, F; P) \geq (M_i - m_i)(F(x_i) - F(x_{i-1})) \geq cc' > 0.$$

Here $c > 0$ equals the difference between left and right limits of F at x', and since F is increasing it follows that $F(x_i) - F(x_{i-1}) \geq c$. Also $M_i - m_i \geq c' > 0$, with c' defined as the difference between the one-sided limits of $g(x)$. Thus by Proposition 4.12, $\int_a^b g(x)dF$ does not exist.

If $x' = a$ then:

$$U(g, F; P) - L(g, F; P) \geq (M_1 - m_1)(F(x_1) - F(a)) \geq cc' > 0,$$

where c is the difference between $F(a)$ and the right limit of $F(x)$ at a, and c' is similarly defined with respect to $g(x)$. The case $x' = b$ is similar. ∎

Exercise 4.16 ($\int_a^b g(x)dF$ exists $\Rightarrow$ No common discontinuities) *Prove that Proposition 4.15 generalizes to state that bounded $g(x)$ and increasing $F(x)$ can also have no common one-sided discontinuity points. Hint: For example, if $x' \in (a, b)$ is a common left discontinuity, consider all partitions with $x' = x_i$ for some i.*

The following proposition consolidates insights from Example 4.13 on the existence of $\int_a^b g(x)dF$, and applies Proposition 4.14 for a new result in item 3.

Note that part 2 of Example 4.13 shows that item 2 below is a sufficient but not necessary condition for the existence of this integral.

Proposition 4.17 (Existence of $\int_a^b g(x)dF$) *Let $g(x)$ and $F(x)$ be defined on $[a, b]$. Then $\int_a^b g(x)dF$ exists in the following cases:*

1. *$g(x)$ is continuous and $F(x)$ is increasing.*

2. *$g(x)$ is of bounded variation, and $F(x)$ is increasing and continuous.*

3. *$g(x)$ is increasing and continuous, and $F(x)$ is of bounded variation.*

Proof. *For items 1 and 2, let $\epsilon > 0$ be given. To prove existence in each case we identify a partition so that (4.18) is satisfied. In both cases we assume that $F(b) > F(a)$, since if $F(b) = F(a)$ the integral exists and equals 0 by definition.*

1. If $g(x)$ is continuous, it is uniformly continuous on compact $[a, b]$ by Exercise 1.14, and thus there is a δ so that $|g(x_i) - g(x_{i-1})| < \epsilon/[F(b) - F(a)]$ if $|x_i - x_{i-1}| < \delta$. Choosing any partition P with mess size $\mu \leq \delta$:

$$
\begin{aligned}
U(g, F; P) - L(g, F; P) &= \sum_{i=1}^{n} [M_i - m_i] [F(x_i) - F(x_{i-1})] \\
&\leq \frac{\epsilon}{F(b) - F(a)} \sum_{i=1}^{n} [F(x_i) - F(x_{i-1})] \\
&= \epsilon.
\end{aligned}
$$

2. If $g(x)$ is of bounded variation with total variation $T_a^b(g)$ over $[a, b]$ as in Definition 3.23, given n choose a partition P so that $F(x_i) - F(x_{i-1}) = [F(b) - F(a)]/n$ for all i. This is possible since $F(x)$ is increasing and continuous, and thus by the intermediate value theorem of Exercise 1.13, for any c with $F(a) \leq c \leq F(b)$ there is an x with $F(x) = c$. Then:

$$
\begin{aligned}
U(g, F; P) - L(g, F; P) &= \sum_{i=1}^{n} [M_i - m_i] (F(x_i) - F(x_{i-1})) \\
&= \frac{F(b) - F(a)}{n} \sum_{i=1}^{n} [M_i - m_i] \\
&\leq T_a^b(g) [F(b) - F(a)]/n.
\end{aligned}
$$

This bound can be made less than ϵ by choosing n large.

3. Item 3 follows from item 2 by Proposition 4.14, that $\int_a^b g(x) dF$ exists if and only if $\int_a^b F(x) dg$ exists. ∎

Remark 4.18 *By the symmetry afforded by Proposition 4.14, it may seem surprising that we only got one additional result in item 3 from its application to item 2. Applying Proposition 4.14 to item 1 provides an existence result for $F(x)$ continuous and $g(x)$ increasing. But this is a special case of item 2 since such $g(x)$ is necessarily of bounded variation.*

The following exercise generalizes item 3 of the above result, and thus item 2 as well by Proposition 4.14. See item 2 of Proposition 4.25 for a generalization of (4.20) for general integrable $g(x)$, and Proposition 4.95 for an n-dimensional version of this result.

Exercise 4.19 *($g(x)$ **continuous**, $F(x) \in B.V.$) Prove directly with Definition 4.3 that if $g(x)$ is continuous and $F(x)$ is of bounded variation, then $\int_a^b g(x) dF$ exists and:*

$$
\left| \int_a^b g(x) dF \right| \leq M_g T_a^b(F). \tag{4.20}
$$

Here $M_g \equiv \sup_{[a,b]} |g(x)|$ and $T_a^b(F)$ denote the total variation of F on $[a, b]$ as in Definition 3.23.

Thus if $F(x)$ is increasing, then by Example 3.26:

$$
\left| \int_a^b g(x) dF \right| \leq M_g [F(b) - F(a)].
$$

Hint: Recall $F(x) = I_1(x) - I_2(x)$ *for monotonically increasing real-valued functions* $I_1(x)$ *and* $I_2(x)$ *by Proposition 3.29, and so:*

$$R(g, F; P) \equiv \sum_{i=1}^{n} g(\widetilde{x}_i)(F(x_i) - F(x_{i-1}))$$
$$= \sum_{i=1}^{n} g(\widetilde{x}_i)(I_1(x_i) - I_1(x_{i-1})) - \sum_{i=1}^{n} g(\widetilde{x}_i)(I_2(x_i) - I_2(x_{i-1})).$$

Justify how this proves existence, then derive (4.20).

Remark 4.20 *Exercise 4.19 proves existence and provides a bound for* $\int_a^b g(x)dF$, *and then* **suggests** *that when* $F(x)$ *is of bounded variation:*

$$\int_a^b g(x)dF = \int_a^b g(x)dI_1 - \int_a^b g(x)dI_2. \tag{1}$$

We are not quite ready to assert this, since the decomposition $F(x) = I_1(x) - I_2(x)$ *is not unique, and in fact there are infinitely many such decompositions.*

To assert (1) requires a proof that $\int_a^b g(x)dF$ *so defined is independent of this decomposition. See Proposition 4.23.*

The next result establishes criteria for the existence of improper Riemann-Stieltjes integrals, and is immediately applicable to the case when $F(x)$ is a distribution function and $0 \leq F(x) \leq 1$.

Proposition 4.21 (Improper Riemann-Stieltjes integrals) *Let* $g(x)$ *be continuous and* $F(x)$ *increasing on* $[a, \infty)$. *If both functions are* **bounded** *on* $[a, \infty)$, *the improper integral:*

$$\int_a^{\infty} g(x)dF \equiv \lim_{b \to \infty} \int_a^b g(x)dF, \tag{4.21}$$

is well defined and finite.

If these assumptions apply on $\mathbb{R}$, *the improper integral:*

$$\int_{-\infty}^{\infty} g(x)dF \equiv \lim_{a \to -\infty, b \to \infty} \int_a^b g(x)dF, \tag{4.22}$$

is well defined and finite.

Proof. *Since* F *is bounded and increasing, given* $\epsilon > 0$ *there exists* b *so that:*

$$\sup_{[a,\infty)} F(x) - \sup_{[a,b]} F(x) < \epsilon/M_g,$$

with $M_g = \max_{[a,\infty)} |g(x)|$. *As* F *is increasing,*

$$\sup_{[a,b]} F(x) = F(b), \quad \sup_{[a,\infty)} F(x) = \sup_{[b,\infty)} F(x),$$

and this obtains:

$$\sup_{[b,\infty)} F(x) - F(b) < \epsilon/M_g. \tag{1}$$

Let $c > b$ *be arbitrary. Applying Proposition 4.24 and Exercise 4.19 to integrals which we know to exist by Proposition 4.17:*

$$\left| \int_a^c g(x)dF - \int_a^b g(x)dF \right| \leq \int_b^c |g(x)|\, dF \leq M_g\left[F(c) - F(b)\right].$$

By (1) it follows that $F(c) - F(b) < \epsilon/M_g$ *for all c, and so:*

$$\int_a^b g(x)dF - \epsilon \leq \int_a^c g(x)dF \leq \int_a^b g(x)dF + \epsilon.$$

Thus $\liminf_{c\to\infty} \int_a^c g(x)dF$ *and* $\limsup_{c\to\infty} \int_a^c g(x)dF$ *satisfy these same bounds, and this obtains:*

$$\limsup_{c\to\infty} \int_a^c g(x)dF - \liminf_{c\to\infty} \int_a^c g(x)dF < 2\epsilon.$$

Since ϵ *is arbitrary,* $\lim_{b\to\infty} \int_a^b g(x)dF$ *is well defined by Corollary I.3.46.*

The proof of (4.22) is similar and left as an exercise. ∎

Remark 4.22 (Existence generalizations) *A very general existence result was proved by* **L. C. (Laurence Chisholm)Young** *(1905–2000) in a 1936 paper using an inequality he developed with* **E. R. (Eric Russell) Love** *(1912–2001), and now called the* **Love-Young inequality**.

*To understand this result, we first define a generalization of total variation called p-**variation**. Given* $p > 0$, *a function* $g(x)$ *is said to be of* **bounded p-variation on an interval** $[a,b]$ *if:*

$$V_p(g) \equiv \sup_\Pi \left(\sum_{i=1}^n |g(x_i) - g(x_{i-1})|^p \right)^{1/p} < \infty, \tag{4.23}$$

where $\sup_\Pi$ *is defined is over all partitions* Π *of the interval* $[a,b]$. *Thus* $V_1(g) \equiv T$, **the total variation** *of Definition 3.23, while* $V_2(g)$ *is called* **quadratic variation** *and will be seen to play a prominent role in the study of stochastic processes and their integrals in Books VII–IX.*

Young's result on Riemann-Stieltjes integration is as follows:

If $g(x)$ *has bounded p-variation and* $F(x)$ *has bounded q-variation on* $[a,b]$ *with* $1/p + 1/q > 1$, *and,* $g(x)$ *and* $F(x)$ *have no common discontinuities, then* $\int_a^b g(x)dF$ *exists.*

The problem of common discontinuities was illustrated in Example 4.13. More generally for increasing integrators, the existence of the integral assures no common discontinuities as proved in Proposition 4.15, and assures no common one-sided discontinuities as proved in Exercise 4.16.

One corollary of Young's result is that if $F(x)$ *is of bounded variation, which is p-variation with* $p = 1$, *then* $\int_a^b g(x)dF$ *exists for any function* $g(x)$ *of bounded q-variation for any* $q > 0$, *again subject to the no common discontinuities constraint.*

Another corollary is that if $\int_a^b g(x)dF$ *exists by Young's p/q-variation criterion, then so too does* $\int_a^b F(x)dg$, *since this variation criterion does not distinguish integrand from integrator. This provides another perspective on the symmetry result in Proposition 4.14.*

See also Propositions 4.23 and 4.62 below for other general results on existence on $\mathbb{R}$.

4.2.4   Integrators of Bounded Variation

We return to a question that was introduced in Remark 4.20. If a Riemann-Stieltjes integral $\int_a^b g(x)dF$ exists with $F(x)$ of bounded variation, meaning exists in the sense of Definition 4.3, can this integral can be expressed as a difference of Riemann-Stieltjes integrals with increasing integrators?

By Proposition 3.29, any such $F(x)$ can be expressed:

$$F(x) = I_1(x) - I_2(x),$$

with monotonically increasing real-valued functions $I_1(x)$ and $I_2(x)$. And thus it seems obvious to assert that the integral with respect to $F(x)$ equals the difference of integrals with respect to $I_1(x)$ and $I_2(x)$:

$$\int_a^b g(x)dF = \int_a^b g(x)dI_1 - \int_a^b g(x)dI_2.$$

However, this proposition proposed two such decompositions, and there are in fact infinitely many as can be obtained by adding any increasing function to both $I_1(x)$ and $I_2(x)$. Questions that must be answered to justify the above decomposition:

1. **Existence:** If $\int_a^b g(x)dF$ exists by Definition 4.3, does this imply that the integrals with respect to $I_1(x)$ and $I_2(x)$ exist for some decomposition? Do they exist for all decompositions?

2. **Well-definedness:** Will any two compositions yield the same value for $\int_a^b g(x)dF$?

The following result addresses these questions.

Proposition 4.23 ($\int_a^b g(x)dF = \int_a^b g(x)dI_1 - \int_a^b g(x)dI_2$; $F \in B.V.$) *Assume that $\int_a^b g(x) dF$ exists for bounded $g(x)$ and $F(x)$ of bounded variation, and let $F(x) = I_1(x) - I_2(x)$ be either decomposition of Proposition 3.29 in terms of increasing functions.*
Then $\int_a^b g(x)dI_1$ and $\int_a^b g(x)dI_2$ exist and:

$$\int_a^b g(x)dF = \int_a^b g(x)dI_1 - \int_a^b g(x)dI_2. \tag{4.24}$$

Further, if $F(x) = F_1(x) - F_2(x)$ is an arbitrary decomposition of $F(x)$ in terms of increasing functions, and if the integrals of $g(x)$ with respect to $F_1(x)$ and $F_2(x)$ exist, then:

$$\int_a^b g(x)dF = \int_a^b g(x)dF_1 - \int_a^b g(x)dF_2. \tag{4.25}$$

If $g(x)$ is continuous, then (4.25) is valid for any such decomposition of $F(x)$. If $g(x)$ is of bounded variation, (4.25) is valid for any decomposition of $F(x)$ with continuous $F_i(x)$.
Proof. *Assume that we have proved existence of the I_j-integrals and (4.24). Then existence of the integrals with respect to increasing $F_1(x)$ and $F_2(x)$ assures by item 5 of Proposition 4.24 that the integrals of $g(x)$ exist with respect to both $I_1(x) + F_2(x)$ and $I_2(x) + F_1(x)$. Further, each of these integrals equals the sum of the associated integrals, for example:*

$$\int_a^b g(x)d\,(I_1 + F_2) = \int_a^b g(x)dI_1 + \int_a^b g(x)dF_2.$$

But then $I_1(x) + F_2(x) = I_2(x) + F_1(x)$ obtains:

$$\int_a^b g(x)dI_1 + \int_a^b g(x)dF_2 = \int_a^b g(x)dF_1 + \int_a^b g(x)dI_2,$$

and (4.25) follows from (4.24).
For (4.24), any tagged partition of $[a, b]$ obtains that:

$$\sum_{i=1}^n g(\widetilde{x}_i)(F(x_i) - F(x_{i-1}))$$
$$= \sum_{i=1}^n g(\widetilde{x}_i)(I_1(x_i) - I_1(x_{i-1})) - \sum_{i=1}^n g(\widetilde{x}_i)(I_2(x_i) - I_2(x_{i-1})). \qquad \bullet$$

By assumption, the first summation converges to $\int_a^b g(x)dF$ as $\mu \to 0$ in the sense of Definition 4.3. Thus if it can be proved that one of the I_j-summations similarly converges, then the other must converge by default and this then proves (4.24).

For this demonstration, Proposition 3.29 provides two definitional choices, but a proof with either choice will imply the other. For example, choose $I_1(x) \equiv F(a) + T_a^x$ in (3.22), recalling that T_a^x is the total variation of F over $[a, x]$. If proved for this $I_1(x)$, this then proves the existence with respect to the associated $I_2(x) \equiv 2N_a^x$ as just noted. But this then obtains the existence for the alternative $I_2(x) \equiv N_a^x$ in (3.21), which in turn now proves existence for the alternative $I_1(x) \equiv F(a) + P_a^x$. The reverse implication is identical.

So let $I_1(x) \equiv F(a) + T_a^x$. Given $\epsilon > 0$, choose δ so that for every partition $P \equiv \{x_i\}_{i=0}^n$ with mesh size $\mu \leq \delta$:

$$\left| \sum_{i=1}^n \left(g(\widetilde{x}_i) - g(\widetilde{y}_i) \right) \left(F(x_i) - F(x_{i-1}) \right) \right| < \epsilon, \ all \ \widetilde{x}_i, \widetilde{y}_i \in [x_{i-1}, x_i], \tag{1}$$

and,

$$T_a^b < \sum_{i=1}^n |F(x_i) - F(x_{i-1})| + \epsilon. \tag{2}$$

Given such ϵ the first bound exists for all partitions P_1 of mesh size $\mu_1 \leq \delta_1$ by the definition of the integral $\int_a^b g(x)dF$. The second follows from the definition of T_a^b as the supremum over all partitions, and Exercise 3.25, that proved that if $T_a^b(\mu)$ is defined as the supremum over all partitions of mesh size μ, then $T_a^b = \lim_{\mu \to 0} T_a^b(\mu)$. Thus given such ϵ there exists δ_2 so that $T_a^b < T_a^b(\mu_2) + \epsilon$ for any partition P_2 with mesh size $\mu_2 \leq \delta_2$. Now choose $\delta = \min(\delta_1, \delta_2)$.

We now prove the existence of $\int_a^b g(x)dI_1$ by the Cauchy criterion by showing that for every partition $P \equiv \{x_i\}_{i=0}^n$ with mesh size $\mu \leq \delta$:

$$0 \leq U(g, I_1; P) - L(g, I_1; P) < C\epsilon, \tag{3}$$

for some constant C independent of ϵ. Equivalently, we prove that with $\Delta I_i \equiv I_1(x_i) - I_1(x_{i-1})$ and $\Delta F_i \equiv F(x_i) - F(x_{i-1})$, that both:

$$\sum_{i=1}^n [M_i - m_i] [\Delta I_i - |\Delta F_i|] < C\epsilon/2, \tag{4}$$

$$\sum_{i=1}^n [M_i - m_i] |\Delta F_i| < C\epsilon/2. \tag{5}$$

Since $g(x)$ is bounded, say $|g(x)| \leq M$ on $[a, b]$, and by definition of $I_1(x)$, statement (4) is obtained from (2):

$$\begin{aligned}
\sum_{i=1}^n [M_i - m_i] [\Delta I_i - |\Delta F_i|] &\leq 2M \sum_{i=1}^n [\Delta I_i - |\Delta F_i|] \\
&= 2M \left[T_a^b - \sum_{i=1}^n |\Delta F_i| \right] \\
&< 2M\epsilon.
\end{aligned}$$

For (5), define $\widetilde{x}_i, \widetilde{y}_i \in [x_{i-1}, x_i]$ so that $M_i - m_i \leq [g(\widetilde{x}_i) - g(\widetilde{y}_i)] + \epsilon$, and this is possible by the definition of M_i and m_i. Then from (1):

$$\begin{aligned}
\sum_{i=1}^n [M_i - m_i] |\Delta F_i| &\leq \sum_{i=1}^n [g(\widetilde{x}_i) - g(\widetilde{y}_i)] |\Delta F_i| + \epsilon \sum_{i=1}^n |\Delta F_i| \\
&\leq \left| \sum_{i=1}^n [g(\widetilde{x}_i') - g(\widetilde{y}_i')] \Delta F_i \right| + \epsilon T_a^b \\
&< \epsilon(1 + T_a^b).
\end{aligned}$$

In the second step, let $(\widetilde{x}_i', \widetilde{y}_i') = (\widetilde{x}_i, \widetilde{y}_i)$ if $\Delta F_i \geq 0$ and $(\widetilde{x}_i', \widetilde{y}_i') = (\widetilde{y}_i, \widetilde{x}_i)$ if $\Delta F_i < 0$. Thus (3) is satisfied with $C \equiv \max(2M, (1 + T_a^b))$.

Combining the pieces proves (3).

If $g(x)$ is continuous, then the integrals on the right in (4.25) exist for any increasing integrators $F_j(x)$ by Proposition 4.17, and similarly for $g(x)$ of bounded variation with continuous $F_i(x)$. ∎

4.2.5 Properties of the Integral

Properties of the Riemann-Stieltjes integral are summarized in the next few results, beginning with increasing integrators. We then derive an integration to the limit result.

Proposition 4.24 (Properties of the integral, Increasing $F(x)$) *Let $F(x)$ be an increasing function on $[a, b]$.*

1. *If $g_1(x)$ and $g_2(x)$ are integrable with respect to F, then so too is $cg_1(x) + eg_2(x)$ for real c, e, and:*

$$\int_a^b (cg_1(x) + eg_2(x))\, dF = c \int_a^b g_1(x)dF + e \int_a^b g_2(x)dF.$$

2. *If $g_1(x)$ and $g_2(x)$ are integrable with respect to F and $g_1(x) \leq g_2(x)$, then:*

$$\int_a^b g_1(x)dF \leq \int_a^b g_2(x)dF.$$

3. *If $g(x)$ is integrable with respect to F, then so too is $|g(x)|$ and:*

$$\left| \int_a^b g(x)dF \right| \leq \int_a^b |g(x)|\, dF.$$

4. *If $a < c < b$ and $g(x)$ is integrable with respect to F on $[a, b]$, then $g(x)$ is integrable on $[a, c]$ and $[c, b]$ and:*

$$\int_a^b g(x)dF = \int_a^c g(x)dF + \int_c^b g(x)dF.$$

5. *If $g(x)$ is integrable with respect to both increasing $F_1(x)$ and $F_2(x)$, then $g(x)$ is integrable with respect to $F(x) \equiv cF_1(x) + eF_2(x)$ for real c, e, and:*

$$\int_a^b g(x)dF = c \int_a^b g(x)dF_1 + e \int_a^b g(x)dF_2.$$

6. *If $g(x)$ is integrable with respect to $F(x)$, then $g(x)$ is integrable with respect to $\bar{F}(x) \equiv F(x) + c$ for any $c \in \mathbb{R}$, and:*

$$\int_a^b g(x)dF = \int_a^b g(x)d\bar{F}.$$

Proof. *We address these in turn, leaving some details as exercises. For this, we will freely use the various criteria afforded by Proposition 4.12 for the existence of the integral.*

1. *Integrability of $cg_1(x)$ and $eg_2(x)$ follow by integrability of $g_1(x)$ and $g_2(x)$ by (4.6), as does $\int_a^b cg_1(x)dF = c\int_a^b g_1(x)dF$. Details are left as an exercise.*

Given $\epsilon > 0$, let P_1 and P_2 be partitions defined for $g_1(x)$ and $g_2(x)$, respectively, with mesh size μ_j and that satisfy (4.18), and let P denote the common refinement of P_1 and P_2. By (4.12), and since $\inf cg_1 + \inf eg_2 \leq \inf(cg_1 + eg_2)$ and $\sup(cg_1 + eg_2) \leq \sup cg_1 + \sup eg_2$:

$$
\begin{aligned}
L(cg_1, F; P_1) + L(eg_2, F; P_2) &\leq L(cg_1, F; P) + L(eg_2, F; P) \\
&\leq L(cg_1 + eg_2, F; P) \\
&\leq U(cg_1 + eg_2, F; P) \\
&\leq U(cg_1, F; P) + U(eg_2, F; P) \\
&\leq U(cg_1, F; P_1) + U(eg_2, F; P_2) \\
&< L(cg_1, F; P_1) + L(eg_2, F; P_2) + 2\epsilon.
\end{aligned}
$$

Thus:

$$U(cg_1 + eg_2, F; P) - L(cg_1 + eg_2, F; P) < 2\epsilon,$$

and $cg_1(x) + eg_2(x)$ is integrable by Proposition 4.12.

This set of inequalities also assures that the limits of both $L(cg_1 + eg_2; P, F)$ and $U(cg_1 + eg_2; P, F)$ are within 2ϵ of $\int_a^b cg_1(x)dF + \int_a^b eg_2(x)dF$, and the proof is complete.

2. *For any partition P, $L(g_1, F; P) \leq L(g_2, F; P)$ and $U(g_1, F; P) \leq U(g_2, F; P)$. If $I_1 > I_2$, with I_j shorthand for the associated Riemann-Stieltjes integrals, let $\epsilon \equiv I_1 - I_2$. Choose partitions P_1 and P_2 so that (4.18) is satisfied with Darboux sums underlying I_1 and I_2, respectively, and $\epsilon/4$. That is, for $j = 1, 2$:*

$$U(g_j, F; P_j) - L(g_j, F; P_j) < \epsilon/4.$$

It then follows from (4.12) that the same inequalities are satisfied with P, the common refinement of P_1 and P_2. But then by construction and the definition of ϵ:

$$L(g_2, F; P) \leq I_2 \leq U(g_2, F; P) < L(g_1, F; P) \leq I_1 \leq U(g_1, F; P).$$

The middle inequality is a contradiction and hence $I_1 \leq I_2$.

3. *This result follows from items 1 and 2 once $|g(x)|$ is proved integrable, since then $g(x) \leq |g(x)|$ and $-g(x) \leq |g(x)|$ imply $\pm \int_a^b g(x)dF \leq \int_a^b |g(x)|\, dF$.*

That $|g(x)|$ is integrable is implied by (4.18) by noting that given any partition, the inequality $||x| - |y|| \leq |x - y|$ for real x, y yields:

$$U(|g|, F; P) - L(|g|, F; P) \leq U(g, F; P) - L(g, F; P).$$

4. *Let P be a partition of $[a, b]$ so that (4.18) is satisfied with given $\epsilon > 0$. Then it follows by (4.12) that (4.18) is also satisfied with P' defined to add point c to P if it is not already included. Define $P_1' = P' \bigcap [a, c]$ and $P_2' = P' \bigcap [c, b]$. Then $F(x_j) - F(x_{j-1}) \geq 0$ since $F(x)$ is increasing, and so by (4.18):*

$$
\begin{aligned}
\epsilon &> \sum_{P'} (M_j - m_j)(F(x_j) - F(x_{j-1})) \\
&\geq \max_{j=1,2} \sum_{P_j'} (M_j - m_j)(F(x_j) - F(x_{j-1})).
\end{aligned}
$$

Thus (4.18) is satisfied with ϵ for either of these subinterval sums and thus both subinterval integrals exist.

That the value of the integral over $[a,b]$ equals the sum of the subinterval integrals is implied by part 1, defining $g_1(x) = g(x)$ on $I_1 \equiv [a,c]$ and 0 elsewhere, and $g_2(x) = g(x)$ on $I_2 \equiv [c,b]$ and 0 elsewhere, and showing that the integral of $g(x)$ over I_j equals the integral of $g_j(x)$ over I_j. This last step is an exercise.

5. *For any partition P of $[a,b]$:*

$$R(g, F; P) = cR(g, F_1; P) + eR(g, F_2; P).$$

Since $g(x)$ is integrable with respect to both $F_1(x)$ and $F_2(x)$, we obtain that:

$$\lim_{\mu \to 0} R(g, F; P) = c \int_a^b g(x)dF_1 + e \int_a^b g(x)dF_2.$$

Thus $g(x)$ is integrable with respect to $F(x)$ by Definition 4.3, and the equality of integrals is proved.

6. *This result follows from part 5 noting that $\int_a^b g(x)dc = 0$ by (4.6).*

∎

We now summarize properties for integrators of bounded variation.

Proposition 4.25 (Properties of the integral, $F(x) \in B.V.$) *Let $F(x)$ be of bounded variation on $[a,b]$.*

1. *If $g_1(x)$ and $g_2(x)$ are integrable with respect to F, then so too is $cg_1(x) + eg_2(x)$ for real c, e, and:*

$$\int_a^b \left(cg_1(x) + eg_2(x) \right) dF = c \int_a^b g_1(x)dF + e \int_a^b g_2(x)dF.$$

2. *If $g(x)$ is integrable with respect to F, then so too is $|g(x)|$ and:*

$$\left| \int_a^b g(x)dF \right| \leq \int_a^b |g(x)| \, dT,$$

where $T(x) \equiv T_a^x$, the (increasing) total variation function of F over $[a,x]$.

3. *If $a < c < b$ and $g(x)$ is integrable with respect to F on $[a,b]$, then $g(x)$ is integrable on $[a,c]$ and $[c,b]$ and:*

$$\int_a^b g(x)dF = \int_a^c g(x)dF + \int_c^b g(x)dF.$$

4. *If $g(x)$ is integrable with respect to both bounded variation functions $F_1(x)$ and $F_2(x)$, then $g(x)$ is integrable with respect to $F(x) \equiv cF_1(x) + eF_2(x)$ for real c, d, and:*

$$\int_a^b g(x)dF = c \int_a^b g(x)dF_1 + e \int_a^b g(x)dF_2.$$

Proof. *Items 1, 3, and 4 are corollaries to items 1, 4, and 5 of Proposition 4.24, using (4.24).*

For item 2, it follows from (4.24) that:

$$\left| \int_a^b g(x)dF \right| \le \left| \int_a^b g(x)dI_1 \right| + \left| \int_a^b g(x)dI_2 \right|.$$

An application of items 3 then 5 of Proposition 4.24 then obtains that $|g(x)|$ is integrable with respect to $I_1(x) + I_2(x)$ and:

$$\left| \int_a^b g(x)dF \right| \le \int_a^b |g(x)| \, d\,(I_1 + I_2).$$

Now let $I_1(x) \equiv F(a) + P_a^x$ and $I_2(x) \equiv N_a^x$ by (3.21). Since $I_1(x) + I_2(x) = F(a) + T_a^x$ by Proposition 3.27, it follows from item 6 of Proposition 4.24:

$$\left| \int_a^b g(x)dF \right| \le \int_a^b |g(x)| \, dT.$$

■

Finally, we consider arbitrary integrators $F(x)$, meaning not necessarily increasing or of bounded variation. We identify two properties. It will be noted that in this general case, we can again deduce integrability of the linear combination in item 1 as above, but must assume subinterval integrability in item 2.

Proposition 4.26 (Properties of the integral, General $F(x)$) *Let $F(x)$ be a function on $[a, b]$.*

1. *If $g_1(x)$ and $g_2(x)$ are integrable with respect to F, then so too is $cg_1(x) + eg_2(x)$ for real c, e, and:*

$$\int_a^b (cg_1(x) + eg_2(x))\, dF = c \int_a^b g_1(x)dF + e \int_a^b g_2(x)dF.$$

2. *If $a < c < b$ and $g(x)$ is integrable with respect to F on $[a, b]$, $[a, c]$, and $[c, b]$, then:*

$$\int_a^b g(x)dF = \int_a^c g(x)dF + \int_c^b g(x)dF.$$

Proof. *For item 1, the case $c = e = 0$ needs no discussion so we assume that $|c| + |e| > 0$. Definition 4.3 obtains that given $\epsilon > 0$ there exists δ_j so that $|R(g_j, F; P) - I_j| < \epsilon/\,(|c| + |e|)$ for all tagged partitions P of mesh size $\mu \le \delta_j$, where $I_j \equiv \int_a^b g_j(x)dF$. Thus for all tagged partitions with $\mu \le \min \delta_j$, since $R(g, F; P)$ is linear in g by definition:*

$$|R(cg_1 + eg_2, F; P) - (cI_1 + eI_2)| \quad < \quad |c|\,|R(g_1, F; P) - I_1| + |e|\,|R(g_2, F; P) - I_2|$$
$$< \quad \epsilon.$$

Hence, $cg_1(x) + eg_2(x)$ is integrable and has value as stated.

For item 2, given $\epsilon > 0$ there exists δ, so that:

$$|R(g, F; P_{[a,b]}) - I| < \epsilon/2, \tag{1}$$

for all tagged partitions $P_{[a,b]}$ of $[a,b]$ of mesh size $\mu \leq \delta$, where $I \equiv \int_a^b g(x)dF$. Similarly, there exists δ_1 so that $\left| R(g,F;P_{[a,c]}) - I_1 \right| < \epsilon/4$, and δ_2 so that $\left| R(g,F;P_{[c,b]}) - I_2 \right| < \epsilon/4$, for all tagged partitions of the respective intervals with respective mesh sizes, where I_j are the associated integrals.

If $P'_{[a,b]}$ is any tagged partition of $[a,b]$ that contains $\{c\}$ as a partition point and with mesh size $\mu \leq \min(\delta, \delta_1, \delta_2)$, then this induces tagged partitions $P_{[a,c]}$ and $P_{[c,b]}$ with the same or smaller mesh size, and thus by splitting the Riemann-Stieltjes sum:

$$\left| R(g,F;P'_{[a,b]}) - (I_1 + I_2) \right| \leq \left| R(g,F;P_{[a,c]}) - I_1 \right| + \left| R(g,F;P_{[c,b]}) - I_2 \right| < \epsilon/2. \quad (2)$$

Thus by (1) and (2) and the triangle inequality:

$$\begin{aligned} |I - (I_1 + I_2)| &\leq \left| R(g,F;P'_{[a,b]}) - (I_1 + I_2) \right| \\ &\quad + \left| R(g,F;P'_{[a,b]}) - I \right| \\ &< \epsilon. \end{aligned}$$

As this is true for all $\epsilon > 0$, the result follows. ∎

We end the section with a result on integration to the limit. While stated for a bounded variation integrator, this result also applies to increasing integrators by Example 3.26.

Proposition 4.27 (On continuous $g_m(x) \to g(x)$, $F(x) \in B.V.[a,b]$) If $\{g_m(x)\}_{m=1}^{\infty}$ and $g(x)$ are continuous with $g_m(x) \to g(x)$ uniformly on $[a,b]$, and $F(x)$ is of bounded variation on this interval, then:

$$\int_a^b g_m(x)dF \to \int_a^b g(x)dF,$$

and:

$$\int_a^b |g_m(x) - g(x)| \, dF \to 0.$$

Proof. *Existence of all integrals is assured by Proposition 4.17. Then using the properties of Proposition 4.25 and (4.20):*

$$\left| \int_a^b g_m(x)dF - \int_a^b g(x)dF \right| = \left| \int_a^b (g_m(x) - g(x)) \, dF \right| \leq M_{(g_m - g)} T_a^b(F),$$

where $M_{(g_m - g)} \equiv \sup_{[a,b]} |g_m(x) - g(x)|$. Uniform convergence of $g_m(x) \to g(x)$ implies $M_{(g_m - g)} \to 0$, and the convergence of integrals is verified.

Similarly:

$$\left| \int_a^b |g_m(x) - g(x)| \, dF \right| \leq M_{(g_m - g)} T_a^b(F),$$

completing the proof. ∎

4.2.6 Evaluating Riemann-Stieltjes Integrals

In Chapter IV.1, a complete characterization of probability distribution functions on $\mathbb{R}$ will be developed. There it will be seen that the most general distribution function on $\mathbb{R}$ is a sum of up to three component parts:

1. An **increasing absolutely continuous** function: Studied in Chapter 3, this class of functions contains, but is not equivalent to, the class of continuously differentiable functions. This latter collection plays a prominent role as the distribution functions of continuous probability theory, with derivatives that are the continuous and thus Riemann integrable density functions. More generally, an absolutely continuous function is differentiable almost everywhere, and this derivative is Lebesgue measurable and thus serves as a density function with respect to Lebesgue integration (Proposition 3.62).

2. An **increasing saltus** function: Though these will initially be defined more generally, for this decomposition, saltus functions are the distribution functions of discrete probability theory.

3. A **singular** function: Again studied in Chapter 3, these are continuous, increasing functions with $f'(x) = 0$ almost everywhere.

The next result addresses the evaluation of Riemann-Stieltjes integrals in $\mathbb{R}$ in the two special cases of $F(x)$ most commonly encountered as a distribution function in probability theory and discussed in Books IV and VI. The first case addresses increasing, continuously differentiable integrators of item 1 above. The second case reflects item 2, and $F(x)$ is defined to be right continuous to be consistent with distribution functions in discrete probability theory.

The first part of case 1 is stated in the more general context of integrators of bounded variation, but this class includes increasing functions by Example 3.26.

Proposition 4.28 ($F(x) \in B.V.[a,b]$, $F'(x)$ **continuous, or,** $F(x)$ **a step function**) *Let $g(x)$ be continuous on $[a,b]$.*

1. *If $F(x)$ is of bounded variation on $[a,b]$ with $F'(x) = f(x)$ continuous on $[a,b]$, then:*

$$\int_a^b g(x)dF = (\mathcal{R}) \int_a^b g(x)f(x)dx, \tag{4.26}$$

where $(\mathcal{R})$ denotes that this is a Riemann integral.

If $F(x)$ is increasing and bounded, and $g(x)$ is bounded, then (4.26) remains true for $a = -\infty$ and/or $b = \infty$.

If $F(x)$ is increasing and bounded, and $g(x)$ is unbounded, then (4.26) remains true for $a = -\infty$ and/or $b = \infty$ if the improper Riemann integral exists for $a = -\infty$ and/or $b = \infty$.

2. *Let $\{y_j\}_{j=1}^N \subset \mathbb{R}$ with $N \leq \infty$, where it is assumed that $\{y_j\}_{j=1}^N$ has no accumulation points if $N = \infty$. Given $f(x)$ with nonnegative $\{f(y_j)\}_{j=1}^N$, assume that the right continuous, increasing step function:*

$$F(x) = \sum_{y_j \leq x} f(y_j),$$

is finite for all x.

Then:

$$\int_a^b g(x)dF = \sum_{y_j \in (a,b]} g(y_j)f(y_j). \tag{4.27}$$

If $g(x)$ and $F(x)$ are bounded, then (4.27) remains true for $a = -\infty$ and/or $b = \infty$.

For unbounded $g(x)$, if the summation in (4.27) is absolutely convergent for $a = -\infty$ and/or $b = \infty$, then the associated improper Riemann-Stieltjes integral is well defined.

Proof. *In both cases, the existence of $\int_a^b g(x)dF$ is assured by Proposition 4.17 for bounded $[a, b]$.*

For item 1, let $\epsilon > 0$ be given. By Definition 4.3 and (4.7), choose δ so that for any $\mu \le \delta$ and arbitrary $\widetilde{x}_i \in [x_i, x_{i-1}]$:

$$\left| \sum_{i=1}^n g(\widetilde{x}_i)(F(x_i) - F(x_{i-1})) - \int_a^b g(x)dF \right| < \epsilon/2.$$

Since $F(x)$ is continuously differentiable with $F'(x) = f(x)$, the mean value theorem obtains $F(x_i) - F(x_{i-1}) = f(x_i')(x_i - x_{i-1})$ for some $x_i' \in (x_i, x_{i-1})$. Since $\widetilde{x}_i$ can be arbitrarily chosen, choose $\widetilde{x}_i = x_i'$ and thus:

$$\left| \sum_{i=1}^n g(x_i')f(x_i')(x_i - x_{i-1}) - \int_a^b g(x)dF \right| < \epsilon/2.$$

Now $g(x)f(x)$ is continuous and thus Riemann integrable by Proposition 1.15, so given this ϵ there exists δ' by Definition 1.2 so that for any partition with $\mu \le \delta'$ and arbitrary $\widetilde{x}_i \in (x_i, x_{i-1})$:

$$\left| \sum_{i=1}^n g(\widetilde{x}_i)f(\widetilde{x}_i)(x_i - x_{i-1}) - (\mathcal{R})\int_a^b g(x)f(x)dx \right| < \epsilon/2.$$

Once again $\widetilde{x}_i$ here can be chosen to equal the above x_i' from the mean value theorem.

Thus for $\mu \le \min(\delta, \delta')$, the triangle inequality yields:

$$\left| \int_a^b g(x)dF - (\mathcal{R})\int_a^b g(x)f(x)dx \right| < \epsilon,$$

which is (4.27).

When $g(x)$ and $F(x)$ are bounded, $\int_a^b g(x)dF$ exists for $a = -\infty$ and/or $b = \infty$ by Proposition 4.21. Thus, for example in the general case $a = -\infty$ and $b = \infty$:

$$\lim_{\substack{a \to -\infty \\ b \to \infty}} \int_a^b g(x)dF \equiv \int_{-\infty}^{\infty} g(x)dF,$$

so this limit exists and is finite. But then by (4.26), the limit of Riemann integrals exists and is finite:

$$\lim_{\substack{a \to -\infty \\ b \to \infty}} (\mathcal{R})\int_a^b g(x)f(x)dx \equiv (\mathcal{R})\int_{-\infty}^{\infty} g(x)f(x)dx.$$

For unbounded $g(x)$, if the Riemann integral is well defined as an improper integral, then the same logic assures the existence of the improper Riemann-Stieltjes integral.

For item 2, again by Definition 4.3 and (4.7), for any $\epsilon > 0$ there exists δ so that for any $\mu \le \delta$ and arbitrary $\widetilde{x}_i \in [x_i, x_{i-1}]$:

$$\left| \sum_{i=1}^n g(\widetilde{x}_i)(F(x_i) - F(x_{i-1})) - \int_a^b g(x)dF \right| < \epsilon.$$

The terms of this summation are nonzero only for intervals with $x_{i-1} < y_j < x_i$ for some j. Choosing μ small enough assures that there is at most one such y_j in each partition *interval since* $\{y_j\}_{j=1}^{N}$ *has no accumulation points. Then choosing* $\widetilde{x}_i = y_j$, *and noting that* $F(x_i) - F(x_{i-1}) = f(y_j)$ *completes the proof of (4.27).*

The statements on improper integrals only need discussion when $N = \infty$. *Then* $F(x)$ *bounded implies* $\sum_{j=1}^{\infty} f(y_j) < \infty$. *Hence, if* $g(x)$ *is bounded, then* $\sum_{y_j \in (a,b]} g(y_j) f(y_j)$ *converges absolutely in the cases* $a = -\infty$ *and/or* $b = \infty$. *Thus the validity of (4.27) over bounded intervals, plus the convergence of both expressions as* $a \to -\infty$ *and/or* $b \to \infty$, *completes the proof for bounded* $g(x)$. *The case for unbounded* $g(x)$ *is similar.* ∎

Remark 4.29 (Integration by parts) *Recall Proposition 4.14, which stated that if a bounded function* $g(x)$ *is Riemann-Stieltjes integrable with respect to* $f(x)$ *over* $[a, b]$, *then* $f(x)$ *is Riemann-Stieltjes integrable with respect to* $g(x)$, *and:*

$$\int_a^b g(x)df + \int_a^b f(x)dg = f(b)g(b) - f(a)g(a).$$

This was called a result in integration by parts, although this was not apparent in this formulation.

When $g(x)$ *and* $f(x)$ *are continuously differentiable, this result can be expressed using (4.26):*

$$(\mathcal{R}) \int_a^b g(x)f'(x)dx + (\mathcal{R}) \int_a^b f(x)g'(x)dx = f(b)g(b) - f(a)g(a),$$

which is (1.30).

We end this section with a result related to the fundamental theorem of calculus, Version II.

Corollary 4.30 (FTC, Version II) *Let* $F(x)$ *be an increasing, continuous function on* $[a, b]$ *that is continuously differentiable on* (a, b). *If* $g(x)$ *is continuous, define:*

$$G(x) \equiv G(a) + \int_a^x g(y)dF,$$

for arbitrary $G(a)$. *Then* $G(x)$ *is differentiable on* (a, b) *and:*

$$G'(x) = g(x)F'(x). \tag{4.28}$$

Further, if $g(x)$ *and* $F(x)$ *are bounded, then (4.28) remains true for* $G(x)$ *defined with* $a = -\infty$.

Proof. *For bounded* $[a, b]$ *this is Proposition 1.33 applied to (4.26) with* $b = x$.

For $a = -\infty$, *Proposition 1.33 requires that* $F'(x)$ *be Riemann integrable. This follows from the (4.26) with* $g(x) \equiv 1$, *since if* $F(x)$ *is bounded this implies* $\int_{-\infty}^{\infty} dF$ *is finite by Proposition 4.21.* ∎

4.3   A Result on R-S Integrators

In Remark 4.22, the existence result of L. C. Young was noted:

The integral $\int_a^b g(x)dF$ *exists if* $g(x)$ *has bounded p-variation and* $F(x)$ *has bounded q-variation on* $[a, b]$ *with* $1/p + 1/q > 1$, *and,* $g(x)$ *and* $F(x)$ *have no common discontinuities.*

When $q = 1$, meaning that $F(x)$ is of bounded variation, this integral therefore exists for all $g(x)$ of bounded p-variation for any $p > 0$, subject to the no common discontinuities constraint.

Exercise 4.19 obtains a result on existence when $F(x)$ is of bounded variation, that $\int_a^b g(x)dF$ exists for all continuous $g(x)$.

Given Young's result, it is natural to wonder if this integral can exist for all continuous $g(x)$ using integrators $F(x)$ of bounded q-variation with $q \neq 1$? In other words, is it necessary for $F(x)$ to be of bounded variation in order for this integral to be defined for all continuous $g(x)$? Perhaps surprisingly, Proposition 4.62 proves that integrators of bounded variation are in fact necessary to integrate all continuous functions.

In detail, Proposition 4.62 states:

Given bounded $F(x)$ on $[a, b]$, if $\int_a^b g(x)dF$ exists for all continuous $g(x)$, then $F(x)$ is of bounded variation.

This result will play an instrumental role in motivating the need for a new definition of "integral" when the integrator is a Brownian motion, which is not of bounded variation. See Books VII–VIII for details.

While the formal proof of this result is quite subtle, and indeed somewhat long, the intuition behind this conclusion is easy to obtain. Given a partition $P_n \equiv \left\{x_i^{(n)}\right\}_{i=0}^{m_n}$ of $[a, b]$, define a continuous function $g_n(x)$ on this interval with $|g_n(x)| \leq 1$, and for tags $\widetilde{x}_i^{(n)} \in \left(x_{i-1}^{(n)}, x_i^{(n)}\right)$ define $g_n(\widetilde{x}_i^{(n)}) = 1$ if $F(x_i^{(n)}) - F(x_{i-1}^{(n)}) > 0$, and $g_n(\widetilde{x}_i^{(n)}) = -1$ if $F(x_i^{(n)}) - F(x_{i-1}^{(n)}) < 0$. Extending $g_n(x)$ to be continuous on $[a, b]$:

$$\sum_{i=1}^{m_n} g(\widetilde{x}_i^{(n)})(F(x_i^{(n)}) - F(x_{i-1}^{(n)})) = \sum_{i=1}^{m_n} \left| F(x_i^{(n)}) - F(x_{i-1}^{(n)}) \right| \equiv t_n. \tag{1}$$

Because every $g_n(x)$ is continuous, the integral $\int_a^b g_n(x)dF$ exists by assumption. Since t_n in (1) is a Riemann-Stieltjes sum, it follows that $t_n < \infty$ for all n.

Unfortunately, this falls short of compelling the conclusion that $F(x)$ must be of bounded variation, meaning that $T \equiv \sup t_n < \infty$. The problem is that we have not constructed a **single integrand** $g(x)$ for which these sums obtain such t_n, and by which it could be concluded from integrability that $\sup t_n < \infty$. To prove this important result in fact requires a general theory that will in the current context appear to be a case of mathematical overkill.

The proof will require the following steps. Collectively, these results will provide important details on a space of continuous functions denoted $C([a, b], \|\cdot\|_\infty)$, representing the space of integrands, and on bounded linear functionals on this space, noting that a Riemann-Stieltjes sum is such a functional.

The subsections needed for this result are as follows:

- **Banach Spaces** introduces various notions on **vector spaces** V. This includes **norms** $\|\cdot\|$ on these spaces that generalize $|x|$ for $x \in \mathbb{R}^n$, and the notion of **completeness**, which assures that if $\{x_n\}_{n=1}^\infty \subset V$ converges by the **Cauchy criterion**, then this sequence converges to a point in V. A **Banach space** is then defined as a complete, normed, vector space.

  Simple examples include $\mathbb{R}$ and $\mathbb{R}^n$, as well as $C([a, b], \|\cdot\|_\infty)$ above, where the norm $\|\cdot\|_\infty$ is defined by $\|f\|_\infty \equiv \sup_{x \in [a,b]} |f(x)|$.

- **Baire's category theorem** introduces topological concepts on normed vector spaces such as **open** and **closed** sets, as well as **dense** and **nowhere dense** sets. A dense set is analogous to the rationals, in the sense that rationals are arbitrarily close to all elements of $\mathbb{R}$. A nowhere dense set is intuitively sparse, so sparse in fact, that even if you add all limit points, it still contains no open set. The integers $\mathbb{Z}$ are countable and nowhere dense in $\mathbb{R}$. The Cantor set is uncountable and nowhere dense in $[0, 1]$.

Baire's famous category theorem states that a complete metric space cannot equal a countable union of nowhere dense sets. Thus $\mathbb{R}$ cannot equal a countable union of sets like $\mathbb{Z}$, and $[0, 1]$ cannot equal a countable union of Cantor-like sets.

- **Banach-Steinhaus theorem** begins with a study of **linear transformations** on normed vector spaces, and properties of such transformations such as **boundedness,** and introduces the **norm of a transformation**. It is then proved that given any tagged partition $P \equiv (\{x_i\}_{i=0}^n, \{\widetilde{x}_i\}_{i=1}^n)$ of $[a, b]$, that we can define a linear transformation T_P on $C([a, b], \|\cdot\|_\infty)$ by the Riemann-Stieltjes summation of (4.5), with range in the normed space $(\mathbb{R}, |\cdot|)$. This is in fact a bounded linear transformation, and its norm $\|T_P\|$ is given by the variation t of the integrator $F(x)$ from Definition 3.23 using the partition $\{x_i\}_{i=0}^n$. Further, for any $g \in C([a, b], \|\cdot\|_\infty)$, the set $\{|T_P g|\}_P$ is bounded where P ranges over all tagged partitions of $[a, b]$.

  The **Banach-Steinhaus theorem** states that given any family of bounded linear transformations $\{T_\alpha\}_{\alpha \in I}$, that the collection of norms $\{\|T_\alpha\|\}_{\alpha \in I}$ is bounded if and only if the collection $\{\|T_\alpha g\|\}_{\alpha \in I}$ is bounded for every g in the domain vector space. Here $\|T_\alpha g\|$ is defined using the norm in the range space, so in the above Riemann-Stieltjes result, $\|T_\alpha g\| \equiv |T_P g|$. Prominent in the proof of this result is Baire's category theorem.

- **Final Result on R-S integrators** assembles the pieces. Defining I as the collection of all tagged partitions of $[a, b]$, it has been shown in the prior step that $\{|T_\alpha g|\}_{\alpha \in I}$ is bounded for every $g \in C([a, b], \|\cdot\|_\infty)$. The Banach-Steinhaus theorem then assures that $\{\|T_\alpha\|\}_{\alpha \in I}$ is bounded. Recalling that $\|T_\alpha\|$ equals the variation t of the integrator $F(x)$ from Definition 3.23 with tagged partition indexed by α, it follows that $\{\|T_\alpha\|\}_{\alpha \in I}$ is bounded if and only if $\sup_\Pi t < \infty$, and this is the definition that $F \in B.V.$

Given this summary, the reader may well choose to skip the remaining details of this section until their interest in Book VIII and stochastic integration motivates a deeper curiosity in the key result ultimately developed below. However, the reader is encouraged to at least scan the logic flow of the development, as there are many useful concepts that will arise again in later books.

4.3.1   Banach Spaces

The notion of a **Banach space**, is named after **Stefan Banach** (1892–1945) who defined this structure and systematically studied its properties. It will reappear in Book V in a very different context, a small testament to the generality of this notion. In essence, a Banach space is a complete, normed vector space, a simple example of which is the vector space $\mathbb{R}^n$ endowed with the standard **Euclidean norm:**

$$|x| \equiv \left[\sum\nolimits_{j=1}^n x_j^2 \right]^{1/2}. \tag{4.29}$$

Sometimes in $\mathbb{R}^n$, and nearly always in other Banach spaces, the norm of an element x is denoted $\|x\|$, and then often with a subscript for added definitional detail.

References for this subject in order of increasing completeness and abstraction are **Reitano** (2010), **Royden** (1971), **Rudin** (1974), and **Hewitt and Stromberg** (1965). These references also provide other collections of Banach spaces based on norms defined in terms of integrals, a topic addressed in Book V.

For completeness, we begin with a collection of definitions. There is a lot of formality in the notion of a vector space, but the goal is to abstract all the operations one takes for granted in the familiar vector space of $\mathbb{R}^n$, which is formally a vector space over the real field $\mathbb{R}$. The reader may find it useful to translate these abstract definitions to the familiar context of $\mathbb{R}^n$ to provide concreteness and familiarity.

The underlying field $\mathcal{F}$ is notationally general in the definition, but in these books, $\mathcal{F}$ is always $\mathbb{R}$ or $\mathbb{C}$. The general definition will identify familiar properties of these special fields.

Definition 4.31 (Field) *A field $\mathcal{F}$ is a collection of elements on which is defined an **addition**, denoted $+$, and a **multiplication**, denoted $\cdot$ or simply by juxtaposition, with the following properties:*

1. **Closure:** *If $\alpha, \beta \in \mathcal{F}$ then $\alpha + \beta \in \mathcal{F}$ and $\alpha\beta \in \mathcal{F}$.*

2. **Commutativity:** *If $\alpha, \beta \in \mathcal{F}$ then $\alpha + \beta = \beta + \alpha$ and $\alpha\beta = \beta\alpha$.*

3. **Units:** *There exists $0, 1 \in \mathcal{F}$ so that for all $\alpha \in \mathcal{F}$, $\alpha + 0 = \alpha$ and $1\alpha = \alpha$.*

4. **Inverses:** *For all $\alpha \in \mathcal{F}$ there exists $-\alpha$ so that $-\alpha + \alpha = 0$; For $\alpha \neq 0$ there exists α^{-1} so that $\alpha^{-1}\alpha = 1$.*

5. **Associativity:** *If $\alpha, \beta, \gamma \in \mathcal{F}$ then $(\alpha + \beta) + \gamma = \alpha + (\beta + \gamma)$ and $(\alpha\beta)\gamma = \alpha(\beta\gamma)$.*

6. **Distributivity:** *If $\alpha, \beta, \gamma \in \mathcal{F}$ then $\alpha(\beta + \gamma) = \alpha\beta + \alpha\gamma$.*

Exercise 4.32 (Fields $\mathbb{Q}$ and $\mathbb{C}$, not $\mathbb{Z}$) *Besides $\mathbb{R}$, for which the above definition is seen to apply, the reader should confirm that both the rationals $\mathbb{Q}$, and the complex numbers $\mathbb{C}$, are indeed fields, while the integers $\mathbb{Z}$ are not.*

In the same way that the notion of a field abstracts the familiar properties of $\mathbb{R}$, a vector space abstracts the familiar properties of $\mathbb{R}^n$, when one identifies n-tuples $(x_1, ..., x_n)$ with the vector x. This is the view of $\mathbb{R}^n$ that is prominent in the subject of linear algebra.

Definition 4.33 (Vector space) *A space V is called a **vector space over a field** $\mathcal{F}$, and also called a **linear space over a field** $\mathcal{F}$, if addition of elements of V is defined, as is multiplication of elements of V by elements of $\mathcal{F}$, called **scalar multiplication**, and these operations satisfy the following properties:*

1. **Zero Vector:** *There is an element $\theta \in V$ so that $x + \theta = x$ for all $x \in V$. Often θ is denoted 0 and this rarely creates ambiguity.*

2. *For all $x, y, z \in V$:*

    a. **Closure:** *$x + y \in V$;*

    b. **Commutativity:** *$x + y = y + x$;*

    c. **Associativity:** *$(x + y) + z = y + (x + z)$.*

3. **Scalar Multiplication:** *For all $x, y \in V$, $\alpha, \beta \in \mathcal{F}$:*

    a. *$\alpha x \in V$;*

    b. *$\alpha(x + y) = \alpha x + \alpha y$;*

    c. *$(\alpha + \beta)x = \alpha x + \beta x$;*

    d. *$\alpha(\beta x) = (\alpha\beta)x$.*

4. For all $x \in V$:

    a. $0x = \theta$, where "0" denotes the additive unit in $\mathcal{F}$;

    b. $1x = x$, where "1" denotes the multiplicative unit in $\mathcal{F}$.

Remark 4.34 (On additive inverses; zero vector) *Vector spaces contain additive inverses: $-x \equiv -1x$, where -1 is the additive inverse of 1 in $\mathcal{F}$. That $x + (-x) = \theta$, the zero vector, follows from 3.c and 4.a.*

    In addition, the zero vector θ is unique since if also $x = x + \theta'$ for all x, then with $x = \theta$ obtains $\theta = \theta + \theta' = \theta'$ by 1.

Exercise 4.35 (Vector spaces $\mathbb{R}^n$, $\mathbb{C}^n$, and maybe $\mathbb{Q}^n$) *Check that $\mathbb{R}^n$ is a vector space over the field $\mathbb{R}$ but not over the field $\mathbb{C}$, and $\mathbb{C}^n$ is a vector space over $\mathbb{C}$ or $\mathbb{R}$.*

    Is $\mathbb{Q}^n$, the space of n-tuples of rationals, a vector space over any of the fields: $\mathbb{Q}$, $\mathbb{R}$, or $\mathbb{C}$?

While a space like $\mathbb{R}^n$ is indeed a vector space, it is so much more. For one thing, it is a vector space where one has a notion of the size or norm of a vector, as well as of the distance between vectors.

Definition 4.36 (Normed vector space) *A vector space V over $\mathcal{F} \in \{\mathbb{R}, \mathbb{C}\}$ is called a **normed vector space** or **normed linear space** or **normed space**, if there exists a functional denoted $\|\cdot\| : V \to \mathbb{R}^+$, so that:*

1. $\|x\| = 0$ if and only if $x = 0$.

*2. **Homogeneity:** $\|\alpha x\| = |\alpha| \, \|x\|$ for all $x \in V$ and $\alpha \in \mathcal{F}$, where $|\cdot|$ denotes the norm in $\mathbb{R}$ or $\mathbb{C}$.*

*3. **Triangle inequality:** $\|x + y\| \leq \|x\| + \|y\|$ for all $x, y \in V$.*

Exercise 4.37 (Norms on $\mathbb{R}^n$, $\mathbb{C}^n$) *Prove that $\mathbb{R}^n$ is a normed space with respect to (4.29), while $\mathbb{C}^n$ is a normed space relative to:*

$$|z| \equiv \left[\sum\nolimits_{j=1}^{n} z_j \bar{z}_j \right]^{1/2}. \tag{4.30}$$

*Here, if $z = a + bi$, then $\bar{z}$ denotes the **complex conjugate of** z:*

$$\bar{z} = a - bi.$$

    Then check that with (4.31), that $d(x,y)$ so defined is the familiar distance function on these spaces.

Remark 4.38 (Normed space $\Rightarrow$ Metric space) *A simple but important observation is that a normed vector space $(V, \|\cdot\|)$ is also a **metric space** (V, d) with **distance function** or **metric** $d(x, y)$ **induced by** $\|\cdot\|$. This distance function is defined by:*

$$d(x, y) \equiv \|x - y\|. \tag{4.31}$$

Definition 4.39 (Metric; Induced metric) *For the induced $d(x,y)$ in (4.31), or any functional $d : V \times V \to \mathbb{R}$ to be a **metric** on V, requires four properties:*

*1. **Nonnegativity:** $d(x, y) \geq 0$.*

*2. **Positivity:** $d(x, y) = 0$ if and only if $x = y$.*

3. Symmetry: $d(x, y) = d(y, x)$.

4. Triangle inequality: *For any* x, y, z:

$$d(x, y) \leq d(x, z) + d(z, y). \tag{4.32}$$

Example 4.40 (Induced metric is a metric) *Unsurprisingly, the induced metric of (4.31) is indeed a metric.*

1. *The range of* $\|\cdot\|$ *is* $\mathbb{R}^+$ *which assures this.*

2. *This is norm property 1.*

3. *This is norm property 2 with* $\alpha = -1$.

4. *This is norm property 3 with* $x' = x - z$, $y' = z - y$ *and thus* $x' + y' = x - y$.

Though not required of a metric in general, it is the case that the **induced metric** *or* **distance function** *on a normed vector space in (4.31) also satisfies:*

5. **Homogeneity:** $d(\alpha x, \alpha y) = |\alpha|\, d(x, y)$ *for* $\alpha \in \mathcal{F}$.
   This is norm property 2.

6. **Translation invariance:** *For any* x, y, z:

$$d(x + z, y + z) = d(x, y).$$

This is true by definition in (4.31).

Remark 4.41 (Metric space $\not\Rightarrow$ Normed space) *For completeness, we note that while every normed vector space is a metric vector space using (4.31), not all metric vector spaces are normed vector spaces. Thus a metric vector space is the more general concept.*

One can define a metric on any collection of points, not just on vector spaces. A norm on a space V *requires that there are arithmetic operations on* V *by definition.*

Exercise 4.42 *Check that a metric on a vector space obtains a normed vector space if the metric* d *is homogeneous and translation invariant as defined above. Then* $\|x\| \equiv d(x, 0)$ *is a norm.*

In a normed vector space, one can consider the notion of **convergence of a sequence** $s_n \to s$, using the standard definition of convergence in $\mathbb{R}$, and substituting $\|s_n - s\|$ for $|s_n - s|$.

Definition 4.43 (Complete Normed space) *Given a normed linear space* $(V, \|\cdot\|)$, *a sequence* $\{s_n\}_{n=1}^{\infty} \subset V$ *is called a* **Cauchy sequence** *if it satisfies the* **Cauchy Criterion:**
For any $\epsilon > 0$ *there is an* $N \in \mathbb{N}$ *so that* $\|s_n - s_m\| < \epsilon$ *for all* $n, m \geq N$.

A normed linear space is **complete** *if given any Cauchy sequence* $\{s_n\}_{n=1}^{\infty} \subset V$, *there is an* $s \in V$ *so that* $\|s_n - s\| \to 0$. *In other words, for any* $\epsilon > 0$ *there is an* $N \in \mathbb{N}$ *so that* $\|s_n - s\| < \epsilon$ *for all* $n \geq N$.

This is then denoted as $s_n \to s$ *as well as* $s = \lim_{n \to \infty} s_n$.

The notion of a Cauchy sequence is named after **Augustin-Louis Cauchy** (1789–1857) and is useful because it provides a criterion for convergence that does not depend on knowing the limit of the sequence. The reader is invited to rethink the Cauchy criteria in the various integration results from this point of view, Namely, that this criterion proved that the sequences of Riemann or Riemann-Stieltjes sums and Darboux sums converged, without knowing to what they converged.

Thus in a complete space, proving that a sequence satisfies the Cauchy criterion provides a proof of the **existence** of a limit in that space.

Putting all these ingredients together, we arrive at the **Banach space.** We leave it as an exercise to convince yourself that $\mathbb{R}^n$ is a real Banach space, and $\mathbb{C}^n$ is a complex Banach space, using the above references as needed.

Definition 4.44 (Banach space) *A **Banach space** is a complete normed linear space. A **real Banach space** is a complete normed linear space over $\mathcal{F} = \mathbb{R}$, and analogously, a **complex Banach space** is a complete normed linear space over $\mathcal{F} = \mathbb{C}$.*

We now focus on the first step of the application in hand. Recall from the above discussion that $C([a,b], \|\cdot\|_\infty)$ will be the space of integrands for the Riemann-Stieltjes integrals we are investigating.

Proposition 4.45 $(C([a,b], \|\cdot\|_\infty)$ **is a real Banach space)** *Let* $C([a,b], \|\cdot\|_\infty)$ *denote the collection of real-valued continuous functions on* $[a,b]$, *with the functional* $\|\cdot\|_\infty$ *defined on* $f \in C([a,b], \|\cdot\|_\infty)$ *by:*

$$\|f\|_\infty \equiv \sup_{x \in [a,b]} |f(x)|. \tag{4.33}$$

Then $\|\cdot\|_\infty$ *is a norm and* $C([a,b], \|\cdot\|_\infty)$ *is a **real Banach space.***
Proof. *That* $C([a,b], \|\cdot\|_\infty)$ *is a vector space over* $\mathbb{R}$ *is left as an exercise.*

For the definition of a norm, only the triangle inequality needs investigation. Given $f, g \in C([a,b], \|\cdot\|_\infty)$:

$$
\begin{aligned}
\|f + g\|_\infty &\equiv \sup_{x \in [a,b]} |f(x) + g(x)| \\
&\leq \sup_{x \in [a,b]} [|f(x)| + |g(x)|] \\
&\leq \sup_{x \in [a,b]} |f(x)| + \sup_{x \in [a,b]} |g(x)| \\
&= \|f\|_\infty + \|g\|_\infty.
\end{aligned}
$$

The first inequality is the triangle inequality on $\mathbb{R}$, *while the second follows because the supremum of* $|f|$ *and* $|g|$ *may be obtained at different* x.

For completeness, assume that $\{f_n\}_{n=1}^\infty$ *is a Cauchy sequence. Since* $\mathbb{R}$ *is complete and* $\{f_n(x)\}_{n=1}^\infty$ *is a real Cauchy sequence for each* $x \in [a,b]$, *define* $f(x)$ *to be the limit of this sequence for each* x. *We prove that* $\|f_n - f\|_\infty \to 0$, *and thus as the uniform limit of continuous functions on compact* $[a,b]$, *it follows from Exercise 1.44 that* $f \in C([a,b], \|\cdot\|_\infty)$.

To this end, given $\epsilon > 0$ *choose* N *so that* $\|f_n - f_m\|_\infty < \epsilon$ *for* $n, m \geq N$. *Then for any* x, $|f_n(x) - f_m(x)| \leq \|f_n - f_m\|_\infty$, *and thus for* $n, m \geq N$:

$$|f_n(x) - f(x)| = \lim_{m \to \infty} |f_n(x) - f_m(x)| \leq \limsup_{m \to \infty} \|f_n - f_m\|_\infty \leq \epsilon.$$

Taking supremum over $x \in [a,b]$ *obtains* $\|f_n - f\|_\infty \leq \epsilon$, *and so* $f_n \to f$ *uniformly on* $[a,b]$. ■

4.3.2 Baire's Category Theorem

A normed linear space has a great deal of structure, even more than may at first be apparent. In this section, we turn to a few notions from topology that will allow an investigation into dense sets that will lead to the important category theorem of **René-Louis Baire** (1874–1932). A classic reference for topology is **Dugundji** (1970).

In the following, $(V, \|\cdot\|)$ is a normed linear space. The reader is invited to translate these notions to $C([a, b], \|\cdot\|_\infty)$, the subject space of this section, as an exercise in making these notions less abstract. See also Exercise 4.54.

1. **Open/closed balls:** Define $B(x, r)$, **the open ball of radius** r **about** $x \in V$, by:

$$B_r(x) = \{y|\, \|x - y\| < r\},$$

and analogously the **closed ball of radius** r **about** x by:

$$\bar{B}_r(x) = \{y|\, \|x - y\| \leq r\}.$$

These balls are sometimes denoted $B(x, r)$ and $\bar{B}(x, r)$.

2. **Open/closed sets:** A normed space $(V, \|\cdot\|)$ has a **natural topology** in which the notions of **open** and **closed** are defined:

   a. A set $A \subset V$ is **open** if for all $x \in A$ there exists $r > 0$ so that $B(x, r) \subset A$;

   b. A set $A \subset V$ is **closed** if $\widetilde{A} \equiv V - A$ is open.

Note that "open" and "closed" sets are proper subcollections of all sets, and many sets are neither open nor closed as is readily verified in $\mathbb{R}$.

Exercise 4.46 *Verify that $\mathcal{T} \equiv \{A \subset V | A \text{ is open}\}$ is a topology on V in the sense of Definition I.2.15. Then show that finite unions and arbitrary intersections of closed sets are closed. By "arbitrary", it is meant that the collection need not be countable.*

Finally, prove that if A is closed, $\{x_n\}_{n=1}^\infty \subset A$ and $x_n \to x$ for some $x \in V$, then $x \in A$. Hint: By Definition 4.43, $x_n \to x$ is equivalent to $\|x_n - x\| \to 0$.

3. **Closure/interior of a set:** Given $A \subset V$:

   a. The **closure of** A, denoted $\bar{A}$, is defined:

$$\bar{A} \equiv \left\{x \in V | B_r(x) \bigcap A \neq \emptyset, \text{ for all } r > 0 \right\}.$$

   The set $\bar{A}$ can also be defined as the collection of **limit points** of A:

$$\bar{A} \equiv \{x \in V|\ \|x_n - x\| \to 0 \text{ for some } \{x_n\}_{n=1}^\infty \subset A\}.$$

   Exercise 4.47 *Prove that these definitions are equivalent; that $\bar{A}$ is closed; and that A is closed if and only if $A = \bar{A}$.*

   b. The **interior of** A, denoted $int(A)$ or $\mathring{A}$, is defined as:

$$\mathring{A} \equiv \{x \in A | B_r(x) \subset A \text{ for some } r > 0\}.$$

   Exercise 4.48 *Prove that $\mathring{A}$ is open and that A is open if and only if $A = \mathring{A}$.*

4. **Dense set:** A set $D \subset V$ is **dense** in V if for any $x \in V$ and any $r > 0$, $B_r(x) \cap D \neq \emptyset$.

Note that a set is dense if this statement is true for $x \in V - D$.

5. Nowhere dense set: A set $A \subset V$ is **nowhere dense** in V if $\bar{A}$, the closure of A, contains no non-empty open sets. That is, A is nowhere dense if $int\left(\bar{A}\right) = \emptyset$. A closed set A is nowhere dense if $\mathring{A} = \emptyset$.

Remark 4.49 (On dense sets) *A dense set can be neither open nor closed, as the rationals $\mathbb{Q} \subset \mathbb{R}$ illustrates.*

If a dense set $D \subset V$ is closed, then of necessity $D = V$. This follows because D closed means $\tilde{D}$ is open, so $x \in \tilde{D}$ implies $B_r(x) \subset \tilde{D}$ for some r, and so $B_r(x) \cap D = \emptyset$. This contradicts that D is dense, and thus $\tilde{D} = \emptyset$ and $D = V$.

On the other hand, a dense set D can be open and have $D \subsetneq V$. For example, let $\{r_j\}_{j=1}^{\infty}$ be an enumeration of the rationals in $\mathbb{R}$ and consider $D \equiv \bigcup_{j=1}^{\infty}(r_j - \epsilon/2^{j+1}, r_j + \epsilon/2^{j+1})$. Then D is open as a countable union of open sets, and also dense in $\mathbb{R}$ since it contains the rationals that are dense. But by subadditivity of Lebesgue measure, $m(D) \le \epsilon$ and thus $D \subsetneq \mathbb{R}$. Indeed, "most" points are outside D. The same construction applies in $\mathbb{R}^n$ using balls with centers x with rational coordinates and Lebesgue measure $\epsilon/2^j$.

We next introduce **Baire's Category theorem,** named after **René-Louis Baire** (1874–1932), who developed these notions in his 1899 doctoral thesis. This theorem has many applications in functional analysis, and applies to Banach spaces and more generally to complete metric spaces.

For his result, Baire introduced the notions of **first category** and **second category** sets:

Definition 4.50 (First/second category sets) *A set $E \subset V$ is:*

- *A **set of the first category** if it is a countable union of nowhere dense sets;*

- *A **set of the second category** is any set not of the first category, meaning a set that cannot be expressed as a countable union of nowhere dense sets.*

The first part of Baire's result states that in a complete metric space, the intersection of a countable collection of **open** dense sets must be dense. Being open is key here. This intersection need not be open, and formally it is a $\mathcal{G}_\delta$-set (Notation I.2.16). Baire's result states that this $\mathcal{G}_\delta$-set must be dense.

The second part of Baire's result implies that a non-empty complete metric space is of the second category. This is the key result needed in the next section. See Corollary 4.53.

Proposition 4.51 (Baire's Category theorem) *Let (X, d) be a complete metric space.*

1. *If $\{A_j\}_{j=1}^{\infty} \subset X$ is a sequence of open dense sets, then $D \equiv \bigcap_{j=1}^{\infty} A_j$ is dense in X.*

2. *If $X \ne \emptyset$, and $\{C_j\}_{j=1}^{\infty} \subset X$ is a sequence of closed sets with $X = \bigcup_{j=1}^{\infty} C_j$, then there exists C_j that is not nowhere dense. In other words, there exists C_j, $x \in C_j$, and $r > 0$, so that $B_r(x) \subset C_j$.*

Proof. *For item 1, let $x \in X$ and $r > 0$ be given, and we claim that $D \bigcap B(x, r) \ne \emptyset$. Since A_1 is dense, there exists $x_1 \in A_1 \bigcap B(x, r)$, and since both are open there exists $r_1 \le 2^{-1}$ so that $\bar{B}(x_1, r_1) \subset A_1 \bigcap B(x, r)$. Since A_2 is dense there exists $x_2 \in A_2 \bigcap B(x_1, r_1)$ and $r_2 \le 2^{-2}$ so that $\bar{B}(x_2, r_2) \subset A_2 \bigcap B(x_1, r_1)$. Continuing in this way, sequences $\{x_j\}_{j=1}^{\infty} \subset X$ and $\{r_j\}_{j=1}^{\infty} \subset \mathbb{R}$ are produced for which $r_j \le 2^{-j}$, and*

$$\bar{B}(x_{j+1}, r_{j+1}) \subset A_{j+1} \bigcap B(x_j, r_j).$$

Thus $d(x_{j+1}, x_j) < r_j \leq 2^{-j}$, and so $\{x_j\}_{j=1}^{\infty}$ is a Cauchy sequence. Specifically, $d(x_n, x_m) < \epsilon$ for $n, m \geq N$ when $2^{-(N-1)} < \epsilon$. Since X is complete, let $x' \in X$ with $d(x_n, x') \to 0$. Since:

$$\bar{B}(x_1, r_1) \supset \bar{B}(x_2, r_2) \supset \bar{B}(x_3, r_3) \supset \cdots,$$

it follows that $x_k \in \bar{B}(x_j, r_j)$ for $j \leq k$ and thus $x' \in \bigcap_{j=1}^{\infty} \bar{B}(x_j, r_j)$. But then $\bar{B}(x_j, r_j) \subset A_j \bigcap B(x_{j-1}, r_j)$ obtains $x' \in \bigcap_{j=1}^{\infty} A_j = D$, while $\bigcap_{j=1}^{\infty} \bar{B}(x_j, r_j) \subset B(x, r)$ obtains $x' \in B(x, r)$. Thus $D \bigcap B(x, r) \neq \emptyset$.

The proof of item 2 is by contradiction. Assume that $X = \bigcup_{j=1}^{\infty} C_j$ and each C_j is nowhere dense. Since each C_j is closed by assumption, $A_j \equiv X - C_j$ is open. Also, each A_j is dense. Otherwise, there exists $x \in X$ and $r > 0$ so that $B(x, r) \bigcap A_j = 0$. This would imply $B(x, r) \subset C_j$, contradicting that C_j is nowhere dense.

Combining, if $X = \bigcup_{j=1}^{\infty} C_j$ then $\bigcap_{j=1}^{\infty} A_j = \tilde{X}$, with $\{A_j\}_{j=1}^{\infty}$ a sequence of open, dense sets. But by assumption $X \neq \emptyset$, and thus $\bigcap_{j=1}^{\infty} A_j = \tilde{X} = \emptyset$. This contradicts part 1, and thus some C_j is not nowhere dense. ∎

Remark 4.52 *A couple of clarifying comments on Baire's result:*

1. *The assumption that each A_j is open is essential for part 1. On $\mathbb{R}$ define $A_j = \mathbb{Q} - \{r_j\}$, with r_j denoting the jth rational. Then $\{A_j\}_{j=1}^{\infty}$ is a countable collection of dense sets and $\bigcap_j A_j = \emptyset$.*

2. *The assumption that $X \neq \emptyset$ is essential for part 2 since otherwise, all $C_j = \emptyset$ contradicts the conclusion.*

Corollary 4.53 (Baire's Category theorem) *Let (X, d) be a non-empty, complete metric space. Then X is of the second category.*
Proof. *By contradiction, if $X = \bigcup_{j=1}^{\infty} A_j$, where each A_j is nowhere dense, then by definition no $\bar{A}_j$ contains a non-empty open set. Thus with $C_j \equiv A_j$ we have $X = \bigcup_{j=1}^{\infty} C_j$, and all C_j are closed and nowhere dense, in contradiction to item 2 of Proposition 4.51.* ∎

Exercise 4.54 (On $C([a,b], \|\cdot\|_{\infty})$) *The space $C([a,b], \|\cdot\|_{\infty})$ is a Banach space by Proposition 4.45, and thus a complete metric space by Remark 4.38. By Corollary 4.53, $C([a,b], \|\cdot\|_{\infty})$ is then of the second category. The reader is asked to provide a detailed description of the following sets in this space:*

1. *Given $f \in C([a,b], \|\cdot\|_{\infty})$, what does $g \in B_r(f)$ say about the function g for $r > 0$? Characterize the functions in $B_r(f)$; in $\bar{B}_r(f)$.*

2. *Explain what it means that $\{f_j\}_{j=1}^{\infty} \subset C([a,b], \|\cdot\|_{\infty})$ is a Cauchy sequence.*

3. *What does it mean that $A \subset C([a,b], \|\cdot\|_{\infty})$ is an open set? What functions would be added to A to create $\bar{A}$? Or removed for $\mathring{A}$?*

4. *What does it mean that a set $D \subset C([a,b], \|\cdot\|_{\infty})$ is a dense set? A nowhere dense set?*

5. *Restate item 2 of Baire's result in the context of $C([a,b], \|\cdot\|_{\infty})$.*

4.3.3   Banach-Steinhaus Theorem

Proposition 4.45 placed $C([a, b], \|\cdot\|_\infty)$, the collection of continuous functions on $[a, b]$, within the framework of Banach spaces. The next step is to reframe how Riemann-Stieltjes sums act on this space. In particular, we will see that such sums can be identified with **bounded linear transformations** on this space, which are also called **bounded linear operators**.

We begin with more definitions, which are stated in generality for completeness. However, our primary investigation will restrict the field $\mathcal{F}$ to $\mathbb{R}$, the domain space of such transformations to the normed linear space $C([a, b], \|\cdot\|_\infty)$, and the range space of such transformations to the normed linear space $(\mathbb{R}, |\cdot|)$.

Definition 4.55 (Transformations between normed linear spaces) *Let* $(V_1, \|\cdot\|_1)$ *and* $(V_2, \|\cdot\|_2)$ *be normed linear spaces over the same field* $\mathcal{F}$.

1. *A* **transformation** $T : V_1 \to V_2$ *is a rule under which* $Tx \in V_2$ *for all* $x \in V_1$.

2. *A* **linear transformation** *is a transformation such that for all* $x, y \in V_1$, $a, b \in \mathcal{F}$:

$$T[ax + by] = aTx + bTy. \tag{4.34}$$

3. *A* **bounded linear transformation** *is a linear transformation for which* $\|T\|$, *the* **norm of** T, *satisfies:*

$$\|T\| \equiv \sup_{x \neq 0} \frac{\|Tx\|_2}{\|x\|_1} < \infty. \tag{4.35}$$

4. *A* **linear transformation** T *is* **continuous at** $x \in V_1$ *if given* $\epsilon > 0$, *there is a* δ, *so that:*

$$\|Tx - Ty\|_2 < \epsilon \text{ if } \|x - y\|_1 < \delta.$$

Equivalently, given $\epsilon > 0$ *there is a* δ, *so that:*

$$T[B_1(x, \delta)] \subset B_2(Tx, \epsilon),$$

where B_j *denotes the open ball defined relative to* $\|\cdot\|_j$.

5. *A* **linear transformation** T *is* **sequentially continuous at** $x \in V_1$ *if* $x_n \to x$ *in* V_1 *implies that* $Tx_n \to Tx$ *in* V_2. *In other words,* $\|Tx_n - Tx\|_2 \to 0$ *if* $\|x_n - x\|_1 \to 0$.

Remark 4.56 (Transformations are "functions") *While the above definition conveys a good deal of abstraction, the concepts described have likely been seen before.*

1. **Transformations** *are simply functions, which provide a rule that assigns a single value in the range space to every value in the domain space. The word function is usually (but not always) reserved for such rules when the domain space is* $\mathbb{R}^n$ *or* $\mathbb{C}^n$ *for some* n, *and range space is* $\mathbb{R}$ *or* $\mathbb{C}$.

2. **Linear transformations** *have likely been encountered in linear algebra. If* A *is an* $n \times m$ *matrix of real numbers, then* A *induces a linear transformation* $A : \mathbb{R}^m \to \mathbb{R}^n$ *by ordinary matrix multiplication. Interpreting* x *and* y *in (4.34) as elements of* $\mathbb{R}^m$ *expressed as column vectors, then* $A[ax + by] = aAx + bAy$ *can be verified by matrix multiplication.*

More generally, any linear transformation $T : \mathbb{R}^m \to \mathbb{R}^n$ *can be so represented by a matrix* A *and matrix multiplication. But such* A *is not unique and reflects the basis used in each space. Recall that the basis used determines the numerical components in the vector representation of a given point.*

3. **Bounded linear transformations** *are easiest to understand in the representation below in (4.36). That is, T is bounded if, when restricted to the unit ball in V_1 defined by $\{y|\ \|y\|_1 = 1\}$, the set $\{Ty\}$ is bounded in the V_2 norm. Proposition 4.58 provides more intuition.*

4. **Continuous linear transformations** *are initially defined exactly like continuous functions $f : \mathbb{R} \to \mathbb{R}$, using the $\epsilon - \delta$ terminology. The only difference here is that we replace the usual absolute value norms on $\mathbb{R}$ with the respective norms in the domain and range spaces. The equivalent formulation comes from the definition of a ball, for example, that $\|x - y\|_1 < \delta$ if and only if $y \in B_1(x, \delta)$.*

5. **Sequentially continuous transformations** *are defined exactly like sequentially continuous functions. As for continuity, one only needs to replace the usual absolute value norms on $\mathbb{R}$ with the respective norms in the domain and range spaces.*

We start with a few results that reflect linearity of T.

Proposition 4.57 (On $\|T\|$) *For a bounded linear transformation T, the norm $\|T\|$ can be equivalently defined:*

$$\|T\| \equiv \sup_{\|y\|_1 = 1} \|Ty\|_2 . \tag{4.36}$$

Then for all x:

$$\|Tx\|_2 \le \|T\| \, \|x\|_1 , \tag{4.37}$$

and T maps balls into balls:

$$T\left[B_1(x, r)\right] \subset B_2(Tx, \|T\| \, r). \tag{4.38}$$

Proof. *Given $x \in V_1$ let $y \equiv x/\|x\|_1$. Then $\|y\|_1 = 1$ by definition, $Ty = Tx/\|x\|_1$ by linearity, and $\|Ty\|_2 = \|Tx\|_2/\|x\|_1$ by item 2 of Definition 4.36, which yield (4.36).*

The inequality in (4.37) follows by definition since $\|T\|$ is the supremum of such ratios. Finally, if $y \in B_1(x, r)$ then $T(y - x) = Ty - Tx$ by linearity, and so by (4.37):

$$\|Tx - Ty\|_2 \le \|T\| \, \|x - y\|_1 < r \, \|T\| .$$

Thus $Ty \in B_2(Tx, r \, \|T\|)$. ∎

The following proposition is quite useful and allows many tools to be applied in proving any result related to bounded linear transformations.

Proposition 4.58 (On boundedness; continuity) *Given a linear transformation $T: V_1 \to V_2$:*

1. *If T is continuous or sequentially continuous at x_0, then T is continuous or sequentially continuous for all x.*

2. *The following are equivalent for T:*

   a. *bounded,*

   b. *continuous,*

   c. *sequentially continuous.*

Proof. *For item 1, assume that T is continuous at x_0, so for any $\epsilon > 0$ there is δ so that $T[B_1(x_0, \delta)] \subset B_2(Tx_0, \epsilon)$. If $x \in V_1$, then $z \in B_1(x_0, \delta)$ if and only if $z + (x - x_0) \in B_1(x, \delta)$. Then by linearity:*

$$T[B_1(x, \delta)] \subset B_2(Tx_0, \epsilon) + T(x - x_0) = B_2(Tx, \epsilon),$$

and T is continuous at y. A similar argument applies for sequential continuity, and is left as an exercise.

For item 2, we first prove $a \Longleftrightarrow b$. If T is bounded then continuity follows from (4.38) with $\delta = \epsilon / \|T\|$. If T is continuous at x, then given ϵ there exists δ so that $\|x - y\|_1 < \delta$ implies $\|Tx - Ty\|_2 < \epsilon$. Then by linearity $\|z\|_1 < \delta$ implies $\|Tz\|_2 < \epsilon$. Letting $y \equiv z / \|z\|_1$ for such z obtains by linearity that $\|Ty\|_2 < \epsilon \delta$, and thus $\|T\| < \epsilon \delta$ and T is bounded.

To prove $b \Longleftrightarrow c$, if T is continuous at x say, then given ϵ there exists δ so that $\|x - y\|_1 < \delta$ implies $\|Tx - Ty\|_2 < \epsilon$. If $x_n \to x$ there exists N so that $\|x_n - x\|_1 < \delta$ for $n \geq N$, and thus by linearity $\|Tx_n - Tx\|_2 < \epsilon$ for $n \geq N$ and T is sequentially continuous. If T is sequentially continuous but not continuous at x, there exists $\epsilon > 0$ so that for no δ is $T[B_1(x, \delta)] \subset B_2(Tx, \epsilon)$. Letting $\delta = 1/m$, there exists $\{x_m\}_{m=1}^{\infty}$ so that $\|x_m - x\|_1 < 1/m$ and $\|Tx_m - Tx\|_2 \geq \epsilon$. Thus $x_m \to x$ and $Tx_m \nrightarrow Tx$, contradicting sequential continuity, and so T is continuous at x. ∎

At last, we are in a position to connect Riemann-Stieltjes integration theory with the formal development above.

Proposition 4.59 (Riemann-Stieltjes sum as bounded linear transformation) *Let $F(x)$ be a bounded integrator so that $\int_a^b g(x) dF$ exists for all continuous $g(x)$.*

*Given any tagged partition $P \equiv (\{x_i\}_{i=0}^n, \{\widetilde{x}_i\}_{i=1}^n)$ of $[a, b]$, meaning $x_0 = a$, $x_i < x_{i+1}$, $x_n = b$, and $\widetilde{x}_i \in [x_{i-1}, x_i]$, define a **transformation:***

$$T_P : C([a, b], \|\cdot\|_\infty) \to (\mathbb{R}, |\cdot|),$$

by:

$$T_P g \equiv \sum_{i=1}^n g(\widetilde{x}_i)(F(x_i) - F(x_{i-1})). \tag{4.39}$$

*Then T_P is a **bounded linear transformation** with norm:*

$$\|T_P\| \equiv \sum_{i=1}^n |F(x_i) - F(x_{i-1})|. \tag{4.40}$$

In other words, $\|T_P\|$ equals the variation t of $F(x)$ from Definition 3.23 using partition $\Pi \equiv \{x_i\}_{i=0}^n$.

Further, given any $g \in C([a, b], \|\cdot\|_\infty)$, there exists $C_g < \infty$, so that:

$$\sup_P |T_P g| \leq C_g, \tag{4.41}$$

where the supremum is over all tagged partitions $P \equiv (\{x_i\}_{i=0}^m, \{\widetilde{x}_i\}_{i=1}^m)$.

Proof. *Linearity of T_P is clear by definition. For boundedness, since $|g(\widetilde{x}_i)| \leq \|g\|_\infty \equiv \sup_{x \in [a,b]} |g(x)|$ for all i:*

$$|T_P g| \leq \|g\|_\infty \sum_{i=1}^n |F(x_i) - F(x_{i-1})|. \tag{1}$$

Thus by (4.36):

$$\|T_P\| \leq \sum_{i=1}^n |F(x_i) - F(x_{i-1})|, \tag{2}$$

so T_P is bounded.

For the norm of T_P, consider continuous $g(x)$ with $\|g\|_\infty = 1$. The upper bound in (2) is achieved with continuous, piecewise linear $g(x)$ with $g(\widetilde{x}_i) = \text{sgn}\,[F(x_i) - F(x_{i-1})]$, and this obtains (4.40).

Fixing $g \in C([a,b], \|\cdot\|_\infty)$, the assumed existence of $I \equiv \int_a^b g(x)dF$ obtains by Definition 4.3 that for any $\epsilon > 0$ there is a δ, so that:

$$|R(g, F; P) - I| < \epsilon,$$

for all tagged partitions P of mesh size $\mu \leq \delta$, where $R(g, F; P)$ is the Riemann-Stieltjes sum in (4.39). Thus:

$$\sup_{\mu \leq \delta} |T_P g| \leq \max |I \pm \epsilon|, \tag{3}$$

where this supremum is over all tagged partitions with mesh size $\mu \leq \delta$.

By assumption $|F| \leq M_F$ on $[a,b]$, so it follows that for any interval $[x_{i-1}, x_i]$:

$$|F(x_i) - F(x_{i-1})| \leq 2M_F. \tag{4}$$

Given a tagged partition with mesh size $\mu \geq \delta$, then $n \leq \frac{b-a}{\delta}$ in (1), and since $|g(x)|$ is bounded by $\|g\|_\infty$, (1) and (4) obtain:

$$\sup_{\mu \geq \delta} |T_P g| \leq 2\|g\|_\infty M_F \frac{b-a}{\delta}.$$

This bound and (3) prove (4.41). ∎

This result illustrates that when $F(x)$ is a bounded integrator such that $\int_a^b g(x)dF$ exists for all continuous $g(x)$, there exists a family $\{T_P\}$ of bounded linear transformations on $C([a,b], \|\cdot\|_\infty)$, indexed by the family of all tagged partitions $P \equiv (\{x_i\}_{i=0}^n, \{\widetilde{x}_i\}_{i=1}^n)$, for which:

1. The norm $\|T_P\|$ of each transformation equals the variation t of $F(x)$ from Definition 3.23 using partition $\Pi \equiv \{x_i\}_{i=0}^n$.

2. As in (4.41), $\sup_P |T_P g| \leq C_g$ for any $g(x) \in C([a,b], \|\cdot\|_\infty)$, where this supremum is defined over all tagged partitions.

Thus $\{|T_P g|\}_P$ is bounded for every $g(x) \in C([a,b], \|\cdot\|_\infty)$.

To prove the key result of this section requires the conclusion that the norms of T_P are uniformly bounded:

$$\sup_P \|T_P\| < \infty.$$

Then from item 1 it would follow that the variations $\{t\}_\Pi$ of $F(x)$ are uniformly bounded over all partitions, and then $F(x)$ will be of bounded variation by definition.

It might well be an unexpected result that for any family of bounded linear operators $\{T_P\}$, the boundedness of $\{|T_P g|\}_P$ for every g is equivalent to the uniform boundedness of norms $\{\|T_P\|\}_P$. But remarkably this is the conclusion of the **Banach-Steinhaus theorem**. This result is named after **Stefan Banach** (1892–1945) and **Hugo Steinhaus** (1887–1972), who coauthored the result in a 1927 paper. It is also known as the **uniform boundedness principle** and was independently discovered by **Hans Hahn** (1879–1934).

The classical proof below relies on Baire's category theory, although there are now proofs of this result that do not use this approach. See for example **Sokal** (2011). The decision to use the classical proof using Baire's result was motivated by the desire to introduce properties of Banach spaces and this category theorem, as these are foundational results of classical analysis that will reappear in later books.

Before stating and proving this theorem, we consider an example where the result is analytically tractable.

Example 4.60 (A simplified Banach-Steinhaus example) *Let* $(X, \|\cdot\|_X)$ *be a finite* n-*dimensional Banach space over the field* $\mathbb{R}$ *with basis* $\{x_i\}_{i=1}^n$, *and* $(Y, \|\cdot\|_Y)$ *a normed linear space. For more specificity, these spaces can be taken to be* $(\mathbb{R}^n, |\cdot|)$ *with* $|\cdot|$ *the standard Euclidean norm in (4.29).*

Given any family of bounded linear transformations $\{T_\alpha\}_{\alpha \in I}$ *with* $T_\alpha \colon X \to Y$, *the Banach-Steinhaus theorem asserts that:*

$$\sup_{\alpha \in I} \|T_\alpha\| < \infty, \text{ if and only if, } \sup_{\alpha \in I} \|T_\alpha x\|_Y < \infty \text{ for all } x.$$

Now $\sup_{\alpha \in I} \|T_\alpha\| < \infty$ *implies* $\sup_{\alpha \in I} \|T_\alpha x\|_Y < \infty$ *for all* x *by (4.37), so we focus on the reverse implication.*

If $x \in X$, *then* $x = \sum_{i=1}^n a_i x_i$ *for* $\{a_i\}_{i=1}^n \subset \mathbb{R}$. *Then by the triangle inequality:*

$$\|T_\alpha x\|_Y \leq \sum_{i=1}^n |a_i| \|T_\alpha x_i\|_Y \leq \sum_{i=1}^n |a_i| M_i,$$

where $M_i \equiv \sup_{\alpha \in I} \|T_\alpha x_i\|_Y$ *is finite by assumption. Thus:*

$$\sup_{\alpha \in I} \|T_\alpha\| \leq \sup_{\|x\|_X = 1} \sum_{i=1}^n |a_i| M_i, \tag{1}$$

where the supremum on the right is defined over all x *with* $\|x\|_X = 1$ *by (4.36).*

Without loss of generality we can assume that the basis is normalized to have $\|x_i\|_X = 1$ *for all* i, *and then by the triangle inequality:*

$$\|x\|_X \leq \sum_{i=1}^n |a_i|.$$

It then follows from (1) that:

$$\sup_{\alpha \in I} \|T_\alpha\| \leq \sup \sum_{i=1}^n |a_i| M_i, \tag{2}$$

where the supremum is now over all $\{a_i\}_{i=1}^n \subset \mathbb{R}$ *with* $\sum_{i=1}^n |a_i| \leq 1$.

Since $f(a_1, ..., a_n) \equiv \sum_{i=1}^n |a_i| M_i$ *is continuous on* $\mathbb{R}^n$, *and* $A \equiv \{\sum_{i=1}^n |a_i| \leq 1\}$ *is compact, we obtain from Exercise 1.13 that:*

$$\sup_A f(a_1, ..., a_n) < \infty.$$

Thus $\sup_{\alpha \in I} \|T_\alpha\| < \infty$, *completing the proof.*

For this example, the assumption of finite dimensionality of X, and thus the existence of a finite basis, provided an enormous simplification to the problem by allowing the direct application of tools from analysis. For the more general result, a more powerful tool is required, here in the form of Baire's category theorem.

The following result is most often stated in terms of families of continuous linear transformations, but by Proposition 4.58 can be equivalently stated in terms of bounded linear transformations.

Proposition 4.61 (Banach-Steinhaus theorem) *Let* $(X, \|\cdot\|_X)$ *be a Banach space where* $X \neq \emptyset$, *and* $(Y, \|\cdot\|_Y)$ *a normed linear space.*

Given a family of continuous linear transformations $\{T_\alpha\}_{\alpha \in I}$ *with* $T_\alpha \colon X \to Y$:

$$\sup_{\alpha \in I} \|T_\alpha\| < \infty, \text{ if and only if, } \sup_{\alpha \in I} \|T_\alpha x\|_Y < \infty \text{ for all } x. \tag{4.42}$$

Proof. *If* $\sup_{\alpha \in I} \|T_\alpha\| < \infty$, *then since* $\|T_\alpha x\|_Y \leq \|T_\alpha\| \|x\|_X$ *by (4.37), it follows that for each* x:

$$\sup_{\alpha \in I} \|T_\alpha x\|_Y \leq \|x\|_X \sup_{\alpha \in I} \|T_\alpha\| < \infty.$$

Next, assume that $\sup_{\alpha \in I} \|T_\alpha x\|_Y < \infty$ *for each* x. *Define the sequence* $\{A_n\}_{n=1}^\infty \subset X$ *by:*

$$A_n \equiv \bigcap_{\alpha \in I} A_n^{(\alpha)},$$
$$A_n^{(\alpha)} \equiv \{x \in X | \ \|T_\alpha x\|_Y \le n\}.$$

We prove that A_n *is closed by showing that each set* $A_n^{(\alpha)}$ *is closed, recalling Exercise 4.46, that an arbitrary intersection of closed sets is closed.*

Let $y \in \widetilde{A}_n^{(\alpha)}$, *the complement of* $A_n^{(\alpha)}$, *and so* $\|T_\alpha y\|_Y = n + a$ *with* $a > 0$. *If* $z \in B(y, a/\ (2\ \|T_\alpha\|))$, *then writing* $y = z + (y - z)$ *and applying linearity and the triangle inequality obtains:*

$$
\begin{aligned}
\|T_\alpha y\|_Y &\le \ \|T_\alpha z\|_Y + \|T_\alpha (y - z)\|_Y \\
&\le \ \|T_\alpha z\|_Y + \|T_\alpha\| \ \|y - z\|_Y \\
&< \ \|T_\alpha z\|_Y + a/2.
\end{aligned}
$$

Since $\|T_\alpha y\|_Y = n + a$, *it follows that* $\|T_\alpha z\|_Y > n + a/2$, *and thus* $B(y, a/\ (2\ \|T_\alpha\|)) \subset \widetilde{A}_n^{(\alpha)}$. *This proves that* $\widetilde{A}_n^{(\alpha)}$ *is open, and hence each* $A_n^{(\alpha)}$ *is closed, as is each* A_n *as an intersection of closed sets.*

Now for each $x \in X$, $\sup_{\alpha \in I} \|T_\alpha x\|_Y \le n$ *for some* n, *and it follows that* $x \in A_n$ *and thus* $\bigcup_{n=1}^\infty A_n = X$. *Because* X *is complete and non-empty, and* $\{A_n\}_{n=1}^\infty$ *are closed sets, part 2 of Baire's category theorem of Proposition 4.51 obtains that at least one set has non-empty interior, say* A_N. *Thus there exists* $x \in A_N$ *and* $\epsilon > 0$ *so that* $B(x, 2\epsilon) \subset A_N$. *Then by definition of* A_N, $\|T_\alpha y\|_Y \le N$ *for all* $\alpha \in I$, *for all* $y \in B(x, 2\epsilon)$.

Any $y \in \bar{B}(x, \epsilon) \subset B(x, 2\epsilon)$ *can be expressed as* $y = x + \epsilon z$ *where* $z \in \bar{B}(0, 1)$, *and equivalently, any* $z \in \bar{B}(0, 1)$ *can be expressed as* $z = (y - x)/\epsilon$. *By linearity of* T_α *and the triangle inequality, this obtains that for all* $z \in \bar{B}(0, 1)$ *and all* $\alpha \in I$:

$$\|T_\alpha z\|_Y \le (\|T_\alpha y\|_Y + \|T_\alpha x\|_Y)\ /\epsilon \le 2N/\epsilon.$$

Taking a supremum with respect to $z \in \bar{B}(0, 1)$ *with* $\|z\|_Y = 1$ *yields by (4.36), that for all* α:

$$\|T_\alpha\| \le 2N/\epsilon.$$

Taking a supremum completes the proof. ∎

4.3.4 Final Result on R-S Integrators

With the above machinery from functional analysis in place, we are finally ready to state and prove a fundamental result on bounded Riemann-Stieltjes integrators that integrate all continuous functions.

Proposition 4.62 (Bounded $F(x)$ **integrates** $C([a, b], \|\cdot\|_\infty) \Rightarrow$ $F(x) \in B.V.$ **)** *Given bounded* $F(x)$ *defined on* $[a, b]$, *if* $\int_a^b g(x)dF$ *exists for all continuous* $g(x)$, *then* $F(x)$ *is of bounded variation.*

Proof. *Given any tagged partition* $P \equiv (\{x_i\}_{i=0}^n, \{\widetilde{x}_i\}_{i=1}^n)$ *of* $[a, b]$, *meaning* $x_0 = a$, $x_i < x_{i+1}$, $x_n = b$, *and* $\widetilde{x}_i \in [x_{i-1}, x_i]$, *define a linear transformation* $T_P : C([a, b], \|\cdot\|_\infty) \to (\mathbb{R}, |\cdot|)$ *by:*

$$T_P g \equiv \sum_{i=1}^n g(\widetilde{x}_i)(F(x_i) - F(x_{i-1})).$$

Define the family of all linear transformations $\{T_{P_\alpha}\}$ *analogously with* $I \equiv \{P_\alpha\} \equiv \{\{x_i^{(\alpha)}\}_{i=0}^{m(\alpha)}, \{\widetilde{x}_i^{(\alpha)}\}_{i=0}^{m(\alpha)}\}$, *the family of all tagged partitions of* $[a, b]$.

Then by (4.41), for every $g \in C([a,b], \|\cdot\|_\infty)$:

$$\sup_{P_\alpha} |T_{P_\alpha} g| \leq C_g < \infty.$$

By the Banach-Steinhaus theorem, this implies that $\sup_{P_\alpha} \|T_{P_\alpha}\| < \infty$, and thus by (4.40):

$$\sup_{P_\alpha} \|T_{P_\alpha}\| \equiv \sup_{P_\alpha} \sum_{i=1}^{m(\alpha)} \left| F(x_i^{(\alpha)}) - F(x_{i-1}^{(\alpha)}) \right| < \infty.$$

Thus $F(x)$ is of bounded variation. ∎

4.4   Riemann-Stieltjes Integrals on $\mathbb{R}^n$

4.4.1   Multivariate Integrators

For $g : \mathbb{R}^n \to \mathbb{R}$ and $F : \mathbb{R}^n \to \mathbb{R}$, the development of the Riemann-Stieltjes integral of g over a rectangle $A \equiv \prod_{j=1}^n [a_j, b_j]$ generalizes the above template when $n = 1$ in much the same way as was seen in the Riemann development. But there is one added complication here of appropriately defining the n-dimensional analog of $\Delta F \equiv F(x_i) - F(x_{i-1})$.

Recalling the Riemann notational conventions, let each interval $[a_j, b_j]$ be partitioned:

$$a_j = x_{j,0} < x_{j,1} < \cdots < x_{j,m_j-1} < x_{j,m_j} = b_j. \tag{4.43}$$

The mesh size μ of this partition in defined as in (1.34) by:

$$\mu \equiv \max_{i,j} \{ x_{j,i} - x_{j,i-1} \},$$

where the maximum is defined over $1 \leq i \leq m_j$ and $1 \leq j \leq n$.

Rectangle partitions will be denoted $P \equiv \{ \{ x_{j,i} \}_{i=0}^{m_j} \}_{j=1}^n$ generalizing the 1-dimensional case. These rectangle partitions lead to a decomposition of A into subrectangles:

$$A = \bigcup_{J \in I} A_J,$$

where $A_J \equiv \prod_{j=1}^n A_{j,i_j}$ with $A_{j,i_j} \equiv [x_{j,i_j-1}, x_{j,i_j}]$, and $I = \{ (i_1, i_2, ..., i_n) | 1 \leq i_j \leq m_j \}$ is the index set that identifies the $\prod_{j=1}^n m_j$ subrectangles that are defined by this partition.

For the Riemann integral in $\mathbb{R}^n$, the "volume" $|A_J|$ of the rectangle $A_J \equiv \prod_{j=1}^n [x_{j,i_j-1}, x_{j,i_j}]$ is defined in (1.37) by the Lebesgue measure:

$$|A_J| \equiv \prod_{j=1}^n (x_{j,i_j} - x_{j,i_j-1}).$$

For the Riemann-Stieltjes integral, $|A_J| \equiv m(A_J)$ must be generalized to an F-volume $|A_J|_F \equiv \Delta F_J$, in an analogous way as the 1-dimensional interval length $x_i - x_{i-1}$ in the Riemann integral was generalized to F-length $\Delta F \equiv F(x_i) - F(x_{i-1})$. The notion of F-length was introduced in Section I.5.2.

A necessary property of such $\Delta F_J \equiv |A_J|_F$ will be **finite additivity**. If $A \equiv \prod_{j=1}^n [a_j, b_j]$ and $A = \bigcup_{J \in I} A_J$, where $\{A_J\}$ is a collection of closed rectangles induced by a partition, then we want:

$$|A|_F = \sum_{J \in I} |A_J|_F,$$

or in terms of the function F:

$$\Delta F(A) = \sum_{J \in I} \Delta F(A_J).$$

The development of Chapter I.8 provides an insight into a definition of $\Delta F(A)$ for certain $F : \mathbb{R}^n \to \mathbb{R}$, which was there proved to be finitely additive in Proposition I.8.13.

Definition 4.63 (F-content) *Given* $F : \mathbb{R}^n \to \mathbb{R}$ *and a bounded rectangle* $A \equiv \prod_{i=1}^{n}[a_i, b_i]$, *define the* F-***content of*** A, *or* F-***volume of*** A, *denoted* $\Delta F(A) \equiv |A|_F$, *by:*

$$\Delta F(A) = \sum_x sgn(x) F(x), \tag{4.44}$$

where each $x = (x_1, ..., x_n)$ *is one of the* 2^n *vertices of* A, *so* $x_i = a_i$ *or* $x_i = b_i$, *and* $sgn(x)$ *is defined as* -1 *if the number of* a_i-*components of* x *is odd, and* $+1$ *otherwise.*

Exercise 4.64 (Lebesgue measure) *Prove that if* $F(x) = \prod_{j=1}^{n} x_j$ *and* $A \equiv \prod_{j=1}^{n}[a_j, b_j]$, *that the Lebesgue measure of* A *as given in (1.37):*

$$m(A) \equiv \prod_{j=1}^{n}(b_j - a_j),$$

is equivalent to $\Delta F(A)$ *in (4.44).*

Exercise 4.65 (ΔF on degenerate A) *Call the rectangle* $A \equiv \prod_{i=1}^{n}[a_i, b_i]$ ***degenerate*** *if* $a_i = b_i$ *for one or more* i. *Prove that as defined in (4.44), that* $\Delta F(A) = 0$. *Hint: Assume unique* $i = 1$ *with this property, split* $\sum_x$ *into two identical sums, and investigate* $sgn(x)$ *in each. Then generalize.*

Remark 4.66 (ΔF as a "volume"; finite additivity) *It may not be apparent that* $\Delta F(A)$ *as defined in (4.44) reflects a "volume" or measure of the rectangle* A, *since without some restrictions on* $F(x)$, ΔF *could in theory be negative. Allowing* $\Delta F(A) < 0$ *would indeed be an illogical result if our goal was to create a Borel measure as in Chapter I.8.*

But, as was seen in Section 4.2 for the Riemann-Stieltjes integral on $\mathbb{R}$, *we did not require that* $\Delta F \equiv F(x_i) - F(x_{i-1}) \geq 0$ *for all integrators. For increasing functions* $\Delta F \geq 0$, *and these functions were largely consistent with those of Chapter I.5 that induced Borel measures.*

But we also investigated as integrators $F \in B.V.$, *bounded variation functions, for which nonnegativity of* ΔF *was in general not expected. In both cases, the critical property for such integrators was finite additivity over partitions of intervals, and this was true by definition of* ΔF.

The same general development will be seen in this section. We will not require that $\Delta F \geq 0$, *but we will require finite additivity. So the outstanding question here:*

If $\Delta F(A)$ *is defined on bounded rectangles* $A = \prod_{j=1}^{n}[a_j, b_j]$ *by (4.44), is this set function finitely additive when applied to partitions?*

Recall that Proposition I.8.9 derived that if μ *is a finite Borel measure on* $\mathbb{R}^n$, *and* $F(x)$ *is defined:*

$$F(x) = \mu[A_x],$$

where $A_x \equiv \prod_{j=1}^{n}(-\infty, x_j]$, *then with the notational conventions of (4.44):*

$$\mu\left[\prod_{j=1}^{n}(a_j, b_j]\right] = \sum_x sgn(x) F(x). \tag{1}$$

Proposition I.8.12 derived the same result for functions $F(x)$ *induced by general Borel measures. Thus functions induced by Borel measures provide nonnegative and finitely additive set functions defined on rectangles* $\prod_{j=1}^{n}(a_j, b_j]$ *as in (1).*

Conversely, it was proved in Proposition I.8.13 that the set function defined on $\prod_{j=1}^{n}(a_j, b_j]$ *in (1) is finitely additive and nonnegative under the assumption that* F *is* ***continuous from above*** *(Definition I.8.5) and* n-***increasing*** *(Definition I.8.7). However,* *neither of these assumptions was needed for the **finite additivity** portion of this proof, as can be verified. Further, this proof did not rely on the geometry of rectangles having the form $\prod_{j=1}^{n}(a_j, b_j]$, and the proof would be identical if we deemed the formula in (1) to represent a set function defined on $\prod_{i=1}^{n}[a_i, b_i]$.*

For that proof, only the functional form of $\sum_x sgn(x)F(x)$ was used, and it was shown that finite additivity resulted from the cancellation of F-valuations at internal points, just as is more easily seen in the 1-dimensional case with $\Delta F \equiv F(x_i) - F(x_{i-1})$.

We simply record the result here.

Proposition 4.67 (Finite additivity of ΔF) *Given $F : \mathbb{R}^n \to \mathbb{R}$, define ΔF on $A \equiv \prod_{i=1}^{n}[a_i, b_i]$ as in (4.44) by:*

$$\Delta F(A) \equiv \sum_x sgn(x)F(x),$$

where each $x = (x_1, ..., x_n)$ is one of the 2^n vertices of A, so $x_i = a_i$ or $x_i = b_i$, and $sgn(x)$ is defined as -1 if the number of a_i-components of x is odd, and $+1$ otherwise.

Then ΔF is finitely additive. If $A \equiv \prod_{j=1}^{n}[a_j, b_j]$ and $A = \bigcup_{J \in I} A_J$, where $\{A_J\}_{J \in I}$ is a collection of closed rectangles induced by a partition P, then:

$$\Delta F(A) = \sum_{J \in I} \Delta F(A_J), \tag{4.45}$$

where $\Delta F(A_J)$ is defined as in (4.44) applied to A_J.

*Further, ΔF so defined is **linear**, meaning that if $F(x) = cF_1(x) + dF_2(x)$, then for all A:*

$$\Delta F(A) = c\Delta F_1(A) + d\Delta F_2(A). \tag{4.46}$$

Proof. *As noted above, finite additivity is proved in parts 1 and 2 of the proof of Proposition I.8.13, while linearity follows from the definition.* ∎

Example 4.68 ($\frac{\partial^n F}{\partial x_1 \partial x_2 \cdots \partial x_n}$ continuous) *To get an intuition for the formula in (4.44), let $F(x)$ be a differentiable function defined on $x = (x_1, ..., x_n)$ with continuous $f(x)$ defined by:*

$$f(x) \equiv \frac{\partial^n F}{\partial x_1 \partial x_2 \cdots \partial x_n}.$$

Then ΔF defined on $A \equiv \prod_{j=1}^{n}[a_j, b_j]$ in (4.44) is given by:

$$\Delta F(A) = (\mathcal{R}) \int_A f(x)dx,$$

which exists by Proposition 1.63 given continuity of $f(x)$.

To see this, note that this integral can be expressed as an iterated Riemann integral by Corollary 1.77, so we seek to prove:

$$\Delta F(A) = \int_{a_1}^{b_1} \int_{a_2}^{b_2} \cdots \int_{a_n}^{b_n} f(x)dx_n dx_{n-1} \cdots dx_1. \tag{4.47}$$

Since $f(x)$ is a continuous function of x_n for any $x_1, x_2, ...x_{n-1}$, the fundamental theorem of calculus in (1.24) obtains with $\partial^{n-1} \equiv \frac{\partial^{n-1}}{\partial x_1 \partial x_2 \cdots \partial x_{n-1}}$:

$$\int_{a_n}^{b_n} f(x)dx_n = \partial^{n-1}F(x_1, x_2, ..., b_n) - \partial^{n-1}F(x_1, x_2, ..., a_n).$$

For any $x_1, x_2, \ldots x_{n-2}$, this is then a continuous function of x_{n-1} and thus the dx_{n-1}-integral can be similarly evaluated.

It is then verified with a little algebra that after n such applications of the fundamental theorem of calculus that the summation in (4.44) is derived, where the coefficient of $F(x)$ depends on the parity of the number of a_j-components.

In probability theory, if $F(x)$ is the joint distribution function of a random vector X defined on a probability space $(\mathcal{S}, \mathcal{E}, \mu)$, then $f(x)$ defined above is the density function of this random vector when it exists, and in so-called "continuous" probability theory, this function is continuous. In the general case, $\Delta F(A)$ represents the probability that $X \in A$, which is defined:

$$\Pr[X \in A] \equiv \mu\left[X^{-1}(A)\right].$$

When the density function $f(x)$ exists and is integrable, this probability can be recovered by integrating the density function over A.

With ΔF now defined, we next generalize the 1-dimensional notions of an integrator function $F(x)$ being **increasing** or **of bounded variation** and will see that these play an important role in the existence results for the Riemann-Stieltjes integral in $\mathbb{R}^n$.

We use the terminology n-**increasing**, as noted above and consistent with Book I, and **of Vitali bounded variation**, also called **of bounded variation in the sense of Vitali**, and named after **Giuseppe Vitali** (1875–1932). It can be checked that when $n = 1$, these notions reduce to the earlier definitions.

The reader is alerted that there are other notions of bounded variation for functions defined on $\mathbb{R}^n$.

Definition 4.69 (*n-increasing; Vitali B.V.*) *Given $F : \mathbb{R}^n \to \mathbb{R}$ and bounded rectangle $A \equiv \prod_{i=1}^n [a_i, b_i]$, define $\Delta F(A)$ as in (4.44).*

Then F is said to be n-increasing if for all such A:

$$\Delta F(A) \geq 0. \tag{4.48}$$

*Further, F is said to be **of Vitali bounded variation** on a rectangle $A \equiv \prod_{j=1}^n [a_j, b_j]$, or **of bounded variation** on A in the sense of Vitali, if:*

$$T_F(A) \equiv \sup_P \sum_{J \in I} |\Delta F(A_J)| < \infty, \tag{4.49}$$

where the supremum is over all partitions $P \equiv \{\{x_{j,i}\}_{i=0}^{m_j}\}_{j=1}^n$ of A, so $A = \bigcup_{J \in I} A_J$.

*The quantity $T_F(A)$ is called the **Vitali total variation** of F on A.*

Exercise 4.70 (*On $\mu \to 0$*) *Generalizing Exercise 3.25, let P_μ denote a partition of mesh size μ as defined in (1.34). Prove that the supremum in the definition of $T_F(A)$ is necessarily obtained as $\mu \to 0$ by proving that:*

$$T = \lim_{\mu \to 0} \sup_{P_\mu} \sum_{J \in I} |\Delta F(A_J)|,$$

where $\sup_{P_\mu}$ denotes the supremum over all partitions of mesh size μ. Hint: Given any two partitions $P_1 = \{\{x_{j,i}\}_{i=0}^{m_j}\}_{j=1}^n$ and $P_2 = \{\{x'_{j,i}\}_{i=0}^{m'_j}\}_{j=1}^n$, show that with $P = \{\{x_{j,i}\}_{i=0}^{m_j}\}_{j=1}^n \bigcup \{\{x'_{j,i}\}_{i=0}^{m'_j}\}_{j=1}^n$ and obvious notation:

$$\max\left\{\sum_{J \in I(P_1)} |\Delta F(A_J)|, \sum_{J \in I(P_2)} |\Delta F(A_J)|\right\} \leq \sum_{J \in I(P)} |\Delta F(A_J)|.$$

*The partition P is called the **common refinement** of P_1 and P_2.*

Exercise 4.71 (n-increasing $\Rightarrow$ Vitali B.V.) *Verify that generalizing the 1-dimensional case, every function F that is n-increasing is of Vitali bounded variation, and further, $T_F(A) = \Delta F(A)$.*

Exercise 4.72 (T_F on degenerate A) *Recalling Exercise 4.65, prove that if A is degenerate then $T_F(A) = 0$.*

Remark 4.73 (n-increasing; Vitali B.V. on subrectangles) *If F is of Vitali bounded variation on a rectangle $A \equiv \prod_{i=1}^{n}[a_i, b_i]$, then it is an exercise to check that it enjoys this property on every closed sub rectangle of A.*

On the other hand, knowing that $\Delta F \geq 0$ on such a rectangle does not imply that this property is satisfied on sub rectangles. This is why the definition of n-increasing explicitly demands that this property holds on all rectangles.

The following exercise would appear to be a simple corollary of Proposition 4.76 below, using Proposition 4.67. However, this result requires an independent proof because the proof of Proposition 4.76 will require this property.

Exercise 4.74 (Finite additivity of $T_F(A)$) *Prove that given A with $T_F(A) < \infty$, and any partition $A = \bigcup_{J \in I} A_J$ into subrectangles, that:*

$$T_F(A) = \sum_{J \in I} T_F(A_J).$$

Thus $T_F(A)$ is finitely additive on partitions of rectangles on which it is finite. Hint: Justify first that by induction you can assume that the I-partition contains two rectangles. Now relate partitions of A to collections of partitions of $\{A_J\}_{J \in I}$, and apply Exercise 4.70.

The next result characterizes $T_F(A')$ for $A' \subset A$ using the **Vitali total variation function** defined on A. This function is well defined by Remark 4.73.

Definition 4.75 (Vitali total variation function on A) *If $F(x)$ is of Vitali bounded variation on the rectangle $A \equiv \prod_{i=1}^{n}[a_i, b_i]$, define the **Vitali total variation function** $T_{F,A}(x)$ for $x \in A$ by:*

$$T_{F,A}(x) \equiv T_F(A_x), \tag{4.50}$$

where $A_x \equiv \prod_{i=1}^{n}[a_i, x_i]$, and $T_F(A_x)$ is defined in (4.49).

Proposition 4.76 ($T_F(A') = \Delta T_{F,A}(A')$) *If $F(x)$ is of Vitali bounded variation on the rectangle $A \equiv \prod_{i=1}^{n}[a_i, b_i]$, then for any rectangle $A' \subset A$:*

$$T_F(A') = \Delta T_{F,A}(A'), \tag{4.51}$$

where $\Delta T_{F,A}(A')$ is defined in (4.44).
Proof. *We start with a simple example, which follows from the general result below, but is a good warm-up. If $A' = A_x \equiv \prod_{j=1}^{n}[a_j, x_j]$, then by (4.44) and (4.50):*

$$\Delta T_{F,A}(A_x) \equiv \sum_y sgn(y) T_F(A_y) = T_F(A_x).$$

This follows because if $y \neq x$ is a vertex of A_x, then some $y_i = x_i$ by definition. Thus A_y is degenerate and $T_F(A_y) = 0$ by Exercise 4.72. For $y = x$, $sgn(x) = 1$ and the result in (4.51) follows for $A' = A_x$.

More generally, if $A' = \prod_{j=1}^{n}[a'_j, b'_j] \subset A$, we prove as in (4.51):

$$T_F(A') = \sum_y sgn(y) T_{F,A}(y), \tag{1}$$

where the sum is over all vertices of A', and $sgn(y)$ is defined as above.

For this proof, we use the **inclusion-exclusion formula** of Proposition I.8.8, attributed to **Abraham de Moivre** *(1667–1754). While the result was notationally stated there as a property of measures, it can be seen from the proof that the only property used there for the given set function μ was finite additivity. Thus this result applies to the set function T_F by Exercise 4.74.*

Now as in the proof of Proposition I.8.9, note that:

$$A' = A_{(b'_1,\ldots,b'_n)} - \bigcup_{i=1}^n A_{x^{(i)}},$$

where $x_j^{(i)} = b'_j$ for $j \neq i$ and $x_i^{(i)} = a'_i$. This follows because if $x \in A_{(b'_1,\ldots,b'_n)} - A'$, then $x_j \leq a'_j$ for at least one $j = i$ and thus $x \in A_{x^{(i)}}$. Now $\bigcup_{i=1}^n A_{x^{(i)}} \subset A_{(b'_1,\ldots,b'_n)}$, so it follows by finite additivity of T_F that:

$$T_F(A') = T_F\left(A_{(b'_1,\ldots,b'_n)}\right) - T_F\left(\bigcup_{i=1}^n A_{x^{(i)}}\right). \tag{2}$$

Note that $T_F\left(A_{(b'_1,\ldots,b'_n)}\right) = sgn(y)T_{F,A}(y)$ for the term in (1) with no a_i components.

Then by the inclusion-exclusion formula of I.(8.6):

$$
\begin{aligned}
-T_F\left[\bigcup_{i=1}^n A_{x^{(i)}}\right] = \;& -\sum_{i=1}^n T_F[A_{x^{(i)}}] + \sum_{i<j} T_F\left[A_{x^{(i)}} \bigcap A_{x^{(j)}}\right] \\
& -\sum_{i<j<k} T_F\left[A_{x^{(i)}} \bigcap A_{x^{(j)}} \bigcap A_{x^{(k)}}\right] \pm \\
& \cdots + (-1)^n T_F\left[\bigcap_{j=1}^n A_{x^{(j)}}\right].
\end{aligned}
\tag{4.52}
$$

For the first sum, using (4.50):

$$-\sum_{i=1}^n T_F[A_{x^{(i)}}] = \sum_{i=1}^n sgn\left(x^{(i)}\right) T_{F,A}\left(x^{(i)}\right),$$

which equals the sum in (1) over all vertices with a single a_i-component, and so $sgn\left(x^{(i)}\right) = -1$ by definition.

For the second summation, $A_{x^{(i)}} \bigcap A_{x^{(j)}} \equiv A_{x^{(i,j)}}$ defined by $x_l^{(i,j)} = b'_l$ for $l \neq i,j$ and $x_l^{(i,j)} = a'_l$ for $l = i,j$. Thus:

$$\sum_{i<j} T_{F,A}\left(x^{(i)} \bigcap A_{x^{(j)}}\right) = \sum_{i<j} sgn\left(x_l^{(i,j)}\right) T_{F,A}\left(x^{(i,j)}\right),$$

which equals the sum in (1) over all vertices with two a_i-components, and $sgn\left(x_l^{(i,j)}\right) = 1$ by definition.

Similarly:

$$-\sum_{i<j<k} T_F\left[A_{x^{(i)}} \bigcap A_{x^{(j)}} \bigcap A_{x^{(k)}}\right] = sgn\left(x_l^{(i,j,k)}\right) \sum_{i<j<k} T_{F,A}\left(x^{(i,j,k)}\right),$$

is the sum over all vertices with three a_i-components, and $sgn\left(x_l^{(i,j)}\right) = -1$ by definition. The same result is obtained for the other terms in (4.52).

Combining these identifications with (2) obtains (1) and the proof is complete. ∎

The next result demonstrates that there are many functions that satisfy these definitions.

Proposition 4.77 (Examples of n-increasing; Vitali B.V.) *Given functions $\{F_j(x_j)\}_{j=1}^n$, define:*

$$F(x) = \prod_{j=1}^n F_j(x_j).$$

1. *If $\{F_j\}_{j=1}^n$ are increasing, then F is n-increasing.*

2. *If $\{F_j\}_{j=1}^n$ are of bounded variation on intervals $\{[a_j, b_j]\}_{j=1}^n$, then F is of Vitali bounded variation on $A \equiv \prod_{j=1}^n [a_j, b_j]$.*

Proof. *First note that given any rectangle $B \equiv \prod_{j=1}^n [c_j, d_j]$, that:*

$$\Delta F(B) \equiv \sum_x sgn(x) F(x) = \prod_{j=1}^n [F_j(d_j) - F_j(c_j)]. \tag{1}$$

This follows because when expanded, the product becomes a summation over all vertices of B, with the coefficient of each term reflecting the parity of the number of c_j-factors as in the definition of $sgn(x)$. Item 1 now follows from (1).

For item 2, let P be a partition of A using the notational convention above. Then:

$$\begin{aligned} \sum_{J \in I} |\Delta F_J| &= \sum_{J \in I} \prod_{j=1}^n |F_j(x_{j,i_j}) - F_j(x_{j,i_j-1})| \\ &= \prod_{j=1}^n \left[\sum_{i_j=1}^{m_j} |F_j(x_{j,i_j}) - F_j(x_{j,i_j-1})| \right]. \end{aligned}$$

This follows by recalling that $I = \{(i_1, i_2, ..., i_n) | 1 \le i_j \le m_j\}$.

Taking a supremum over all such P, and denoting by P_j the induced partition of component $[a_j, b_j]$:

$$\begin{aligned} T_F(A) &= \sup_P \prod_{j=1}^n \left[\sum_{i_j=1}^{m_j} |F_j(x_{j,i_j}) - F_j(x_{j,i_j-1})| \right] \\ &\le \prod_{j=1}^n \sup_{P_j} \left[\sum_{i_j=1}^{m_j} |F_j(x_{j,i_j}) - F_j(x_{j,i_j-1})| \right] \\ &= \prod_{j=1}^n T_{a_j}^{b_j}[F_j], \end{aligned}$$

where $T_{a_j}^{b_j}[F_j]$ denotes the total variation of F_j on $[a_j, b_j]$. ∎

The final result for this study of Riemann-Stieltjes integrators generalizes Proposition 3.29, which stated that a function $f(x)$ defined on $[a, b]$ is of bounded variation if and only if

$$f(x) = I_1(x) - I_2(x),$$

for increasing, real-valued functions $I_1(x)$ and $I_2(x)$.

As noted in Remark 3.30, the implication for Riemann-Stieltjes integration was that for all $x \in [a, b]$:

$$\Delta f = \Delta I_1 - \Delta I_2,$$

where $\Delta f \equiv f(x) - f(a)$ and ΔI_1 and ΔI_2 are defined analogously. In other words, the interval set function Δf can be decomposed into a difference of nonnegative interval set functions, since $\Delta I_1 \ge 0$ and $\Delta I_2 \ge 0$. Two explicit examples were given. This result led to Proposition 4.23, which connected the associated Riemann-Stieltjes integrals.

For Vitali bounded variation functions on a rectangle $A = \prod_{j=1}^n [a_j, b_j]$, a similar decomposition is true for $\Delta F(A)$.

Proposition 4.78 (Characterization of Vitali B.V.) *A function $F(x)$ is of Vitali bounded variation on $A = \prod_{j=1}^n [a_j, b_j]$ if and only if there exists nonnegative, finitely additive set functions G_1 and G_2, so that for all rectangles $A' \subset A$:*

$$\Delta F(A') = G_1(A') - G_2(A'), \tag{4.53}$$

where ΔF is defined as in (4.44).

In particular, such a decomposition is given by:

$$G_1(A') = \frac{1}{2}\left[T_F(A') + \Delta F(A')\right], \quad G_2(A') = \frac{1}{2}\left[T_F(A') - \Delta F(A')\right], \qquad (4.54)$$

where $T_F(A')$ denotes the Vitali total variation of F on A' as in (4.49).

Proof. *If $F(x)$ is of Vitali bounded variation on A, then it is of Vitali bounded variation on A' for any rectangle $A' \subset A$ by Remark 4.73. Thus the set functions G_1 and G_2 in (4.54) are well-defined, and it is apparent that (4.53) is satisfied. Further, each G_j is finitely additive by Proposition 4.67 and Exercise 4.74.*

For nonnegativity, let $A' = \bigcup_{J \in I} A_J$ be a decomposition of A' into subrectangles. Then $|\Delta F(A_J)| = \max[\pm\Delta F(A_J)]$ for any A_J, and thus $|\Delta F(A_J)| \pm \Delta F(A_J) \geq 0$. It then follows by finite additivity of ΔF that:

$$\sum_{J \in I} |\Delta F(A_J)| \pm \Delta F(A') \geq 0.$$

Taking a supremum over all such partitions obtains $G_j(A') \geq 0$ for all x.

Conversely, given $F(x)$ with decomposition of $\Delta F(A')$ as in (4.53), let $A = \bigcup_{J \in I} A_J$ be a decomposition of A into subrectangles. Then by finite additivity of ΔF and the assumed nonnegativity and finite additivity of G_j, it follows that:

$$\sum_{J \subset I} |\Delta F(A_J)| \leq \sum_{J \in I} G_1(A_J) + \sum_{J \in I} G_2(A_J)$$
$$= G_1(A) + G_2(A).$$

This bound is valid after taking a supremum over all partitions, and thus F is of Vitali bounded variation on A. ∎

The conclusion of Proposition 3.29 that bounded variation $f(x)$ on $[a, b]$ can be expressed as $f(x) = I_1(x) - I_2(x)$ for increasing, real-valued functions $I_1(x)$ and $I_2(x)$, also provides a decomposition of the "measure" or "content" of $[a, x]$ under f. As noted in Remark 3.30, the content $\Delta_x f \equiv f(x) - f(a)$ equals a difference of the nonnegative I_j-lengths of this interval, $\Delta_x I_j \equiv I_j(x) - I_j(a)$ (recall I.(5.4)), using increasing, though not necessarily right continuous functions. This followed because $I_1(a) - I_2(a) = f(a)$ for both representations identified in (3.21) and (3.22).

While the above result generalizes this to $F(x)$ of Vitali bounded variation on A, and characterizes the n-dimensional content $\Delta F(A')$ as a difference of nonnegative, finitely additive set functions $G_1(A')$ and $G_2(A')$, this latter result is not yet equivalent to the former. In particular, this result does not obtain a decomposition of the **function** $F(x)$ into identified n-increasing functions I_1 and I_2 for which G_1 and G_2 in (4.54) are given by ΔI_1 and ΔI_2 as defined in (4.44).

The following result completes the generalization.

Corollary 4.79 (Characterization of Vitali $B.V.$) *A function $F(x)$ is of Vitali bounded variation on $A = \prod_{j=1}^{n}[a_j, b_j]$ if and only if there exists n-increasing functions $I_1(x)$ and $I_2(x)$, so that:*

$$F(x) = I_1(x) - I_2(x). \qquad (4.55)$$

In particular, one such decomposition is given by:

$$I_1(x) = \frac{1}{2}\left[T_{F,A}(x) + F(x)\right], \quad I_2(x) = \frac{1}{2}\left[T_{F,A}(x) - F(x)\right], \qquad (4.56)$$

where $T_{F,A}(x) \equiv T_F(A_x)$ by Definition 4.75.

Proof. *If $F(x)$ is of Vitali bounded variation on A, then $I_1(x)$ and $I_2(x)$ given in (4.56) are well-defined and satisfy (4.55).*

To prove that these functions are n-increasing, we prove that for all rectangles $A' \subset A$:

$$G_1(A') = \Delta I_1(A'), \quad G_2(A') = \Delta I_2(A'), \tag{1}$$

with G_1 and G_2 as in Proposition 4.78. Thus $\Delta I_j(A') \geq 0$ for all A'.

To prove (1), it follows from linearity of Δ in (4.46) and Proposition 4.76:

$$
\begin{aligned}
\Delta I_j(A') &= \frac{1}{2}\left[\Delta T_{F,A}(A') \pm \Delta F(A')\right] \\
&= \frac{1}{2}\left[T_F(A') \pm \Delta F(A')\right] = G_j(A').
\end{aligned}
$$

Conversely, if (4.55) is satisfied on A for n-increasing functions $I_1(x)$ and $I_2(x)$, then by linearity of Δ and nonnegativity of $\Delta I_j(A_J)$ for $A_J \subset A$:

$$|\Delta F(A_J)| \leq \Delta I_1(A_J) + \Delta I_2(A_J).$$

Thus given any partition $A = \bigcup_J A_J$, finite additivity of ΔI_j obtains:

$$\sum_J |\Delta F(A_J)| \leq \Delta I_1(A) + \Delta I_2(A).$$

This is then true for the supremum over all such partitions, and so $F(x)$ is of Vitali bounded variation. ■

4.4.2 Definition of the Integral

Given a bounded function $g(x)$ defined on A, and an appropriate integrator $F(x)$, existence of the Riemann-Stieltjes integral:

$$\int_A g\,dF,$$

will be defined as in the 1-dimensional case. That is, we provide both a Riemann-Stieltjes sums approach suitable for general integrators, as well as the Darboux sums approach that requires n-increasing integrators.

In later sections, we then investigate equivalence for increasing integrators, as well as existence of these integrals, and finally properties and evaluations.

We begin with the definition using the Riemann-Stieltjes sums approach.

Definition 4.80 (Riemann-Stieltjes integral; General F) *Let $g(x)$ be bounded and $F(x)$ defined on a rectangle $A = \prod_{j=1}^n [a_j, b_j]$. Given the partition $P \equiv \left\{ \{x_{j,i_j}\}_{i_j=0}^{m_j} \right\}_{j=1}^n$ of A and tags $\{\widetilde{x}_J\}_{J \in I}$ with $\widetilde{x}_J \in A_J \equiv \prod_{j=1}^n [x_{j,i_j-1}, x_{j,i_j}]$ for $J = (i_1, i_2, ..., i_n)$, define the* **Riemann-Stieltjes summation** *by:*

$$R(g, F; P) \equiv \sum_{J \in I} g(\widetilde{x}_J)\, \Delta F(A_J), \tag{4.57}$$

where $\Delta F(A_J)$ is given in (4.44).
If:

$$\lim_{\mu \to 0} R(g, F; P) = I < \infty, \tag{4.58}$$

meaning that for any $\epsilon > 0$ there is a δ so that for all **tagged partitions** *P of mesh size $\mu \leq \delta$:*

$$|R(g, F; P) - I| < \epsilon,$$

*then define the **Riemann-Stieltjes integral of** $g(x)$ **with respect to** $F(x)$ **over** A:*

$$\int_A g(x)dF = I. \tag{4.59}$$

If (4.58) is not satisfied, we say that Riemann-Stieltjes integral of $g(x)$ with respect to $F(x)$ over A does not exist.

If $g(x)$ is Riemann-Stieltjes integrable with respect to $F(x)$, one sometimes uses the notation $g \in \mathcal{R}(F)$.

Notation 4.81 (Tagged partition) *The collections $\left\{\{x_{j,i_j}\}_{i_j=0}^{m_j}\right\}_{j=1}^n$ and $\{\widetilde{x}_J\}_{J \in I}$ are sometimes referred to as a **tagged partition** and denoted $P \equiv \left(\left\{\{x_{j,i_j}\}_{i_j=0}^{m_j}\right\}_{j=1}^n, \{\widetilde{x}_J\}_{J \in I}\right)$, while the latter set of points are called the **subrectangle tags**.*

For the Darboux sums approach, given bounded $g(x)$ and partition $P \equiv \left\{\{x_{j,i_j}\}_{i_j=0}^{m_j}\right\}_{j=1}^n$ of $A = \prod_{j=1}^n [a_j, b_j]$ as in (4.43), we again suppress the **lower and upper bounding step functions,** and instead introduce lower and upper Darboux sums directly.

Definition 4.82 (Darboux sums) *Given a **bounded function** $g(x)$ and an n-increasing function $F(x)$ defined on $A = \prod_{j=1}^n [a_j, b_j]$ with partition $P \equiv \left\{\{x_{j,i_j}\}_{i_j=0}^{m_j}\right\}_{j=1}^n$, so that $A = \bigcup_{J \in I} A_J$, define $L(g, F; P)$, the **lower Darboux sum,** and $U(g, F; P)$, **the upper Darboux sum,** by:*

$$L(g, F; P) \equiv \sum_{J \in I} m_J \Delta F(A_J), \quad U(g, F; P) \equiv \sum_{J \in I} M_J \Delta F(A_J), \tag{4.60}$$

where $m_J \equiv \inf\{g(x) | x \in A_J\}$, $M_J \equiv \sup\{g(x) | x \in A_J\}$, and $\Delta F(A_J)$ is defined in (4.44).

Because the integrator function $F(x)$ is assumed to be n-increasing, it is apparent by definition that for any given partition P that:

$$L(g, F; P) \le U(g, F; P).$$

This relationship is less apparent but also true when different partitions are used for the upper and lower sums. The following proposition clarifies this with an affirmative result, but first a familiar definition.

Definition 4.83 (Refinement of a partition) *Given $A = \prod_{j=1}^n [a_j, b_j]$, a partition $P' \equiv \left\{\{x'_{j,i_j}\}_{i_j=0}^{m'_j}\right\}_{j=1}^n$ of A is a **refinement** of a partition $P \equiv \left\{\{x_{j,i_j}\}_{i_j=0}^{m_j}\right\}_{j=1}^n$ if $\left\{\{x_{j,i_j}\}_{i_j=0}^{m_j}\right\}_{j=1}^n \subset \left\{\{x'_{j,i_j}\}_{i_j=0}^{m'_j}\right\}_{j=1}^n$.*

Proposition 4.84 $(L(g, F; P_1) \le U(g, F; P_2))$ *If $g(x)$ is bounded and $F(x)$ is n-increasing on A, then for any partitions P_1 and P_2:*

$$L(g, F; P_1) \le U(g, F; P_2).$$

Proof. Let P be a partition of A defined by all the points of both P_1 and P_2, and thus P is a **common refinement** of both P_1 and P_2. We claim that:

$$L(g, F; P_1) \le L(g, F; P) \le U(g, F; P) \le U(g, F; P_2), \tag{4.61}$$

which proves the stated result.

By definition $L(g, F; P) \leq U(g, F; P)$, so the proof will be complete by demonstrating that $L(g, F; P_1) \leq L(g, F; P)$ and $U(g, F; P) \leq U(g, F; P_2)$. By induction, it is enough to prove this assuming that P contains only one more point than P_1 or P_2.

For example, let $j = 1$. If $\{x_{1,i}\}_{i=0}^{m_1}$ defines partition P_1 or P_2, let P be defined with $\{x_{1,i_1}\}_{i_1=0}^{m_1} \bigcup \{z\}$ with $x_{1,0} < z < x_{1,1}$, and assume that P_1, P_2, and P have the same partition points $\{x_{j,i_j}\}_{i_j=0}^{m_j}$ for $j > 1$. Then the first and third inequalities in (4.61) follow from a simple observation. If A_J is any subrectangle of A with first interval component $[x_{1,0}, x_{1,1}]$, and m_J and M_J denote the associated bounds for $g(x)$ on A_J, then:

$$m_J = \min(m'_J, m''_J), \quad \max(M'_J, M''_J) = M_J.$$

Here m'_J and m''_J denote the infima of $g(x)$ on the two rectangles A'_J and A''_J defined by z that union to A_J, and M'_J and M''_J are similarly defined in terms of suprema.

Thus by finite additivity of ΔF, letting I' denote the collection of indexes with first interval component $[x_{1,0}, x_{1,1}]$:

$$
\begin{aligned}
L(g, F; P) - L(g, F; P_1) &= \sum\nolimits_{J \in I'} [m'_J \Delta F(A'_J) + m''_J \Delta F(A''_J) - m_J \Delta F(A_J)] \\
&= \sum\nolimits_{J \in I'} [m'_J - m_J] \Delta F(A'_J) + \sum\nolimits_{J \in I'} [m''_J - m_J] F(A''_J) \\
&\geq 0.
\end{aligned}
$$

Similarly, $U(g, F; P_2) - U(g, F; P) \geq 0$, and the proof is complete. ∎

We now define the lower and upper Darboux integrals as in Definition 4.9.

Definition 4.85 (Upper/lower Darboux integrals) *If $g(x)$ is bounded and $F(x)$ is n-increasing on $A = \prod_{j=1}^{n}[a_j, b_j]$, the **lower Darboux integral of g with respect to F over A** is defined:*

$$\underline{\int_A} g(x) dF \equiv \sup_P L(g, F; P), \tag{4.62}$$

*and the **upper Darboux integral of g with respect to F over A** is defined:*

$$\overline{\int_A} g(x) dF \equiv \inf_P U(g, F; P), \tag{4.63}$$

where the supremum and infimum are defined over all partitions of A.

The lower and upper Darboux integrals always exist and are finite since the associated Darboux sums are monotonic with respect to partition refinements by Proposition 4.84. Further, these sums have an upper bound of $M\Delta F(A)$ and lower bound of $m\Delta F(A)$ if $m \leq g(x) \leq M$.

The following definition states that a bounded function $g(x)$ is Riemann-Stieltjes integrable with respect to n-increasing $F(x)$ over A when the upper and lower Darboux integrals agree.

Definition 4.86 (Riemann-Stieltjes integral: Increasing $F(x)$) *Let $g(x)$ be bounded and $F(x)$ n-increasing on $A = \prod_{j=1}^{n}[a_j, b_j]$ and assume that:*

$$\sup_P L(g, F; P) = \inf_P U(g, F; P) = I < \infty, \tag{4.64}$$

where the supremum and infimum are taken over all partitions P of A.

*Then the **Riemann-Stieltjes integral of $g(x)$ with respect to $F(x)$ over A** is defined by:*

$$\int_A g(x) dF = I. \tag{4.65}$$

If (4.15) is not satisfied, then we say that the Riemann-Stieltjes integral of $g(x)$ with respect to $F(x)$ over A does not exist.

If $g(x)$ is Riemann-Stieltjes integrable with respect to $\dot{F}(x)$, one sometimes uses the notation $g \in \mathcal{R}(F)$.

Remark 4.87 (On $\mu \to 0$) *Recalling Remark 4.11, is it again the case that by virtue of (4.61), the integrability condition of (4.64) can be restated as:*

$$\lim_{\mu \to 0} \sup_{P_\mu} L(g, F; P_\mu) = \lim_{\mu \to 0} \inf_{P_\mu} U(g, F; P_\mu) = I, \qquad (4.66)$$

where P_μ denotes a partition of mesh size μ.

4.4.3 Riemann and Darboux Equivalence

In this section, we prove that the Riemann-Stieltjes sums Definition 4.80 and Darboux sums Definition 4.86 provide equivalent criteria for the existence of the Riemann-Stieltjes integral of a bounded function $g(x)$ for an n-**increasing integrator** $F(x)$. In other words, we prove that for an n-increasing integrator, a bounded function is Riemann-Stieltjes integrable in terms of Riemann-Stieltjes sums if and only if it is so integrable in terms of Darboux sums, and further that the integrals agree.

We do this by showing that each is equivalent to (4.67), which is the **Cauchy criterion for the existence of the Riemann-Stieltjes integral** and named after **Augustin-Louis Cauchy** (1789–1857). The Cauchy criterion here is identical to that stated for the Riemann integral in (1.16) and (1.50), and for the Riemann-Stieltjes integral in (4.18).

Proposition 4.88 (Cauchy criterion for the Riemann-Stieltjes integral) *A bounded function $g(x)$ is Riemann-Stieltjes integrable with respect to n-increasing $F(x)$ on $A = \prod_{j=1}^{n}[a_j, b_j]$ by Definition 4.80 if and only if it is Riemann-Stieltjes integrable by Definition 4.86, and then the integrals agree.*

In either case, $g(x)$ is so integrable if and only for any $\epsilon > 0$ there is a partition P, so that:

$$0 \leq U(g, F; P) - L(g, F; P) < \epsilon, \qquad (4.67)$$

with $U(g, F; P)$ and $L(g, F; P)$ given in (4.60).

Further, if (4.67) is satisfied for P, it is also satisfied for every refinement of P.

Proof. *As noted above, we show that the requirement of either definition of integrability is equivalent to the above Cauchy criterion, and thus each is equivalent to the other.*

Given $\epsilon > 0$, assume that (4.67) is satisfied. It follows from (4.61) and the definition of infimum and supremum that for such P:

$$L(g, F; P) \leq \sup_{P'} L(g, F; P') \leq \inf_{P'} U(g, F; P') \leq U(g, F; P).$$

Thus:

$$\inf_{P'} U(g, F; P') - \sup_{P'} L(g, F; P') < \epsilon. \qquad (1)$$

Since this is true for all ϵ, g is Riemann-Stieltjes integrable with respect to n-increasing $F(x)$ over A by the Darboux sums criterion of Definition 4.86.

Now (4.66) implies that there exists μ' so that for every partition P_1 with mesh size $\mu_1 \leq \mu'$:

$$\sup_{P'} L(g, F; P') \leq L(g, F; P_1) + \epsilon/2.$$

Similarly, there exists μ'' so that for every partition P_2 with mesh size $\mu_2 \leq \mu''$:

$$\inf_{P'} U(g, F; P') \geq U(g, F; P_2) - \epsilon/2.$$

By (4.67) and (1), given any partition P_3 with mesh size $\mu_3 \leq \min\{\mu', \mu''\}$:

$$U(g, F; P_3) - L(g, F; P_3) < 2\epsilon.$$

Thus given any such partition, and any interval tags $\{\widetilde{x}_J\}_{J \in I}$:

$$L(g, F; P_3) \leq \sum_{J \in I} g(\widetilde{x}_J) \Delta F(A_J) \leq L(g, F; P_3) + 2\epsilon. \tag{2}$$

As (2) is satisfied for every partition P_3 with mesh size $\mu_3 \leq \min\{\mu', \mu''\}$, it is thus true taking a limit as $\mu_3 \to 0$. By (4.66), with I denoting the integral using Darboux sums:

$$I \leq \lim_{\mu \to 0} \sum_{J \in I} g(\widetilde{x}_J) \Delta F(A_J) \leq I + 2\epsilon. \tag{3}$$

Since ϵ is arbitrary, g is Riemann-Stieltjes integrable with respect to n-increasing $F(x)$ on A by the Riemann-Stieltjes sums criterion of Definition 4.80.

Further by (3), the Riemann-Stieltjes integrals of these definitions agree.

Conversely, if g is Riemann-Stieltjes integrable with respect to n-increasing $F(x)$ on A by the Darboux sums criterion of Definition 4.86 and $\epsilon > 0$ is given, then by definition of extrema there exists partitions P_1 and P_2, so that:

$$0 \leq U(g, F; P_1) - \int_A g(x) dF < \epsilon/2,$$

$$0 \leq \int_A g(x) dF - L(g, F; P_2) < \epsilon/2.$$

Thus:

$$0 \leq U(g, F; P_1) - L(g, F; P_2) < \epsilon. \tag{4}$$

It then follows from (4.61) that (4) is satisfied with $U(g, F; P)$ and $L(g, F; P)$ defined relative to the common refinement P of these partitions, and this is (4.67).

Similarly, assume that g is Riemann-Stieltjes integrable with respect to n-increasing $F(x)$ on A by the Riemann-Stieltjes sums criterion of Definition 4.80 and $\epsilon > 0$ is given. Then there is a δ so that for all tagged partitions $P \equiv \left(\left\{ \{x_{j,i_j}\}_{i_j=0}^{m_j} \right\}_{j=1}^{n}, \{\widetilde{x}_J\}_{J \in I} \right)$ of mesh size $\mu \leq \delta$:

$$\left| \sum_{J \in I} g(\widetilde{x}_J) \Delta F(A_J) - \int_A g(x) dF \right| < \epsilon/3.$$

Thus for any such partition and any two sets of tags $\{\widetilde{x}_J\}_{J \in I}$ and $\{\widetilde{x}'_J\}_{J \in I}$:

$$\left| \sum_{J \in I} g(\widetilde{x}_J) \Delta F(A_J) - \sum_{J \in I} g(\widetilde{x}'_J) \Delta F(A_J) \right| < 2\epsilon/3.$$

By definition of M_J and m_J, we can choose these tags so that these Riemann-Stieltjes sums are within $\epsilon/6$ of $U(g, F; P)$ and $L(g, F; P)$, respectively, and this obtains (4.67).

Finally, if (4.67) is satisfied for a given partition, it is also satisfied for every refinement by (4.61). ∎

4.4.4 On Existence of the Integral

The primary result of this section is the existence theorem addressed next. This result is stated in terms of an integrator function F of Vitali bounded variation, but then applies to n-increasing integrator functions by Exercise 4.71. This latter result is stated in Corollary 4.90 and applies when F is a joint distribution function of a random vector defined on a probability space by Proposition II.6.9.

We then investigate improper integrals, and the decomposition of integrals with Vitali bounded variation integrators.

Proposition 4.89 ($g(x)$ **continuous,** $F(x)$ **Vitali B.V.**) *If* $F(x)$ *is of Vitali bounded variation and* $g(x)$ *is continuous on a bounded rectangle* $A \equiv \prod_{j=1}^{n}[a_j, b_j]$, *then* $\int_A g(x)dF$ *exists.*

Further, with $M_g \equiv \sup_A |g(x)|$ *and* $T_F(A)$ *defined in (4.49):*

$$\left| \int_A g(x)dF \right| \leq M_g T_F(A). \tag{4.68}$$

Proof. *If* $T_F(A) = 0$, *then* $\Delta F(A_J) = 0$ *for every subrectangle* A_J *induced by any partition* $P = \{\{x_{j,i}\}_{i=0}^{m_j}\}_{j=1}^{n}$ *of* A *by (4.49). Thus* $\int_A g(x)dF = 0$, *and (4.68) is satisfied.*

Assume therefore that $T_F(A) > 0$. *Since* $g(x)$ *is continuous on compact* A, *it is bounded and uniformly continuous by Exercise 1.14. Thus for any* $\epsilon > 0$ *there is* δ_0 *so that if* $|x - x'| < \delta_0$ *then:*

$$|g(x) - g(x')| < \epsilon / [2T_F(A)]. \tag{1}$$

We now show that if $\delta \equiv \delta_0/\sqrt{n}$, *then given partitions* P_1 *and* P_2 *with respective mesh sizes* $\mu_1, \mu_2 < \delta$ *and any tags, that:*

$$|R(g; P_1, F) - R(g; P_2, F)| < \epsilon. \tag{2}$$

This then implies that Riemann-Stieltjes summations can be made arbitrarily close as $\mu_j \to 0$, *and thus* $\lim_{\mu \to 0} R(g; P, F)$ *is well-defined.*

To prove (2), we prove that with P *the common refinement of these partitions, that:*

$$|R(g; P_1, F) - R(g; P, F)| < \epsilon/2, \qquad |R(g; P_2, F) - R(g; P, F)| < \epsilon/2, \tag{3}$$

and apply the triangle inequality.

With notation as above, and I_1 *and* I *denoting the index collections for partitions* P_1 *and* P, *respectively, then for any tags:*

$$R(g; P_1, F) - R(g; P, F) = \sum_{J_1 \in I_1} g(x_{J_1}) \Delta F(A_{J_1}) - \sum_{J \in I} g(x_J) \Delta F(A_J).$$

For every sub rectangle A_J *defined by* P *that is contained in a sub rectangle* A_{J_1} *defined by* P_1, *define* $x'_J = x_{J_1}$. *Then by finite additivity in (4.45):*

$$\sum_{A_J \subset A_{J_1}} g(x'_J) \Delta F(A_J) = g(x_{J_1}) \Delta F(A_{J_1}),$$

and thus:

$$R(g; P_1, F) - R(g; P, F) = \sum_{J \in I} [g(x'_J) - g(x_J)] \Delta F_J. \tag{4}$$

By construction, every pair (x'_J, x_J) *in (4) is in the same sub rectangle of* P_1. *As the mesh size of* P_1 *is less than* $\delta_0/\sqrt{n}$, *it follows that* $|x'_J - x_J| < \delta$ *and then* $|g(x'_J) - g(x_J)| < \epsilon / [2T_F(A)]$ *by (1). Hence:*

$$|R(g; P_1, F) - R(g; P, F)| \leq \epsilon / [2T_F(A)] \sum_{J \in I} |\Delta F_{J'}| \leq \epsilon/2,$$

which is the first result in (3). *The same construction applies to the second estimate in* (3), *and the proof of* (2), *and the well-definedness of* $\lim_{\mu \to 0} R(g; P, F)$, *is complete.*

To prove that this limit is finite and thus equal to $\int_A g(x)dF$, let mesh size μ and an associated P be given. Then:

$$\begin{aligned} |R(g; P, F)| &\leq \sum_{J \in I} |g(x_J)| |\Delta F_J| \\ &\leq M_g T_F(A). \end{aligned}$$

As the same is true for all partitions, this bound is also true for the limit as $\mu \to 0$.

Thus $\int_A g(x)dF$ exists and satisfies (4.68). ■

As noted above, this proposition provides a useful corollary that is worth summarizing.

Corollary 4.90 ($g(x)$ **continuous,** n**-increasing** $F(x)$) *If* $F(x)$ *is* n-*increasing and* $g(x)$ *is continuous on a bounded rectangle* $A \equiv \prod_{j=1}^{n} [a_j, b_j]$, *then* $\int_A g(x)dF$ *exists.*

Further, with $M_g \equiv \sup_A |g(x)|$ *and* $\Delta F(A)$ *defined in* (4.44):

$$\left| \int_A g(x)dF \right| \leq M_g \Delta F(A). \tag{4.69}$$

Proof. *This follows by Exercise 4.71, that* n-*increasing implies of Vitali bounded variation and* $T_F(A) = \Delta F(A)$. ■

Exercise 4.91 *Prove Corollary 4.90 directly using Proposition 4.88.*

The next result establishes criteria for the existence of **improper Riemann-Stieltjes integrals.**

Proposition 4.92 (Improper Riemann-Stieltjes integrals) *Let* $g(x)$ *be continuous and* $F(x)$ n-*increasing on* $A \equiv \prod_{j=1}^{n} [a_j, \infty)$. *If both functions are bounded on* A, *the improper integral:*

$$\int_A g(x)dF \equiv \lim_{b \to \infty} \int_{A_b} g(x)dF, \tag{4.70}$$

is well defined and finite, where $b = (b_1, ..., b_n)$ *and* $A_b \equiv \prod_{j=1}^{n} [a_j, b_j]$.

If both functions are bounded on $\mathbb{R}^n$, *the improper integral:*

$$\int_{\mathbb{R}^n} g(x)dF \equiv \lim_{\substack{a \to -\infty \\ b \to +\infty}} \int_{A_b} g(x)dF, \tag{4.71}$$

is well defined and finite, with $a = (a_1, ..., a_n)$ *and* b *and* A_b *as above.*

More generally, such limits exist for any subset of $a_j \to -\infty$ *and/or* $b_j \to \infty$, *and thus all such improper integrals are well-defined.*

Proof. *Since* F *is bounded and* n-*increasing, so too is* ΔF *bounded on any rectangle. By* (4.44), *if* $|F| \leq M_F$ *then* $\Delta F \leq 2^n M_F$ *on any rectangle by the triangle inequality. Letting* $c > a$ *mean that* $c_j > a_j$ *for all* j, *then for any* $c > a$ *this bound implies that* $\Delta F(A_c) \leq 2^n M_F$ *and thus:*

$$\sup_{c>a} \Delta F(A_c) \leq 2^n M_F.$$

Since this supremum is bounded and $|g| \leq M_g$, *choose* $b > a$, *so that:*

$$\sup_{c>a} \Delta F(A_c) - \Delta F(A_b) < \epsilon/M_g.$$

Then since:

$$\sup_{c>a} \Delta F(A_c) \geq \sup_{c>b} \Delta F(A_c) \geq \Delta F(A_b),$$

it follows that:

$$\sup_{c>b} \Delta F(A_c) - \Delta F(A_b) < \epsilon/M_g. \tag{1}$$

Now with b chosen as above and $c > b$ arbitrary, note that A_c can be expressed as a union of A_b and $2^n - 1$ other rectangles $\{A_k\}_{k=1}^{2^n-1}$:

$$A_c = A_b \bigcup \bigcup_{k=1}^{2^n-1} A_k,$$

by dividing each component interval that defines A_c as follows:

$$[a_j, c_j] = [a_j, b_j] \bigcup [b_j, c_j].$$

Applying Proposition 4.94 below to integrals that exist by Corollary 4.90, and using finite additivity and the decomposition of A_c just constructed:

$$\left| \int_{A_c} g(x)dF - \int_{A_b} g(x)dF \right| \leq \sum_{k=1}^{2^n-1} \int_{A_k} |g(x)|\, dF \leq M_g \left[\Delta F(A_c) - \Delta F(A_b) \right]. \tag{2}$$

Now for any $c > b$:

$$\sup_{c>b} \Delta F(A_c) \geq \Delta F(A_c),$$

and so by (1):

$$\Delta F(A_c) - \Delta F(A_b) < \epsilon/M_g.$$

Thus from (2):

$$\int_{A_b} g(x)dF - \epsilon \leq \int_{A_c} g(x)dF \leq \int_{A_b} g(x)dF + \epsilon. \tag{3}$$

This obtains that $\liminf_{c\to\infty} \int_{A_c} g(x)dF$ and $\limsup_{c\to\infty} \int_{A_c} g(x)dF$ satisfy these same bounds and so:

$$0 \leq \limsup_{c\to\infty} \int_{A_c} g(x)dF - \liminf_{c\to\infty} \int_{A_c} g(x)dF < 2\epsilon.$$

Since ϵ is arbitrary, $\lim_{b_j\to\infty} \int_{A_b} g(x)dF$ is well defined.

The proof of 4.71 is similar and left as an exercise, as is the statement on subsets of $a_j \to -\infty$ and/or $b_j \to \infty$. ∎

The final result of this section generalizes the decomposition result of Proposition 4.23 with the aid of Corollary 4.79.

Proposition 4.93 ($\int_A g(x)dF = \int_A g(x)dI_1 - \int_A g(x)dI_2$; $F \in B.V.$) *Given $A \equiv \prod_{j=1}^{n}[a_j, b_j]$, assume that $\int_A g(x)dF$ exists for bounded $g(x)$ and $F(x)$ of Vitali bounded variation, and let $F(x) = I_1(x) - I_2(x)$ be the decomposition of Corollary 4.79 in terms of n-increasing functions in (4.56).*

Then $\int_A g(x)dI_1$ and $\int_A g(x)dI_2$ exist and:

$$\int_A g(x)dF = \int_A g(x)dI_1 - \int_A g(x)dI_2. \tag{4.72}$$

Further, if $F(x) = F_1(x) - F_2(x)$ is an arbitrary decomposition of $F(x)$ in terms of n-increasing functions, and if the integrals of $g(x)$ with respect to $F_1(x)$ and $F_2(x)$ exist, then:

$$\int_A g(x)dF = \int_A g(x)dF_1 - \int_A g(x)dF_2. \tag{4.73}$$

If $g(x)$ is continuous, then (4.73) is valid for any such decomposition $F(x) = F_1(x) - F_2(x)$.
Proof. *Assume that we have proved existence of the I_j-integrals and (4.72). Then existence of the integrals with respect to increasing $F_1(x)$ and $F_2(x)$ assures by Proposition 4.94 that the integrals of $g(x)$ exist with respect to both $I_1(x) + F_2(x)$ and $I_2(x) + F_1(x)$. Further, each of these integrals equals the sum of the associated integrals, for example:*

$$\int_A g(x)d(I_1 + F_2) = \int_A g(x)dI_1 + \int_A g(x)dF_2.$$

But then $I_1(x) + F_2(x) = I_2(x) + F_1(x)$ obtains:

$$\int_A g(x)dI_1 + \int_A g(x)dF_2 = \int_A g(x)dF_1 + \int_A g(x)dI_2,$$

and (4.73) follows from (4.72).
    For (4.72), any tagged partition of A obtains that:

$$\sum\nolimits_{J \in I} g(\widetilde{x}_J) \Delta F(A_J) = \sum\nolimits_{J \in I} g(\widetilde{x}_J) \Delta I_1(A_J) - \sum\nolimits_{J \in I} g(\widetilde{x}_J) \Delta I_2(A_J).$$

By assumption, the first summation converges to $\int_A g(x)dF$ as $\mu \to 0$ in the sense of Definition 4.80. Thus if it can be proved that one of the I_j-summations similarly converges, then the other must converge by default and this obtains (4.72).
    To this end, given $\epsilon > 0$ choose δ so that for every partition P if A with mesh size $\mu \leq \delta$:

$$\left| \sum\nolimits_{J \in I} [g(\widetilde{x}_J) - g(\widetilde{x}_J)] \Delta F(A_J) \right| < \epsilon, \ \ all \ \widetilde{x}_J, \widetilde{y}_J \in A_J, \tag{1}$$

and,

$$T_F(A) < \sum\nolimits_{J \in I} |\Delta F(A_J)| + \epsilon. \tag{2}$$

    Given such ϵ, the first bound exists for all partitions P_1 of mesh size $\mu_1 \leq \delta_1$ by the Riemann-Stieltjes sums Definition 4.80 of the integral $\int_A g(x)dF$. The second follows from the definition of $T_F(A)$ as the supremum over all partitions, and Exercise 4.70 that proved that if $T_{F,\mu}(A)$ is defined as the supremum over all partitions of mesh size μ, then $T_F(A) = \lim_{\mu \to 0} T_{F,\mu}(A)$. Thus given such ϵ there exists δ_2 so that $T_F(A) < T_{F,\mu_2}(A) + \epsilon$ for any partition P_2 with mesh size $\mu_2 \leq \delta_2$. Now choose $\delta = \min(\delta_1, \delta_2)$.
    Recall that $I_1(x) = \frac{1}{2}[T_{F,A}(x) + F(x)]$. We now prove the existence of $\int_A g(x)dI_1$ by the Cauchy criterion, by showing that for every partition P of A of mesh size $\mu \leq \delta$:

$$0 \leq U(g, I_1; P) - L(g, I_1; P) < C\epsilon, \tag{3}$$

for some constant C independent of ϵ.
    First, by definition of $I_1(x)$ and linearity of Δ from (4.46):

$$U(g, I_1; P) - L(g, I_1; P) = \frac{1}{2}[U(g, T_{F,A}; P) - L(g, T_{F,A}; P)] \tag{4}$$

$$+ \frac{1}{2}[U(g, F; P) - L(g, F; P)].$$

Note that the difference of Darboux sums defined with F need not be nonnegative since $F(x)$ is not assumed to be n-increasing. Thus we bound the absolute value of this term.
    Recalling that $\Delta T_{F,A}(A_J) = T_F(A_J)$ by Proposition 4.76, we prove:

$$\sum\nolimits_{J \in I} [M_J - m_J][T_F(A_J) - |\Delta F(A_J)|] < C_1\epsilon, \tag{5}$$

$$\sum\nolimits_{J \in I} [M_J - m_J]|\Delta F(A_J)| < C_2\epsilon, \tag{6}$$

$$\left| \sum\nolimits_{J \in I} [M_J - m_J]\Delta F(A_J) \right| < C_3\epsilon. \tag{7}$$

It will then follow from these bounds and (4) that (3) is satisfied with $C = (C_1 + C_2 + C_3)/2$.

Since $g(x)$ is bounded, say $|g(x)| \leq M$ on A, statement (5) is obtained from finite additivity of T_F from Exercise 4.74 and (2):

$$\sum_{J \in I} [M_J - m_J] \left[T_F(A_J) - |\Delta F(A_J)| \right] \leq 2M \sum_{J \in I} \left[T_F(A_J) - |\Delta F(A_J)| \right]$$
$$= 2M \left[T_F(A) - \sum_{J \in I} |\Delta F(A_J)| \right]$$
$$< 2M\epsilon.$$

For (6), define $\widetilde{x}_J, \widetilde{y}_J \in A_J$ so that $M_J - m_J \leq [g(\widetilde{x}_J) - g(\widetilde{y}_J)] + \epsilon$, and this is possible by the definition of M_J and m_J. Then from (1):

$$\sum_{J \in I} [M_J - m_J] |\Delta F(A_J)| \leq \sum_{J \in I} [g(\widetilde{x}_J) - g(\widetilde{y}_J)] |\Delta F(A_J)| + \epsilon \sum_{J \in I} |\Delta F(A_J)|$$
$$\leq \left| \sum_{J \in I} [g(\widetilde{x}'_J) - g(\widetilde{y}'_J)] \Delta F_i \right| + \epsilon T_F(A)$$
$$< \epsilon(1 + T_F(A)).$$

In the second step, we let $(\widetilde{x}'_i, \widetilde{y}'_i) = (\widetilde{x}_i, \widetilde{y}_i)$ if $\Delta F_i \geq 0$ and $(\widetilde{x}'_i, \widetilde{y}'_i) = (\widetilde{y}_i, \widetilde{x}_i)$ if $\Delta F_i < 0$.

For (7), defining $\widetilde{x}_J, \widetilde{y}_J \in A_J$ as in the previous step:

$$\left| \sum_{J \in I} [M_J - m_J] \Delta F(A_J) \right| \leq \left| \sum_{J \in I} [g(\widetilde{x}_J) - g(\widetilde{y}_J)] \Delta F(A_J) \right| + \epsilon \sum_{J \in I} |\Delta F(A_J)|$$
$$< \epsilon(1 + T_F(A)).$$

Finally, if $g(x)$ is continuous, then the integrals on the right in (4.73) exist for any n-increasing integrators $F_j(x)$ by Corollary 4.90. ∎

4.4.5 Properties of the Integral

Properties of the Riemann-Stieltjes integral in n-dimensions are summarized first in the case of greatest interest here, and that is for an n-increasing integrator F. Joint distribution functions are important examples of such integrators.

Proposition 4.94 (Properties of the integral, n-increasing $F(x)$) *Let $F(x)$ be an n-increasing function on $A \equiv \prod_{j=1}^{n} [a_j, b_j]$.*

1. If $g_1(x)$ and $g_2(x)$ are integrable, then so too is $cg_1(x) + eg_2(x)$ for real c, e, and:

$$\int_A (cg_1(x) + eg_2(x)) \, dF = c \int_A g_1(x) dF + e \int_A g_2(x) dF.$$

2. If $g_1(x)$ and $g_2(x)$ are integrable and $g_1(x) \leq g_2(x)$, then:

$$\int_A g_1(x) dF \leq \int_A g_2(x) dF.$$

3. If $g(x)$ is integrable, then so too is $|g(x)|$ and:

$$\left| \int_A g(x) dF \right| \leq \int_A |g(x)| \, dF.$$

4. If $A = \bigcup_{J \in I} A_J$, where $\{A_J\}$ is a finite disjoint collection of rectangles, and $g(x)$ is integrable on A, then $g(x)$ is integrable on A_J for all J, and:

$$\int_A g(x) dF = \sum_{J \in I} \int_{A_J} g(x) dF.$$

5. *If $g(x)$ is integrable with respect to both n-increasing functions $F_1(x)$ and $F_2(x)$, then $g(x)$ is integrable with respect to $F(x) \equiv cF_1(x) + eF_2(x)$ for real c, e, and:*

$$\int_A g(x)dF = c\int_A g(x)dF_1 + e\int_A g(x)dF_2.$$

6. *If $g(x)$ is integrable with respect to $F(x)$, then $g(x)$ is integrable with respect to $\bar{F}(x) \equiv F(x) + c$ for any $c \in \mathbb{R}$, and:*

$$\int_A g(x)dF = \int_A g(x)d\bar{F}.$$

Proof. *The proofs here are similar to the proofs of Proposition 4.24, but we include them for completeness.*

1. *Integrability of $cg_1(x)$ and $eg_2(x)$ follow from integrability of $g_1(x)$ and $g_2(x)$ by (4.58), as does $\int_A cg_1(x)dF = c\int_A g_1(x)dF$. Details are left as an exercise.*

Given $\epsilon > 0$, let P_1 and P_2 be partitions defined for $g_1(x)$ and $g_2(x)$, respectively, with mesh size μ_j and which satisfy (4.67), and let P denote the common refinement of P_1 and P_2. By (4.61), and since $\inf cg_1 + \inf eg_2 \leq \inf(cg_1 + eg_2)$ and $\sup(cg_1 + eg_2) \leq \sup cg_1 + \sup eg_2$:

$$
\begin{aligned}
L(cg_1, F; P_1) + L(eg_2, F; P_2) &\leq L(cg_1, F; P) + L(eg_2, F; P) \\
&\leq L(cg_1 + eg_2, F; P) \\
&\leq U(cg_1 + eg_2, F; P) \\
&\leq U(cg_1, F; P) + U(eg_2, F; P) \\
&\leq U(cg_1, F; P_1) + U(eg_2, F; P_2) \\
&< L(cg_1, F; P_1) + L(eg_2, F; P_2) + 2\epsilon.
\end{aligned}
$$

Thus:

$$U(cg_1 + eg_2, F; P) - L(cg_1 + eg_2, F; P) < 2\epsilon,$$

and $cg_1(x) + eg_2(x)$ is integrable by Proposition 4.88.

This set of inequalities also assures that the limits as $\mu \to 0$ of both $L(cg_1 + eg_2; P, F)$ and $U(cg_1 + eg_2; P, F)$ are within 2ϵ of $\int_A cg_1(x)dF + \int_A eg_2(x)dF$, and the proof is complete.

2. *For any partition P, $L(g_1, F; P) \leq L(g_2, F; P)$ and $U(g_1, F; P) \leq U(g_2, F; P)$. If $I_1 > I_2$, with I_j shorthand for the associated Riemann-Stieltjes integral, let $\epsilon \equiv I_1 - I_2$. Choose partitions P_1 and P_2 so that (4.67) is satisfied with Darboux sums underlying I_1 and I_2, respectively, and $\epsilon/4$. That is, for $j = 1, 2$:*

$$U(g_j, F; P_j) - L(g_j, F; P_j) < \epsilon/4.$$

It then follows from (4.61) that the same inequalities are satisfied with P, the common refinement of P_1 and P_2. But then by construction and the definition of ϵ:

$$L(g_2, F; P) \leq I_2 \leq U(g_2, F; P) < L(g_1, F; P) \leq I_1 \leq U(g_1, F; P).$$

The middle inequality is a contradiction and hence $I_1 \leq I_2$.

3. *This result follows from items 1 and 2 once* $|g(x)|$ *is proved integrable, since then* $g(x) \leq |g(x)|$ *and* $-g(x) \leq |g(x)|$ *imply* $\pm \int_A g(x)dF \leq \int_A |g(x)| \, dF$.

That $|g(x)|$ *is integrable is implied by (4.67) by noting that given any partition, the inequality* $||x| - |y|| \leq |x - y|$ *for real* x, y *yields:*

$$U(|g| , F; P) - L(|g| , F; P) \leq U(g, F; P) - L(g, F; P).$$

4. *It is enough to prove this result in the case where* $\{A_J\}_{J \in I}$ *is defined by an arbitrary partition of* A. *Once proved, then given general* $A_J = \prod_{j=1}^n [a_j^J, b_j^J]$, *we form the partition* $P = \{A'_J\}_{J \in I'}$ *of* A *where each* $[a_j, b_j]$ *is partitioned by the ordered collection of points* $\{a_j^J, b_j^J\}_{J \in I}$. *It then follows that for this partition:*

$$\int_A g(x)dF = \sum_{J \in I'} \int_{A'_J} g(x)dF.$$

Then by construction, each of the original A_J-*rectangles is partitioned into a unique subset of the* A'_J-*rectangles, and applying this result again completes the proof.*

By induction, the case of a general partition can be proved by assuming that $A = A_1 \bigcup A_2$, *where for example* $a_1 < c_1 < b_1$, *and:*

$$A_1 = [a_1, c_1] \times \prod_{j=2}^n [a_j, b_j], \quad A_2 = [c_1, b_1] \times \prod_{j=2}^n [a_j, b_j].$$

Then if P *is a partition of* A *so that (4.67) is satisfied with given* ϵ, *then by (4.61) it is also satisfied with refined partition* P', *defined to add point* c_1 *to the partition points of* $[a_1, b_1]$ *if it is not already included.*

Defining $P'_1 = P' \bigcap A_1$ *and* $P'_2 = P' \bigcap A_2$, *the lower Darboux sum splits:*

$$L(g, F; P') = L(g, F; P'_1) + L(g, F; P'_2),$$

where the latter sums are now defined relative to partitions of A_1 *and* A_2. *The same is true for* $U(g, F; P')$:

$$U(g, F; P') = U(g, F; P'_1) + U(g, F; P'_2).$$

The result in (4.67) for $U(g, F; P') - L(g, F; P')$ *now yields that for* $j = 1, 2$:

$$0 \leq U(g, F; P'_j) - L(g, F; P'_j) < \epsilon.$$

Thus both integrals over A_1 *and* A_2 *exist. That the value of the integral over* A *also splits into the subrectangle integrals follows from an application of (4.58).*

5. *For any partition* P *of* A:

$$R(g, F; P) = cR(g, F_1; P) + eR(g, F_2; P).$$

Since $g(x)$ *is integrable with respect to both* $F_1(x)$ *and* $F_2(x)$, *we obtain that:*

$$\lim_{\mu \to 0} R(g, F; P) = c \int_A g(x)dF_1 + e \int_A g(x)dF_2.$$

Thus $g(x)$ *is integrable with respect* $F(x)$ *by Definition 4.80, and the equality of integrals is proved.*

6. *This result follows from part 5 noting that* $\int_A g(x)dc = 0$ *by (4.58).*

■

We now state associated properties for integrals with Vitali bounded variation integrators.

Proposition 4.95 (Properties of the integral, Vitali bounded variation $F(x)$) *Let $F(x)$ be of Vitali bounded variation on $A \equiv \prod_{j=1}^{n}[a_j, b_j]$.*

1. *If $g_1(x)$ and $g_2(x)$ are integrable, then so too is $cg_1(x) + eg_2(x)$ for real c, e, and:*

$$\int_A (cg_1(x) + eg_2(x))\, dF = c\int_A g_1(x)dF + e\int_A g_2(x)dF.$$

2. *If $g(x)$ is integrable, then so too is $|g(x)|$ and:*

$$\left|\int_A g(x)dF\right| \leq \int_A |g(x)|\, dT_{F,A},$$

where $T_{F,A}$ is as given in (4.50).

3. *If $A = \bigcup_{J \in I} A_J$, where $\{A_J\}$ is a finite disjoint collection of rectangles, and $g(x)$ is integrable on A, then $g(x)$ is integrable on A_J for all J, and:*

$$\int_A g(x)dF = \sum_{J \in I}\int_{A_J} g(x)dF.$$

4. *If $g(x)$ is integrable with respect to both Vitali bounded variation functions $F_1(x)$ and $F_2(x)$, then $g(x)$ is integrable with respect to $F(x) \equiv cF_1(x) + eF_2(x)$ for real c, e, and:*

$$\int_A g(x)dF = c\int_A g(x)dF_1 + e\int_A g(x)dF_2.$$

Proof. *Items 1, 3, and 4 are corollaries to items 1, 4, and 5 of Proposition 4.94, using (4.72).*

For item 2, it follows from (4.72) that:

$$\left|\int_A g(x)dF\right| \leq \left|\int_A g(x)dI_1\right| + \left|\int_A g(x)dI_2\right|.$$

An application of items 3 then 5 of Proposition 4.94 obtains that $|g(x)|$ is integrable with respect to $I_1(x) + I_2(x)$ and:

$$\left|\int_a^b g(x)dF\right| \leq \int_a^b |g(x)|\, d\,(I_1 + I_2).$$

Now let $I_1(x) = \frac{1}{2}\left[T_{F,A}(x) + F(x)\right]$ and $I_2(x) = \frac{1}{2}\left[T_{F,A}(x) - F(x)\right]$ as in (4.56). Since $I_1(x) + I_2(x) = T_{F,A}(x)$, it follows from item 6 of Proposition 4.94:

$$\left|\int_a^b g(x)dF\right| \leq \int_a^b |g(x)|\, dT_{F,A}.$$

∎

We end this section with a result on integration to the limit. While stated for a Vitali bounded variation integrator, this result also applies to n-increasing integrators by Exercise 4.71.

Proposition 4.96 (On continuous $g_m(x) \to g(x)$, Vitali bounded variation $F(x)$) *If $\{g_m(x)\}_{m=1}^{\infty}$, $g(x)$ are continuous with $g_m(x) \to g(x)$ uniformly on $A \equiv \prod_{j=1}^{n}[a_j, b_j]$, and $F(x)$ is of Vitali bounded variation on A, then:*

$$\int_A g_m(x)dF \to \int_A g(x)dF,$$

and:

$$\int_A |g_m(x) - g(x)|\, dF \to 0.$$

Proof. *Existence of all integrals is assured by Proposition 4.89. Then using the properties of Proposition 4.95 and (4.68):*

$$\left| \int_A g_m(x)dF - \int_A g(x)dF \right| = \left| \int_A (g_m(x) - g(x))\, dF \right| \leq M_{(g_m-g)} T_F(A),$$

where $M_{(g_m-g)} \equiv \sup_A |g_m(x) - g(x)|$. Uniform convergence of $g_m(x) \to g(x)$ implies $M_{(g_m-g)} \to 0$, and convergence of integrals is proved.

Similarly:

$$\left| \int_A |g_m(x) - g(x)|\, dF \right| \leq M_{(g_m-g)} T_F(A),$$

completing the proof. ∎

4.4.6   Evaluating Riemann-Stieltjes Integrals

This final section addresses the evaluation of Riemann-Stieltjes integrals in $\mathbb{R}^n$ in two special cases of an integrator function $F(x)$, and largely parallels Proposition 4.28. These two cases are commonly encountered in probability theory where $F(x)$ is the joint distribution function of a random vector defined on a probability space. Part 1 of this result is applicable in multivariate continuous probability theory, while part 2 is applicable in multivariate discrete probability theory. These results also apply when $F(x)$ is the distribution function induced by Borel measure on $\mathbb{R}^n$.

To formalize these applications, recall that by Propositions I.8.10 and I.8.12, that distribution functions induced by Borel measures are n-increasing, while in Proposition II.6.9, this conclusion is validated for the joint distribution function of a random vector defined on a probability space. Thus if differentiable, then part 1 applies to these distribution functions; if discrete, then part 2 applies. While part 1 is stated in the more general context of $F(x)$ of Vitali bounded variation, this includes functions that are n-increasing by Exercise 4.71.

For this result, the statement that a given identity "remains true for $a_j = -\infty$ and/or $b_j = \infty$" means it remains true for any and all combinations of such limits. Thus this proposition also implies that all such improper integrals are well defined.

Proposition 4.97 ($F(x)$ **Vitali** *B.V.* **with** $\partial^n F$ **continuous, or,** $F(x)$ **a step function**) *The numerical evaluation of $\int_A g(x)dF$ can be simplified in the following cases.*

1. *Let $F(x)$ be of Vitali bounded variation and assume that:*

$$f(x) \equiv \frac{\partial^n F}{\partial x_1 \partial x_2 \cdots \partial x_n},$$

is a continuous function on $A = \prod_{j=1}^{n}[a_j, b_j]$.

If $g(x)$ is continuous on A:

$$\int_A g(x)dF = (\mathcal{R}) \int_A g(x)f(x)dx, \tag{4.74}$$

where $(\mathcal{R})$ denotes that this is a Riemann integral.

If $F(x)$ is n-increasing and $g(x)$ and $F(x)$ are bounded, then (4.74) remains true for $a_j = -\infty$ and/or $b_j = \infty$.

If $F(x)$ is n-increasing and bounded and $g(x)$ is unbounded, then (4.74) remains true for $a_j = -\infty$ and/or $b_j = \infty$ if the improper Riemann integral exists.

2. *Let $\{y_j\}_{j=1}^N \subset \mathbb{R}^n$ and nonnegative $\{c_j\}_{j=1}^N$ be given, where if $N = \infty$ we assume that $\{y_j\}_{j=1}^\infty$ has no accumulation points and $\sum_{j=1}^\infty c_j < \infty$. Define an increasing step function $F(x)$, by:*

$$F(x) = \sum_{y_j \leq x} c_j,$$

where $y_j \leq x$ is shorthand for $y_{j_k} \leq x_k$ for $1 \leq k \leq n$.

Then $F(x)$ is n-increasing, and with $g(x)$ continuous on $A = \prod_{j=1}^n [a_j, b_j]$:

$$\int_A g(x)dF = \sum_{y_j \in A} g(y_j)c_j. \tag{4.75}$$

If $g(x)$ and $F(x)$ are bounded, then (4.27) remains true for $a_j = -\infty$ and/or $b_j = \infty$.

If $F(x)$ is bounded and $g(x)$ is unbounded, then (4.27) remains true for $a_j = -\infty$ and/or $b_j = \infty$ if $\sum_{y_j \in A} g(y_j)c_j$ is absolutely convergent.

Proof. *In both cases, the existence of $\int_A g(x)dF$ is assured by Proposition 4.89 for bounded A, noting that $F(x)$ of part 2 is n-increasing. This latter conclusion follows from Proposition I.8.10 by defining a finite Borel measure on $\mathbb{R}^n$ by $\mu[y_j] = c_j$, extending by additivity, and noting that $F(x)$ so defined is the distribution function of this measure of I.(8.2).*

For part 1, let $\epsilon > 0$ be given and choose δ' as in Definition 4.80, so that for any partition $P \equiv \left\{ \{x_{j,i_j}\}_{i_j=0}^{m_j} \right\}_{j=1}^n$ of A with mesh size $\mu \leq \delta'$, and arbitrary tags $\{x_J\}_{J \in I}$ with $x_J \in \prod_{j=1}^n [x_{j,i_j-1}, x_{j,i_j}]$:

$$\left| \sum_{J \in I} g(x_J) \Delta F(A_J) - \int_A g(x)dF \right| < \epsilon/2. \tag{1}$$

Recalling (4.47), $\Delta F(A_J)$ is defined on the subrectangle $A_J = \prod_{j=1}^n [x_{j,i_j-1}, x_{j,i_j}]$ by a Riemann integral. In detail, and denoting $A_J \equiv \prod_{j=1}^n [y_j, z_j]$ for simplicity:

$$\Delta F(A_J) = \int_{y_1}^{z_1} \int_{y_2}^{z_2} \cdots \int_{y_n}^{z_n} f(x)dx_n dx_{n-1} \cdots dx_1.$$

By continuity of $f(x)$ on $A_J = \prod_{j=1}^n [x_{j,i_j-1}, x_{j,i_j}]$, the mean value theorem for integrals of Corollary 1.73 yields that for some $x'_J \in \prod_{j=1}^n (x_{j,i_j-1}, x_{j,i_j})$:

$$\Delta F(A_J) = f(x'_J) \prod_{j=1}^n \left(x_{j,i_j} - x_{j,i_j-1} \right).$$

Setting $x_J = x'_J$ in (1), it follows that for any partition of A with mesh size $\mu \le \delta'$, there exist tags $\{x'_J\}_{J \in I}$, so that:

$$\left| \sum_{J \in I} g\left(x'_J\right) f\left(x'_J\right) \prod_{j=1}^{n} \left(x_{j,i_j} - x_{j,i_j-1}\right) - \int_A g(x) dF \right| < \epsilon/2. \tag{2}$$

Now since $g(x)f(x)$ is continuous on A it is Riemann integrable on A by Proposition 1.63, and for this ϵ there exists δ'' so that for any partition of A of mesh size $\mu \le \delta''$ and arbitrary tags $\{x''_J\}_{J \in I'}$:

$$\left| \sum_{J \in I'} g\left(x''_J\right) f\left(x''_J\right) \prod_{j=1}^{n} \left(x_{j,i_j} - x_{j,i_j-1}\right) - (\mathcal{R}) \int_A g(x) f(x) dx \right| < \epsilon/2.$$

Thus for any partition of mesh size $\mu \le \min(\delta', \delta'')$, choose $x''_J = x'_J$ from (2), and the triangle inequality obtains:

$$\left| \int_A g(x) dF - (\mathcal{R}) \int_A g(x) f(x) dx \right| < \epsilon.$$

As ϵ is arbitrary, (4.74) is proved.

When $g(x)$ and $F(x)$ are bounded, $\int_A g(x) dF$ exists for $a_j = -\infty$ and/or $b_j = \infty$ by Proposition 4.92. Then, for example in the general case of all $a_j \to -\infty$ and all $b_j \to \infty$, applying (4.74) for $A \equiv \prod_{j=1}^{n} [a_j, b_j]$:

$$\lim_{\substack{a_j \to -\infty \\ b_j \to \infty}} \int_A g(x) dF = \lim_{\substack{a_j \to -\infty \\ b_j \to \infty}} (\mathcal{R}) \int_A g(x) f(x) dx.$$

Thus both limits exist and are finite, which is (4.74) in this case.

More generally, if any Riemann integral is well defined as an improper integral, then the limit of the Riemann integral on the right above exists. By equality for each bounded set, it follows that the limit of the Riemann-Stieltjes integrals also exists, and so too the associated improper integral.

For part 2, since this integral exists by Definition 4.80 as noted above, given $\epsilon > 0$ there exists δ' so that for any partition $P \equiv \left\{ \{x_{j,i_j}\}_{i_j=0}^{m_j} \right\}_{j=1}^{n}$ of A with mesh size $\mu \le \delta'$, and arbitrary tags $\{x_J\}_{J \in I}$ with $x_J \in \prod_{j=1}^{n} [x_{j,i_j-1}, x_{j,i_j}]$:

$$\left| \sum_{J \in I} g\left(x_J\right) \Delta F\left(A_J\right) - \int_A g(x) dF \right| < \epsilon. \tag{3}$$

For $F(x)$ defined as above as the distribution function of a Borel measure, and $A_J \equiv \prod_{j=1}^{n} [x_{j,i_j-1}, x_{j,i_j}]$:

$$\Delta F\left(A_J\right) = \sum_{y_k \in A_J} c_k.$$

Now either $\{y_j\}_{j=1}^{N}$ is finite, or $N = \infty$ and this set is assumed to have no accumulation points, and thus we can choose μ so that each A_J contains at most one such y_j. Then $\Delta F\left(A_J\right) = c_j$ or $\Delta F_J = 0$ for all J. In the former case, choose $x_J = y_j$ and then by (3):

$$\left| \sum_{y_j \in A} g\left(y_j\right) c_j - \int_A g(x) dF \right| < \epsilon.$$

As ϵ is arbitrary, this is (4.75).

The remaining details are similar to part 1. ∎

References

I have listed below a number of textbook references for the mathematics and finance presented in this series of books. Many provide both theoretical and applied materials in their respective areas that are beyond those developed here and would be worth pursuing by those interested in gaining a greater depth or breadth of knowledge in the given subjects. This list is by no means complete and is intended only as a guide to further study. In addition, a limited number of research papers will be identified in each book if they are referenced therein. A more complete guide to published papers can be found in the references below.

The reader will no doubt observe that the mathematics references are somewhat older than the finance references and upon web searching will find that several of the older texts in each category have been updated to newer editions, sometimes with additional authors. Since I own and use the editions below, I decided to present these editions rather than reference the newer editions that I have not reviewed. As many of these older texts are considered "classics," they are also likely to be found in university and other technical libraries.

That said, there are undoubtedly many very good new texts by both new and established authors with similar titles that are also worth investigating. One that I will at the risk of immodesty recommend for more introductory materials on mathematics, probability theory and finance is:

[1] Reitano, Robert R. *Introduction to Quantitative Finance: A Math Tool Kit.* Cambridge, MA: The MIT Press, 2010.

Topology, Measure, Integration, Linear Algebra

[2] Doob, J. L. *Measure Theory.* New York, NY: Springer-Verlag, 1994.

[3] Dugundji, James. *Topology.* Boston, MA: Allyn and Bacon, 1970.

[4] Edwards, Jr., C. H. *Advanced Calculus of Several Variables.* New York, NY: Academic Press, 1973.

[5] Gemignani, M. C. *Elementary Topology.* Reading, MA: Addison-Wesley Publishing, 1967.

[6] Halmos, Paul R. *Measure Theory.* New York, NY: D. Van Nostrand, 1950.

[7] Hewitt, Edwin, and Karl Stromberg. *Real and Abstract Analysis.* New York, NY: Springer-Verlag, 1965.

[8] Royden, H. L. *Real Analysis,* 2nd Edition. New York, NY: The MacMillan Company, 1971.

[9] Rudin, Walter. *Principals of Mathematical Analysis,* 3rd Edition. New York, NY: McGraw-Hill, 1976.

[10] Rudin, Walter. *Real and Complex Analysis,* 2nd Edition. New York, NY: McGraw-Hill, 1974.

[11] Shilov, G. E., and B. L. Gurevich. *Integral, Measure & Derivative: A Unified Approach.* New York, NY: Dover Publications, 1977.

[12] Strang, Gilbert. *Introduction to Linear Algebra,* 4th Edition. Wellesley, MA: Cambridge Press, 2009.

Probability Theory & Stochastic Processes

[13] Billingsley, Patrick. *Probability and Measure,* 3rd Edition. New York, NY: John Wiley & Sons, 1995.

[14] Chung, K. L., and R. J. Williams. *Introduction to Stochastic Integration.* Boston, MA: Birkhäuser, 1983.

[15] Davidson, James. *Stochastic Limit Theory.* New York, NY: Oxford University Press, 1997.

[16] de Haan, Laurens, and Ana Ferreira. *Extreme Value Theory, An Introduction.* New York, NY: Springer Science, 2006.

[17] Durrett, Richard. *Probability: Theory and Examples,* 2nd Edition. Belmont, CA: Wadsworth Publishing, 1996.

[18] Durrett, Richard. *Stochastic Calculus, A Practical Intriduction.* Boca Raton, FL: CRC Press, 1996.

[19] Feller, William. *An Introduction to Probability Theory and Its Applications,* Volume I. New York, NY: John Wiley & Sons, 1968.

[20] Feller, William. *An Introduction to Probability Theory and Its Applications,* Volume II, 2nd Edition. New York, NY: John Wiley & Sons, 1971.

[21] Friedman, Avner. *Stochastic Differential Equations and Applications, Volumes 1 and 2.* New York, NY: Academic Press, 1975.

[22] Ikeda, Nobuyuki, and Shinzo Watanabe. *Stochastic Differential Equations and Diffusion Processes.* Tokyo, Japan: Kodansha Scientific, 1981.

[23] Karatzas, Ioannis, and Steven E. Shreve. *Brownian Motion and Stochastic Calculus.* New York, NY: Springer-Verlag, 1988.

[24] Kloeden, Peter E., and Eckhard Platen. *Numerical Solution of Stochastic Differential Equations.* New York, NY: Springer-Verlag, 1992.

[25] Lowther, George. *Almost Sure, A Maths Blog on Stochastic Calculus,* https://almostsure.wordpress.com/stochastic-calculus/

[26] Lukacs, Eugene. *Characteristic Functions.* New York, NY: Hafner Publishing, 1960.

[27] Nelson, Roger B. *An Introduction to Copulas,* 2nd Edition. New York, NY: Springer Science, 2006.

[28] Øksendal, Bernt. *Stochastic Differential Equations: An Introduction with Applications,* 5th Edition. New York, NY: Springer-Verlag, 1998.

[29] Protter, Phillip. *Stochastic Integration and Differential Equations: A New Approach.* New York, NY: Springer-Verlag, 1992.

[30] Revuz, Daniel, and Marc Yor. *Continuous Martingales and Brownian Motion,* 3rd Edition. New York, NY: Springer-Verlag, 1991.

[31] Rogers, L. C. G., and D. Williams. *Diffusions, Markov Processes and Martingales,* Volume 1, Foundations, 2nd Edition. Cambridge, UK: Cambridge University Press, 2000.

[32] Rogers, L. C. G., and D. Williams. *Diffusions, Markov Processes and Martingales,* Volume 2, Itô Calculus, 2nd Edition. Cambridge, UK: Cambridge University Press, 2000.

[33] Sato, Ken-Iti. *Lévy Processes and Infinitely Divisible Distributions.* Cambridge University Press, Cambrideg, UK, 1999.

[34] Schilling, René L., and Lothar Partzsch. *Brownian Motion: An Introduction to Stochastic Processes,* 2nd Edition. Berlin/Boston: Walter de Gruyter GmbH, 2014.

[35] Schuss, Zeev. *Theory and Applications of Stochastic Differential Equations.* New York, NY: John Wiley and Sons, 1980.

Finance Applications

[36] Etheridge, Alison. *A Course in Financial Calculus.* Cambridge, UK: Cambridge University Press, 2002.

[37] Embrechts, Paul, Claudia Klüppelberg, and Thomas Mikosch. *Modelling Extremal Events for Insurance and Finance.* New York, NY: Springer-Verlag, 1997.

[38] Hunt, P. J., and J. E. Kennedy. *Financial Derivatives in Theory and Practice,* Revised Edition. Chichester, UK: John Wiley & Sons, 2004.

[39] McLeish, Don L. *Monte Carlo Simulation and Finance.* New York, NY: John Wiley, 2005.

[40] McNeil, Alexander J., Rüdiger Frey, and Paul Embrechts. *Quantitative Risk Management: Concepts, Techniques, and Tools.* Princeton, NJ.: Princeton University Press, 2005.

Research Papers for Book III

[41] Darboux G., Ann. Sci. Ecole Norm. Sup. Ser., Vol. 2 , No. 4: 57–112, 1875.

[42] Lebesgue, Henri. *Leçons sur l'intégration et la recherche des fonctions primitives.* Paris: Gauthier-Villars, 1904.

[43] McCarthy, John., "An Everywhere Continuous Nowhere Differentiable Function," The American Mathematical Monthly, Vol. 60, No. 10: 709, 1953.

[44] Sokal, Alan D., "A Really Simple Elementary Proof of the Uniform Boundedness Theorem," The American Mathematical Monthly, Vol. 118, No. 5: 450–452, 2011.

[45] Riemann, B., "Ueber die Darstellbarkeit einer Function durch eine trigonometrische Reihe," Göttinger Akad. Abh., Vol. 13, 1868.

[46] Stieltjes, Thomas Jan., "Recherches sur les fractions continues." Ann. Fac. Sci. Toulouse, Vol. VIII: 1–122, 1894.

[47] Young, L. C., "An Inequality of the Hölder Type, Connected with Stieltjes Integration," Acta Mathematica, Vol. 67, No. 1: 251–282, 1936.

Index

Taylor & Francis
Taylor & Francis Group
http://taylorandfrancis.com

Printed in the United States
by Baker & Taylor Publisher Services